Quantum Computing

Quantum Computing

Foundations and Practice

Steven Herbert

Great Clarendon Street, Oxford, OX2 6DP,
United Kingdom

Oxford University Press is a department of the University of Oxford.
It furthers the University's objective of excellence in research, scholarship,
and education by publishing worldwide. Oxford is a registered trade mark of
Oxford University Press in the UK and in certain other countries.

© Steven Herbert 2025

The moral rights of the author have been asserted.

All rights reserved. No part of this publication may be reproduced, stored in a retrieval system, transmitted, used for text and data mining, or used for training artificial intelligence, in any form or by any means, without the prior permission in writing of Oxford University Press, or as expressly permitted by law, by licence or under terms agreed with the appropriate reprographics rights organization. Enquiries concerning reproduction outside the scope of the above should be sent to the Rights Department, Oxford University Press, at the address above.

You must not circulate this work in any other form
and you must impose this same condition on any acquirer.

Published in the United States of America by Oxford University Press
198 Madison Avenue, New York, NY 10016, United States of America

British Library Cataloguing in Publication Data

Data available

Library of Congress Control Number: 2025937663

ISBN 9780192868435
ISBN 9780192868442 (pbk.)

DOI: 10.1093/9780191964381.001.0001

The manufacturer's authorised representative in the EU for product safety is
Oxford University Press España S.A. of Parque Empresarial San Fernando de Henares,
Avenida de Castilla, 2 – 28830 Madrid (www.oup.es/en or product.safety@oup.com).
OUP España S.A. also acts as importer into Spain of products made by the manufacturer.).

Links to third party websites are provided by Oxford in good faith and
for information only. Oxford disclaims any responsibility for the materials
contained in any third party website referenced in this work.

Preface

For many years, the Quantum Computing course in Part II of the University of Cambridge Computer Science Tripos consisted of eight packed lectures. After I had the pleasure and privilege of taking over in 2020, the length was increased to 16 lectures, with the scope expanded to cover some more recent results, and in particular to reflect the fact that quantum computing was at last becoming a relevant topic in the so-called 'real world'. As I came to the end of my first term of lecturing, the COVID pandemic hit, and the students were all sent home in great haste (those who could get home, that is). In-person lecturing would not return to Cambridge for nearly two years. In the intervening time, I had to produce high-quality videos of the lectures, and I beefed up the course notes considerably in an attempt to assist students who were left to learn in relative isolation. All of this meant that I had, without particularly intending to, begun to think of the course as a 'product', and so it was an easy decision to say 'yes' when Oxford University Press came calling to see whether I'd like to turn the course into a book.

My own journey into the wonderful field of quantum computation was somewhat unconventional, and I learned the subject by studying Nielsen and Chuang's [2010] iconic textbook. However, in a fast moving field like quantum computing, even a seminal textbook becomes dated, and in the quarter of a century since the original publication of *quantum computation and quantum information*, there have been several major breakthroughs that I have covered in this book. In doing so, I have added considerably to the material from my own lecture course, and my hope is that this textbook will provide the basis for the quantum computing courses that I am sure will spring up in the coming years as the field continues to grow and mature. As I have shown through my own teaching, the course can be condensed into 16 lectures, but I estimate that it would take about 20–30 lectures to cover the material in full.

This book is also born of a deep conviction that quantum computing *is* an accessible subject that is sometimes mystified by those who should know better. Therefore, this textbook is also targeted at information and technology professionals, who in the future will increasingly encounter quantum computing, even if it is not their core area. I believe that many such people will be sufficiently intrigued to invest the time to gain a solid appreciation of *why* and *how* quantum mechanics offers an enhanced computational paradigm, rather than just settling for the disposition of 'operational acceptance' towards quantum computing that many contemporary resources on the subject implicitly adopt. In my experience, any technically-minded person with a working knowledge of linear algebra is capable of following and understanding even the most astonishing results in quantum computing. Broadly speaking, this conviction motivates the 'foundations' aspect of *foundations and practice*, and the 'practice' aspect speaks for itself: as quantum computing does indeed become a 'thing' in the 'real world', I hope that this book will become a valuable resource for practitioners and theoreticians alike.

Steven Herbert

For Oscar and Sienna

Acknowledgement

First and foremost, my thanks to Sathya Subramanian, Adam Ó Conghaile, David Amaro, and Yuta Kikuchi for reviewing various chapters of the book. The book is immeasurably better for the comments, corrections, and gentle criticism I received. Of course, any remaining mistakes are my responsibility alone. I am also indebted to my editor Giulia Lipparini and acquisitions editor Dan Taber-Hewett for their support, patience, and wise counsel, and indeed to Kanimozhi Ramamurthy and the whole production team at Oxford University Press – without whom this book would literally never have happened. Finally, a special mention to Prof. Anuj Dawar, who entrusted me to write and deliver the quantum computing course upon which this book is based, and who has always been a staunch supporter of both me and this project for the past several years.

On a personal note, to my parents for always encouraging my tenacity and the belief that I could grow up to achieve something. To my children for being my inspiration to get up and make the most of every single day. And more than anyone to my wonderful, dear wife: to merely say that you have supported me would be a gross under-statement. For without your unyielding love and belief in me, not to mention your firm but tender encouragement when needed, I would never have been able to accomplish such a feat as writing a book.

Steven Herbert

Contents

1. **Introduction**   1
   1.1 Wave-particle duality   3
   1.2 Entanglement, the measurement problem, and Schrödinger's much-debated cat   5

2. **Linear algebra**   9
   2.1 Complex numbers   9
   2.2 Vector spaces and linear maps   10
      2.2.1 Matrix multiplication   11
      2.2.2 Tensor multiplication   12
   2.3 Vector (quantum state) properties   12
      2.3.1 Inner products, orthogonality, and norms   13
      2.3.2 Outer products and projectors   13
      2.3.3 Vector space bases   14
      2.3.4 Subspaces   17
   2.4 Matrix properties   17
      2.4.1 Transpose and conjugate transpose (adjoint)   17
      2.4.2 Hermitian and unitary matrices   18
      2.4.3 Sparsity and rank   18
      2.4.4 Trace and determinant   19
      2.4.5 Eigenvalues, eigenvectors, and the spectral decomposition   19
      2.4.6 The singular value decomposition   22
      2.4.7 Parameterised matrices and matrix operations   23
      2.4.8 Matrix inversion and pseudo-inversion   24
      2.4.9 Matrix (operator) norms   25
   2.5 Notational miscellany   25

3. **Quantum mechanics and quantum information**   29
   3.1 State space, composition, and evolution   30
      3.1.1 Important single-qubit unitary operations   31
      3.1.2 Evolution of multi-qubit states   32
      3.1.3 Entangling operations and entangled states   33
   3.2 Measurement   33
      3.2.1 Computational basis measurement and the Born rule   34
      3.2.2 Global and relative phase   36
      3.2.3 The Bloch sphere   37
      3.2.4 Distinguishing orthogonal and non-orthogonal quantum states   38
      3.2.5 Measuring the expectation of an observable   41

3.3	Bell inequalities	42
3.4	No-go theorems	46
	3.4.1 The no-signalling principle	46
	3.4.2 The no-cloning theorem	49
	3.4.3 The no-deleting theorem	50

4. Quantum circuits — 53

- 4.1 The essentials of the quantum circuit model — 53
- 4.2 Classical control and deferred and implicit measurement — 58
- 4.3 Rotations and controlled gates — 61
- 4.4 Universality of the quantum circuit model — 66
  - 4.4.1 Computational universality — 66
  - 4.4.2 Quantum universality — 70
- 4.5 Quantum compilation — 78
- 4.6 Post-selection — 80

5. Quantum communication protocols — 83

- 5.1 An entangled pair as a resource for information transfer — 83
  - 5.1.1 Teleportation — 84
  - 5.1.2 Superdense coding — 85
  - 5.1.3 How to think about entanglement as a resource — 86
- 5.2 Quantum key distribution — 87
  - 5.2.1 The one-time pad — 88
  - 5.2.2 The BB84 protocol — 89

6. Quantum advantage — 97

- 6.1 How big is Hilbert space? — 97
  - 6.1.1 The gate complexity of approximately preparing every quantum state — 97
  - 6.1.2 Exploring Hilbert space classically: state vector simulation — 100
- 6.2 Computation theory and computational complexity theory — 100
  - 6.2.1 Classical and quantum models of computation — 102
  - 6.2.2 Computability and the Church–Turing thesis — 106
  - 6.2.3 Decision problems and computational complexity classes — 107
  - 6.2.4 Other types of computational problems — 111
  - 6.2.5 Complexity-theoretic evidence that certain quantum circuits are hard to simulate classically — 112
- 6.3 Quantum circuits that can be efficiently classically simulated — 113
  - 6.3.1 Efficient classical simulation of permutations — 113
  - 6.3.2 Efficient classical simulation of circuits without entanglement — 115
  - 6.3.3 Efficient classical simulation of stabiliser circuits — 116
  - 6.3.4 Generic techniques for quantum circuit simulation — 122
- 6.4 The road to quantum advantage — 123

7. Quantum algorithms for query problems — 127
- 7.1 Deutsch's algorithm — 128
- 7.2 The Deutsch–Jozsa algorithm — 131
- 7.3 The Bernstein–Vazirani algorithm — 133
- 7.4 Simon's algorithm — 135
- 7.5 Conditions for exponential quantum speed-up in query problems — 137

8. Quantum search — 139
- 8.1 Grover's algorithm — 139
  - 8.1.1 The source of quantum advantage in Grover's algorithm — 147
  - 8.1.2 Grover's algorithm for NP-complete problems — 147
- 8.2 Grover's algorithm decides the OR problem — 148
  - 8.2.1 $\Omega(\sqrt{N})$ is a lower-bound for deciding the OR problem — 148

9. Quantum phase estimation and quantum amplitude estimation — 155
- 9.1 The discrete Fourier transform — 155
- 9.2 The quantum Fourier transform — 156
- 9.3 Quantum phase estimation — 160
- 9.4 Quantum amplitude estimation — 163
  - 9.4.1 Quantum amplitude amplification — 163
  - 9.4.2 Estimating the amplitude with phase estimation — 165
  - 9.4.3 Quantum Monte Carlo integration — 168
  - 9.4.4 Amplitude estimation without phase estimation — 169
- 9.5 Quantum counting — 172

10. Order finding, period finding, and quantum factoring — 177
- 10.1 Order finding — 177
  - 10.1.1 Order finding with quantum phase estimation — 178
  - 10.1.2 Preparing the second register in a suitable initial state — 178
  - 10.1.3 Extracting the order from the phase — 179
  - 10.1.4 Efficiently implementing the controlled unitaries — 184
- 10.2 Shor's algorithm — 185
  - 10.2.1 Reduction of factoring to order finding — 186
  - 10.2.2 Shor's algorithm: computational complexity — 187
  - 10.2.3 Shor's algorithm explained in terms of period finding — 188

11. Hamiltonian simulation and ground state energy estimation — 191
- 11.1 Hamiltonian decomposition — 192
- 11.2 Simulating Hamiltonian evolution — 195
  - 11.2.1 Trotterisation — 195
  - 11.2.2 Hamiltonian simulation with the truncated Taylor series — 200
  - 11.2.3 Simulation of sparse Hamiltonian evolution — 205
- 11.3 Estimating the ground state energy of a Hamiltonian — 206
  - 11.3.1 Using quantum phase estimation to find the ground state energy — 207

Contents

    11.3.2 Ground state energy estimation in quantum chemistry   208
  11.4 The variational quantum eigensolver   208
  11.5 Is there a quantum advantage available in Hamiltonian simulation and ground state energy estimation?   213

12. Quantum optimisation   217
  12.1 The quantum adiabatic theorem   217
  12.2 Adiabatic quantum computation   220
    12.2.1 Using adiabatic quantum computation for unstructured search   221
    12.2.2 Using adiabatic quantum computation to solve NP-complete decision problems   223
  12.3 Adiabatic quantum optimisation   226
    12.3.1 NP-hardness and optimisation   226
    12.3.2 Ising models and quadratic unconstrained binary optimisation   227
    12.3.3 Max-cut   227
    12.3.4 Adiabatic quantum algorithms for NP-hard optimisation problems   228
  12.4 Quantum annealing   228
  12.5 The quantum approximate optimisation algorithm   232
  12.6 Other forms of quantum optimisation   233

13. A quantum linear system solver   237
  13.1 Solving linear systems with quantum computers   237
  13.2 Run-time and error analysis   241
  13.3 HHL for non-Hermitian matrix inversion and pseudo-inversion   242
  13.4 HHL: the caveats   246
  13.5 Solving differential equations using HHL   247
  13.6 Quantum linear algebra for machine learning and low rank dequantisation   249

14. Quantum signal processing and the quantum singular value transformation   253
  14.1 Quantum signal processing   253
  14.2 Block encoding and the quantum eigenvalue transformation   257
  14.3 Quantum signal processing as a means of Hamiltonian simulation   260
  14.4 The quantum singular value transformation   262
  14.5 Matrix inversion by the quantum singular value transformation   264
  14.6 Search by the quantum singular value transformation   265
  14.7 A grand unification of quantum algorithms?   270

15. An introduction to quantum error correction   277
  15.1 Classical error correction   277
  15.2 Quantum error correction   279
    15.2.1 The three-qubit bit-flip code   279
    15.2.2 The three-qubit phase-flip code   282

15.3 The Shor code	283
15.3.1 Correcting any single-qubit error with the Shor code	285
15.3.2 Digitisation of errors	288
15.3.3 The depolarising channel	288
15.4 Storing a quantum state indefinitely	289
15.4.1 Correcting noisy 'transversal' logical operations	290
15.5 Fault-tolerant quantum computation	291
15.5.1 Elements of fault tolerance	293
15.5.2 The threshold theorem	293
15.6 Quantum error correction that respects device layout	295
Appendix A: Simulating the Hadamard gate in the quadratic form expansion	297
Answers to chapter problems	301
Bibliography	310
Index	318

1
Introduction

The discovery of physical systems that exhibit what we now call quantum mechanical behaviour was unquestionably one of the most astonishing events of modern physics. No less remarkable was the realisation that quantum mechanics presents a much more powerful paradigm for computation and information processing than its classical counterpart. Indeed, until that point, the theory of computation was largely treated as an abstract mathematical topic, with little consideration given to how the mathematical objects being studied would obtain physical reality. This disposition is largely appropriate when said mathematical objects are (classical) bits and the operations thereupon, but is definitely not satisfactory when considering the computational properties of quantum systems. As one of the founding fathers of quantum computation, David Deutsch, wrote 'The theory of computation has traditionally been studied almost entirely in the abstract, as a topic in pure mathematics. This is to miss the point of it. Computers are physical objects, and computations are physical processes. What computers can or cannot compute is determined by the laws of physics alone, and not by pure mathematics', the essence of which Rolf Landauer, distilled into the pithier 'information is physical'.

Along with Deutsch, Nobel laureate Richard Feynman can too lay claim to the title *founding father of quantum computation*, and his approach was also to put the physical right at the heart of quantum computation, by realising that one of the most compelling reasons to investigate quantum models of computation (and to build quantum computers) would be to simulate physical systems that are governed by the laws of quantum mechanics. Feynman's famous insight – repeated in almost every textbook and popular book on quantum computing – was '...*trying to find a computer simulation of physics seems to me to be an excellent program to follow out... the real use of it would be with quantum mechanics... Nature isn't classical... and if you want to make a simulation of Nature, you'd better make it quantum mechanical, and by golly it's a wonderful problem, because it doesn't look so easy*'.

The quantum circuit model neatly captures the intersection between the physical and the abstract and can help us make sense of each. Quantum circuits completely express the physics of closed quantum systems that are composed of two-level subsystems (qubits), such as most real-world quantum computers. Quantum circuits are also equivalent to an abstractly-defined model of quantum computation, with the important caveat that the circuits for different sizes of the same problem must be uniformly generated, which is met for all of the quantum algorithms studied in this textbook. The opening chapters of this textbook are, therefore, dedicated to developing the quantum circuit model from the postulates of quantum mechanics, including the linear algebraic prerequisites necessary to do so, and also presenting a number of important results that follow directly.

The next several chapters are then concerned with quantum algorithms, a crucial topic for practitioners and theoreticians alike. It is significant that in its 40 year history, the field of quantum algorithm design has undergone a quiet revolution. The original approach of conceptualising idealised problems with queries to unknown functions (and then searching for real-world problems bearing a close resemblance thereto) has largely given way to the modern approach of viewing quantum computers as machines for manipulating the singular values or eigenvalues of very large matrices (and then searching for problems that can be posed in such terms). This textbook aims to pay due homage to both traditions of quantum algorithm design, by presenting a suitable canon of quantum algorithms. Moreover, it is worth highlighting the fact that as well as being fully-fledged algorithms in their own right, these also play the role of algorithmic primitives or subroutines: notably, whilst there has been a recent explosion in the number of proposals for 'new' quantum algorithms, they are typically novel combinations of established primitives, and almost all can trace their roots back to these few core ideas.

The final chapter of the book addresses the fact that quantum mechanics is not quite the right model for describing large systems of qubits in the real world. Whilst the postulates of quantum mechanics describe the behaviour of *closed* quantum systems, all real systems – not least real-world quantum computers – are necessarily *open* quantum systems. However, the implication of this discrepancy can be dealt with by considering the 'ideal' *unitary* behaviour – that is, as dictated by the postulates of quantum mechanics and, therefore, the quantum circuit model – as being perturbed by a well-characterised random process, namely noise. Based on this model, the theory of quantum error correction proves that (under certain circumstances, and with sufficiently mild noise) the ideal unitary model of quantum computation in which all of the preceding results are proven can, in principle, be efficiently simulated on real quantum hardware. All of the theory leading up to this Herculean result can (and does) fill an entire textbook on its own, and so more than in any other topic, a careful curation of what to include was necessary. The guiding philosophy in doing so was to give the minimal amount of content needed to persuade the theorist of the plausibility of the claim, and to describe to the practitioner the right framework in which to think about error-corrected quantum computing. (For example, after reading the final chapter one should be clear that to speak – as many often do – of a *logical* qubit as an isolated entity, that can be defined independently of things like the number of operations in the circuit and the overall acceptable error rate, is nonsensical.)

But before all of that, a more pressing question – indeed *the* question that must be answered to motivate the subject matter of this book – is *why does quantum information and computation exist as a field at all?* To answer this, we begin with the counterintuitive idea of wave-particle duality.

1.1 Wave-particle duality

A famous foundational experiment in quantum mechanics involves a light source being shone at a screen with two narrow slits – the famous double-slit experiment. The theory of electromagnetic wave propagation dictates that the two slits can be treated as two sources from which light waves of the same wavelength emanate, and beyond the first screen, a second screen is placed, this time with no slits. The waves from the two slits then interfere (constructively and destructively) – again as predicted by the theory of electromagnetic wave propagation – to give a light and dark 'interference pattern' when the light hits the second screen.

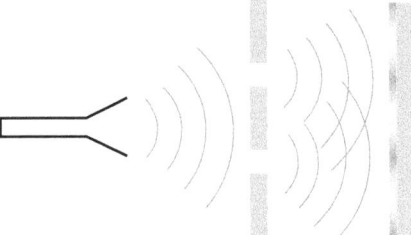

Figure 1.1

If that were the end of the story, the double-slit experiment would amount to nothing more than a confirmation of the wave-like behaviour of light. The clue here, however, is the terminology 'confirmation of the *wave-like* behaviour', rather than a stronger phrase such as 'confirmation that light is a wave', as light has also been observed to have particle-like properties. Indeed, the name 'quantum' derives from the fact that physical entities that were previously thought to exist as perfect fields (and so spread continuously in space) were, in certain experiments, observed to come in localised, finite, quantities – which have a smallest possible size, termed a *quantum*. Photons are the quanta of electromagnetic fields or, if you prefer, light particles.

That light exhibits both wave- and particle-like behaviour is not in and of itself wholly incompatible with classical physics (classical physics/mechanics being the name given to models of physics that do not take into account quantum effects). Consider, for example, the sea. Any surfer will attest to the wave-like behaviour of a mass of water, whilst no chemist has an objection to the claim that water consists of H_2O particles. However, quantum mechanics is most emphatically *not* about a very large number of particles coming together to give rise to wave-like behaviour *en masse*, and the double-slit experiment can show why not in a rather remarkable way.

4   Introduction

Modern experimental apparatus enables the light source to be dimmed to such a low level that single photons are emitted at intervals spaced out in time. Amazingly, even when only a single photon is travelling, the interference pattern remains – the only explanation being that even a single photon can exhibit wave-like behaviour, and 'interfere with itself'. That is to say that the single photon has seemingly passed through *both* slits, or to put it in quantum mechanical parlance, is in a *superposition of two components*, one passing through each of the two slits.

The plot thickens even further when the experiment is adapted by placing a photon detector at each of the slits, which identifies whether a photon passes through the slit, and allows the photon to then continue on its way. In this case, the interference pattern disappears, and each photon is indeed registered at one slit or the other (but never both), and the pattern of light on the second screen is consistent with a single source emanating at whichever slit the photon was detected.

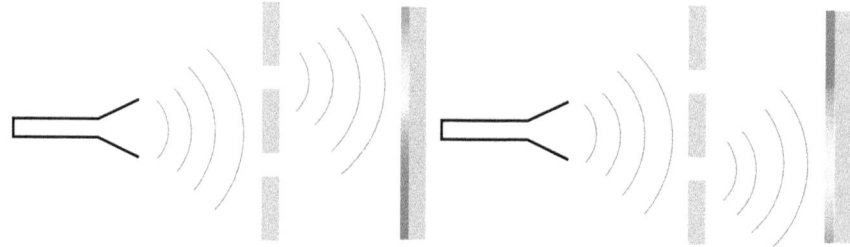

Figure 1.2

This apparently paradoxical behaviour gives rise to the term 'wave-particle duality'. Roughly speaking, wave-particle duality means that a quantum entity will propagate as a wave when it is not encumbered by a measurement, but that when a measurement is taken, the wave will *collapse* and the entity will indeed appear as a particle in a manner that is consistent with one of the possible measurement outcomes.

The double-slit experiment contains many (but not all) of the fundamental quantum phenomena, namely superposition, interference and measurement, and it is interesting to view the double-slit experiment as an instance of a qubit.[1] A qubit has two *computational basis* states, labelled $|0\rangle$ and $|1\rangle$, (which are properly introduced in Chapters 2 and 3), and in the case of the double-slit experiment, the states are the (photon passing through each of the) two slits. The qubit can then optionally be measured, in which case either of the states $|0\rangle$ and $|1\rangle$ is observed, as the measurement apparatus detects through which slit the photon passes. If the experiment is set up to be symmetric, each of $|0\rangle$ and $|1\rangle$ will be obtained as measurement outcomes

---

[1] It is important to appreciate that, whilst analogising the double-slit experiment as a qubit is helpful for attaining a solid, intuitive grasp of how fundamental quantum phenomena propagate into information-theoretic mathematical objects like qubits, it is ultimately just that, an analogy, and the double-slit experiment does not formally meet all of the requirements to be an instance of a qubit.

with 50% probability. If a measurement is not taken, the qubit will be in a superposition of $|0\rangle$ and $|1\rangle$, and its onward evolution can involve the interference of these two components.

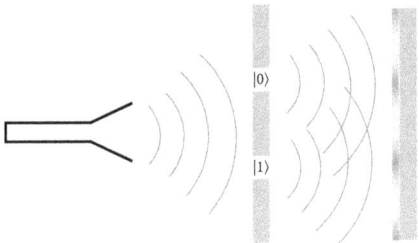

Figure 1.3

Casting the double-slit experiment as a qubit reveals the critical property that a mathematical description of the state of a quantum system must be sufficient to specify the behaviour both: (i) when a measurement is taken (that is, from the description of the state the probability with which each possible measurement outcome occurs should be derivable); *and* (ii) when the state undergoes onward 'wave-like' evolution in the absence of a measurement. The necessity of the second of these means that strictly more *information* is contained in a quantum state than merely an encoding of measurement outcome probabilities, and this point cannot be emphasised enough, since if it were not the case, then quantum computation would be no more powerful than randomised classical computation.

Another attractive feature of the double-slit experiment is that it goes some way towards demystifying measurement. The abstract definition of measurement given in the postulates of quantum mechanics (see Chapter 3) can make it seem somewhat ethereal, but in the double-slit experiment, a measurement is taken by the very tangible physical object that is a photon detector. This also provides a nice way to think about computational bases ($|0\rangle$ and $|1\rangle$), as they, in some sense, represent 'classical' states with an objective reality, and their measurement is simply the process of obtaining and ascertaining this classical reality by collapsing the wave function.

1.2 Entanglement, the measurement problem, and Schrödinger's much-debated cat

One major drawback of the double-slit experiment as an illustration of quantum mechanical behaviour is that it is concerned with a single quantum system, and so another of the fundamental quantum phenomena, *entanglement*, which can occur only in a composition of systems, is missing altogether. From a computer scientist's perspective, entanglement can be thought of as an `if` statement in superposition, and is at the heart of what is almost certainly the most widely discussed thought experiment in quantum mechanics, namely *Schrödinger's cat*.

Erwin Schrödinger proposed that a cat could be hypothetically placed in a box with a flask of deadly poison, which would then be broken (killing the cat) according to some microscopic process that is well-described by the laws of quantum mechanics. That is to say that the flask of poison would exist in a superposition of being broken and unbroken, in the same way that the photon of the double-slit experiment exists in a superposition of two states: one passing through each of the two slits. Notably, this particular flask of poison is not sitting safely in isolation, but has a rather nasty implication for the cat. If the set-up was instead to be described by a classical system, then we would have no issue in writing down the following 'algorithm':

Algorithm 1.1 Schrödinger's cat (classical)

Require: Flask of poison, Cat
1: **if** Flask of poison == broken **then**
2:     Cat ← dead
3: **else**
4:     Cat ← alive
5: **end if**

However, as the system *is* quantum, the flask of poison is not *either* broken or unbroken in the mutually exclusive classical sense required by the if statement, but rather is in a superposition of the two. As the states unbroken and alive can only exist together, and likewise the states broken and dead, the whole system is in a superposition of these two *pairs* of states. Observing (measuring) the system by opening the box and peering inside would then reveal either a broken flask along with a dead cat or an unbroken flask along with an alive cat. Never a dead cat and unbroken flask, or an alive cat and a broken flask. We say that the cats status as dead/alive has become *entangled* with the status of the flask of poison as broken/unbroken. In a format to be properly introduced in Chapter 3, a quantum description of Schrödinger's cat in Dirac notation would be:

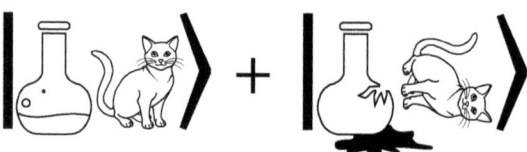

Figure 1.4

Whilst Schrödinger's cat is useful for introducing entanglement, Schrödinger's original motivation for proposing the thought experiment was as a *reductio ad absurdum* to refute the claim that the process of measurement ordains objective ('classical') reality on physical systems: how could a cat be in a superposition of dead and alive until a box is opened? How can conscious observation induce the collapse of a quantum state (such as when the box is opened)? With the debate still raging 90 years later, it is fair to say that Schrödinger did not conclusively shut down the discussion with one absurd example as he had hoped.

In the intervening time, however, the overall ontological deficiency in adequately describing the process of quantum measurement has, at least, been crystallised as the *measurement problem*. In a general sense, the measurement problem comes about because of the lack of clear distinction between a quantum mechanical system that can exist in a superposition of states and must be *measured* in order for the superposition to collapse, and the measurement apparatus itself. Particle physics dictates that even large objects, such as measurement apparatus, are composed of microscopic systems (such as atoms and molecules), which are well-modelled by quantum rather than classical mechanics. Thus, one must ask, at what scale does an object become classical – such that it may take measurements of microscopic *quantum* systems?

One radical *interpretation* of quantum mechanics – the *many-worlds interpretation* – holds (roughly speaking) that there is no such scale, and that our entire universe is really just one of a superposition of states of a quantum multiverse. Whilst the reader is in no way discouraged from deeper contemplation of the philosophical implications of quantum mechanics, for the purposes of studying quantum computation, it suffices to take the pragmatic view that the distinction between quantum systems and measurement apparatus is always clear in the operational settings being studied (a disposition which is largely consistent with the *Copenhagen interpretation*). This is appropriate for the practitioner, as real quantum hardware prepares extremely fragile quantum states that exist only for fleeting amounts of time in laboratory conditions kept extremely close to absolute zero: everything else is classical to all intents and purposes. For the theorist, the postulates of quantum mechanics define an abstract mathematical framework that is independent of the physics – so results are proven to be correct within this framework, and no special consideration of the interpretation of quantum mechanics is needed at all.

Further reading

David Deutsch's quote is from his book *the fabric of reality* [1997], Rolf Landauer's from his eponymous article [1991], whilst the wide-reaching work of Richard Feynman in conceptualising quantum computing has been celebrated by John Preskill [2023]. The double-slit experiment was first proposed by Thomas Young [1804], long before its relationship to quantum physics was discovered; the Stern–Gerlach experiment is another nice way to introduce the essential components of quantum mechanics [1922]. Schrödinger's cat is due to Erwin Schrödinger [1935], and the many-worlds interpretation is due to Hugh Everett [1957]. (By contrast, the Copenhagen interpretation was the accumulation of many conversations and works between the core group of scientists – Niels Bohr, Werner Heisenberg, Max Born, and others – working on the foundations of quantum mechanics in Copenhagen in the 1920s.)

2
Linear algebra

Quantum mechanics is mathematically formulated in terms of a *Hilbert space*, which is a vector space with a defined inner product. The inner product induces a distance function, such that the Hilbert space is a complete metric space, and for the purposes of this book, we restrict our attention to finite-dimensional Hilbert spaces. Linear algebra is the study of vector spaces and the (linear) operations thereon, and so it follows that having a solid grasp of the basics of linear algebra is essential for studying quantum mechanics and, therefore, quantum computation.

2.1 Complex numbers

In general, complex numbers are required to describe quantum states and operations. The set of complex numbers is denoted $\mathbb{C}$, and any $z \in \mathbb{C}$ is of the form $z = a + ib$ for some real numbers a and b, and $i = \sqrt{-1}$. (Note that throughout the book, 'i' is reserved to denote the square root of -1, to keep this distinct from 'i', which is used in various ways, for instance as a summation index.) The complex numbers form a field, with the familiar addition and multiplication operations, along with subtraction and division (respectively) as inverse operations, and $+0$ and $\times 1$ as the identity operations. Additionally, it is useful to define a few further terms and explicitly give some important general properties of complex numbers:

- Each complex number has a *conjugate*, $z^* = a - ib$.
- The modulus of a complex number is given by $|z| = \sqrt{a^2 + b^2} = \sqrt{zz^*}$.
- For two complex numbers, z_1 and z_2, $|z_1 z_2| = |z_1||z_2|$.
- Unit complex numbers lie on the unit circle of the *Argand diagram* (in which complex numbers are plotted on two-dimensional axes with one coordinate representing the real component and the other coordinate the imaginary component) and can be written in the form $e^{i\theta}$. θ is periodic with period 2π in the sense that $e^{i(\theta + 2n\pi)} = e^{i\theta}$ for any integer n. The Argand diagram, with unit circle and key coordinates included, is shown in Fig. 2.1.

10   Linear algebra

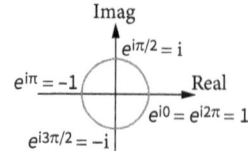

Figure 2.1

2.2 Vector spaces and linear maps

The vector space of all n-tuples of complex numbers is denoted $\mathbb{C}^n$. The elements of a vector space are *vectors* and are often represented in column vector form:

$$\begin{bmatrix} u_1 \\ u_2 \\ \vdots \\ u_n \end{bmatrix} \qquad (2.1)$$

where $u_i \in \mathbb{C}$. A vector of the form $\begin{bmatrix} u_1 & u_2 & \ldots & u_n \end{bmatrix}$ is called a *row vector*.

Two vectors in $\mathbb{C}^n$ can be added by performing 'element-wise' addition, and it is also possible to multiply vectors by a (in general complex) scalar w,

$$\begin{bmatrix} u_1 \\ u_2 \\ \vdots \\ u_n \end{bmatrix} + \begin{bmatrix} v_1 \\ v_2 \\ \vdots \\ v_n \end{bmatrix} = \begin{bmatrix} u_1 + v_1 \\ u_2 + v_2 \\ \vdots \\ u_n + v_n \end{bmatrix} ; \quad w \begin{bmatrix} u_1 \\ u_2 \\ \vdots \\ u_n \end{bmatrix} = \begin{bmatrix} wu_1 \\ wu_2 \\ \vdots \\ wu_n \end{bmatrix} \qquad (2.2)$$

In quantum information and computation, it is usual to express vectors using 'bra-ket' notation (also known as *Dirac* notation, after its inventor, Paul Dirac), which is the convention adopted throughout this book. A 'ket' is a column vector:

$$|u\rangle = \begin{bmatrix} u_1 \\ u_2 \\ \vdots \\ u_n \end{bmatrix} \qquad (2.3)$$

Each 'ket' has a corresponding 'bra', which is its conjugate transpose (see Section 2.4.1), the row vector:

$$\langle u| = \begin{bmatrix} u_1^* & u_2^* & \ldots & u_n^* \end{bmatrix} \qquad (2.4)$$

Linear algebra is the study of *linear* maps between vector spaces; in particular, a linear operator (the terms 'linear operator' and 'linear map' are essentially synonymous for the purposes of this book) between vector spaces U and V is a function

$A: U \to V$ such that:

$$A\left(\sum_i a_i |u_i\rangle\right) = \sum_i a_i A(|u_i\rangle) \tag{2.5}$$

where a_i are scalars, and $|u_i\rangle \in U$, i.e., A is linear in its inputs.

The map may be from a vector space to the same (sized) vector space (i.e., $V = U$), in which case an important operation is the *identity*, I, defined such that $I|u\rangle = |u\rangle$ for all $|u\rangle \in U$. Sometimes the identity will be appended with a subscript indicating the size of the space to which it applies; however, when it is clear and unambiguous from the context, the subscript may be omitted.

Throughout this book, linear operators are invariably defined and analysed in terms of their *matrix representations*. A matrix is an array of (in general) complex numbers:

$$A = \begin{bmatrix} a_{11} & \cdots & a_{1m} \\ \vdots & \ddots & \\ a_{n1} & & a_{nm} \end{bmatrix} \tag{2.6}$$

The elements $a_{11}, a_{22}, \ldots, a_{nn}$ of a matrix are known as its *leading diagonal*.

The addition of two matrices of the same size is defined as the element-wise addition:

$$\begin{bmatrix} a_{11} & \cdots & a_{1m} \\ \vdots & \ddots & \\ a_{n1} & & a_{nm} \end{bmatrix} + \begin{bmatrix} b_{11} & \cdots & b_{1m} \\ \vdots & \ddots & \\ b_{n1} & & b_{nm} \end{bmatrix} = \begin{bmatrix} a_{11}+b_{11} & \cdots & a_{1m}+b_{1m} \\ \vdots & \ddots & \\ a_{n1}+b_{n1} & & a_{nm}+b_{nm} \end{bmatrix} \tag{2.7}$$

Matrices may also be multiplied by a scalar by multiplying each element accordingly:

$$w \begin{bmatrix} a_{11} & \cdots & a_{1m} \\ \vdots & \ddots & \\ a_{n1} & & a_{nm} \end{bmatrix} = \begin{bmatrix} a_{11} & \cdots & a_{1m} \\ \vdots & \ddots & \\ a_{n1} & & a_{nm} \end{bmatrix} w = \begin{bmatrix} wa_{11} & \cdots & wa_{1m} \\ \vdots & \ddots & \\ wa_{n1} & & wa_{nm} \end{bmatrix} \tag{2.8}$$

2.2.1 Matrix multiplication

If A is an $n \times m$ matrix and B is an $m \times l$ matrix, then $C = A \times B$ is the $n \times l$ matrix with entries given by

$$c_{ik} = \sum_{j=1}^{m} a_{ij} b_{jk} \tag{2.9}$$

for all $i = 1, \ldots, n$ and $k = 1, \ldots, l$. This is known as *matrix multiplication* and has the following properties:

- It is associative: $(A \times B) \times C = A \times (B \times C) = ABC$.

- It is distributive: $A(B + C) = AB + AC$; $(A + B)C = AC + BC$.
- It is not (in general) commutative: $AB \neq BA$. Indeed, BA will not even be mathematically meaningful unless $n = l$, and the sizes of the matrices AB and BA will not match (and so they cannot be equal) unless both matrices are *square* and the same size, i.e., $n = m = l$. For special cases where $AB = BA$, the matrices A and B are said to *commute*.

2.2.2 Tensor multiplication

As well as scalar multiplication and matrix multiplication, to describe large quantum states (as are of interest in quantum computation) and the operations thereon, a third form of multiplication of matrices, namely *tensor multiplication*, is also required. Let A and B be matrices of *any dimension*:

$$A \otimes B = \begin{bmatrix} a_{11}B & \cdots & a_{1m}B \\ \vdots & \ddots & \\ a_{n1}B & & a_{nm}B \end{bmatrix} \qquad (2.10)$$

where $\otimes$ denotes the tensor product. For example:

$$\begin{bmatrix} 1 & 0 \\ 0 & 2 \end{bmatrix} \otimes \begin{bmatrix} 1 & 2 & 3 \end{bmatrix} = \begin{bmatrix} 1 & 2 & 3 & 0 & 0 & 0 \\ 0 & 0 & 0 & 2 & 4 & 6 \end{bmatrix} \qquad (2.11)$$

In general, if A is $n \times m$ and B is $n' \times m'$, then $A \otimes B$ is $nn' \times mm'$. The tensor product is associative, so $A \otimes (B \otimes C) = (A \otimes B) \otimes C$.

As an n-element column vector is just an $n \times 1$ matrix, the tensor product applies also to column vectors (and similarly for row vectors). An important property of tensor products when combined with matrix products is as follows: Let A and B be $n \times m$ and $n' \times m'$ matrices, respectively, and $|u\rangle$ and $|v\rangle$ be m and m' dimensional column vectors, respectively:

$$(A \otimes B)(|u\rangle \otimes |v\rangle) = (A|u\rangle) \otimes (B|v\rangle) \qquad (2.12)$$

2.3 Vector (quantum state) properties

In quantum mechanics, complex unit vectors represent quantum states, and so the terms 'vector' and '(quantum) state' are used somewhat interchangeably, depending on the context. It is common (and unambiguous) to drop the tensor product sign when taking the tensor product of vectors, such that the following are all equivalent: $|u\rangle \otimes |v\rangle, |u\rangle|v\rangle, |uv\rangle$ (and similarly for bras). The final form is most commonly used when $|u\rangle$ and $|v\rangle$ are *computational basis states*, as introduced in Section 2.3.3. This notation relies on the associativity of tensor multiplication, such that $(|u\rangle \otimes |v\rangle) \otimes |w\rangle = |u\rangle \otimes (|v\rangle \otimes |w\rangle) = |u\rangle|v\rangle|w\rangle = |uvw\rangle$ (again similarly for bras).

$|\psi\rangle$ is commonly used to denote a general quantum state, and it is worth noting that terms such as $|\psi_0\rangle, |\psi_1\rangle, |\psi_2\rangle, \ldots$ refer to *different* states: the indices do not refer to elements of the vector.

2.3.1 Inner products, orthogonality, and norms

Let $|u\rangle = \begin{bmatrix} u_1 \\ \vdots \\ u_n \end{bmatrix}$ and $|v\rangle = \begin{bmatrix} v_1 \\ \vdots \\ v_n \end{bmatrix}$ be two complex vectors of the same size, n; the *inner product* between $|u\rangle$ and $|v\rangle$ is a function that returns a complex scalar and is defined as follows:

$$\langle u|v\rangle = \langle u| \times |v\rangle = \begin{bmatrix} u_1^* & \cdots & u_n^* \end{bmatrix} \begin{bmatrix} v_1 \\ \vdots \\ v_n \end{bmatrix} = \sum_{i=1}^{n} u_i^* v_i \quad (2.13)$$

Note that the order of $|u\rangle$ and $|v\rangle$ is important, as $\langle u|v\rangle = (\langle v|u\rangle)^*$ and a complex number is only equal to its conjugate if its imaginary part is zero. Furthermore, when each of $|u\rangle$ and $|v\rangle$ have at least one non-zero element:

- If $\langle u|v\rangle = 0$, then $|u\rangle$ and $|v\rangle$ are *orthogonal*.
- $\langle u|u\rangle = \sum_{i=1}^{n} |u_i|^2$, which is a positive real number.

The inner product also provides a way to measure the size of a vector. There are various ways to do this, but the one that is usually relevant in quantum mechanics is the ℓ^2 norm (or simply 'norm'). The norm of $|u\rangle$ is defined as follows:

$$|||u\rangle|| = \sqrt{\langle u|u\rangle} \quad (2.14)$$

Unit vectors have norm = 1. In general, the modulus of the inner product between two unit vectors measures the *overlap* between them (i.e., how well-aligned they are): if they are orthogonal, the inner product is zero, as above, whereas if they are equal, it is one – and between these two extremes there is a continuum of values.

2.3.2 Outer products and projectors

As well as taking the inner product of two vectors, it is also possible to take their outer product, in which case they no longer need to have the same dimension. Continuing with n-element $|u\rangle$, but now letting $|v\rangle$ have some m elements, the outer product is defined as the $n \times m$ complex matrix, $|u\rangle\langle v|$, that is:

$$|u\rangle\langle v| = \begin{bmatrix} u_1 \\ \vdots \\ u_n \end{bmatrix} \begin{bmatrix} v_1^* & \cdots & v_m^* \end{bmatrix} = \begin{bmatrix} u_1 v_1^* & \cdots & u_1 v_m^* \\ \vdots & \ddots & \\ u_n v_1^* & & u_n v_m^* \end{bmatrix} \qquad (2.15)$$

If $|u\rangle$ is a unit vector, then $|u\rangle\langle u|$ is known as a *projector*, as $|u\rangle\langle u|$ is an operator that 'projects' an arbitrary vector (of appropriate dimension) $|v\rangle$ onto $|u\rangle$:

$$(|u\rangle\langle u|)|v\rangle = |u\rangle(\langle u||v\rangle) = (\langle u|v\rangle)|u\rangle \qquad (2.16)$$

The use of parentheses here is not strictly required due to the associativity of matrix multiplication, but has been included as $\langle u|v\rangle$, i.e., the inner product between $|u\rangle$ and $|v\rangle$ is a scalar that measures how well aligned they are, or equivalently the *component* of either in the direction of the other. We can thus see that this is a scalar that multiplies $|u\rangle$; hence, the whole thing has the interpretation of 'the amount of $|v\rangle$ in the direction of $|u\rangle$ times a unit vector in the latter direction', which can be illustrated as in Fig. 2.2.[1]

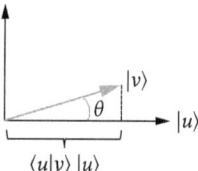

Figure 2.2

Two easy to obtain properties that hold for all projectors and are useful to know are:

- $(|u\rangle\langle u|)^2 = |u\rangle\langle u|u\rangle\langle u| = |u\rangle\langle u|$.
- $(|u\rangle\langle u|)^\dagger = \langle u|^\dagger |u\rangle^\dagger = |u\rangle\langle u|$, i.e., they are self-adjoint (Hermitian).

Indeed, these are often taken definitive of a projector.

2.3.3 Vector space bases

An important property of all vector spaces is the ability to represent any vector therein in terms of any *basis*. In particular, a basis of $\mathbb{C}^n$ is a minimal collection of vectors $|u_1\rangle, |u_2\rangle, \ldots, |u_n\rangle$ such that every vector $|u\rangle \in \mathbb{C}^n$ can be expressed as a linear

---

[1] Note that Fig. 2.2 is something of a simplification, as in general $\langle u|v\rangle$ is a complex value.

combination of these:

$$|u\rangle = a_1|u_1\rangle + a_2|u_2\rangle + \cdots + a_n|u_n\rangle \tag{2.17}$$

where the coefficients $a_i \in \mathbb{C}$. That the basis is a *minimal* collection of vectors means that $|u_1\rangle, |u_2\rangle, \ldots, |u_n\rangle$ are linearly independent: no $|u_i\rangle$ can be expressed as a linear combination of the others. The size of the basis is termed its *dimension*, and it is easy to see that it must always be equal to n. Of particular interest are *orthonormal bases*, in which each basis vector is a unit vector, and the basis vectors are pairwise orthogonal, that is:

$$\langle u_i | u_j \rangle = \begin{cases} 1 & \text{if } i = j \\ 0 & \text{otherwise} \end{cases} \tag{2.18}$$

From this definition, it is immediately apparent that if one or more of the basis vectors of an orthonormal basis are multiplied by (in general different) unit complex numbers, the vectors still constitute an orthonormal basis (also termed an orthonormal *set*).

There is an infinite number of bases for any vector space, and some example bases for $\mathbb{C}^3$ are:

$$\begin{bmatrix} 1 \\ 2 \\ 1 \end{bmatrix}, \begin{bmatrix} 10 \\ 2+i \\ 0 \end{bmatrix}, \begin{bmatrix} 1 \\ 0 \\ 0 \end{bmatrix} \tag{2.19}$$

$$\begin{bmatrix} 0 \\ 1/\sqrt{2} \\ 1/\sqrt{2} \end{bmatrix}, \begin{bmatrix} 0 \\ 1/\sqrt{2} \\ -1/\sqrt{2} \end{bmatrix}, \begin{bmatrix} 1 \\ 0 \\ 0 \end{bmatrix} \tag{2.20}$$

$$\begin{bmatrix} 1 \\ 0 \\ 0 \end{bmatrix}, \begin{bmatrix} 0 \\ 1 \\ 0 \end{bmatrix}, \begin{bmatrix} 0 \\ 0 \\ 1 \end{bmatrix} \tag{2.21}$$

The latter two of these are orthonormal, of which the final one is known as the *computational* basis. The computational basis is particularly important and is often denoted as:

$$|0\rangle = \begin{bmatrix} 1 \\ 0 \\ \vdots \\ 0 \end{bmatrix}, |1\rangle = \begin{bmatrix} 0 \\ 1 \\ \vdots \\ 0 \end{bmatrix}, \ldots, |n-1\rangle = \begin{bmatrix} 0 \\ 0 \\ \vdots \\ 1 \end{bmatrix} \tag{2.22}$$

In the case where n is a power of 2, this is not merely a convenient notation, but actually derives from the previous definition of the tensor product applied to the computational basis states of $\mathbb{C}^2$, using the standard conversion between binary and the positive integers. To see this, consider the first 2^2 computational basis states, first written in binary and then expanded as tensor products according to the rules introduced above:

$$|0\rangle = |00\rangle = |0\rangle \otimes |0\rangle = \begin{bmatrix} 1 \\ 0 \end{bmatrix} \otimes \begin{bmatrix} 1 \\ 0 \end{bmatrix} = \begin{bmatrix} 1 \\ 0 \\ 0 \\ 0 \end{bmatrix} \qquad (2.23)$$

$$|1\rangle = |01\rangle = |0\rangle \otimes |1\rangle = \begin{bmatrix} 1 \\ 0 \end{bmatrix} \otimes \begin{bmatrix} 0 \\ 1 \end{bmatrix} = \begin{bmatrix} 0 \\ 1 \\ 0 \\ 0 \end{bmatrix} \qquad (2.24)$$

$$|2\rangle = |10\rangle = |1\rangle \otimes |0\rangle = \begin{bmatrix} 0 \\ 1 \end{bmatrix} \otimes \begin{bmatrix} 1 \\ 0 \end{bmatrix} = \begin{bmatrix} 0 \\ 0 \\ 1 \\ 0 \end{bmatrix} \qquad (2.25)$$

$$|3\rangle = |11\rangle = |1\rangle \otimes |1\rangle = \begin{bmatrix} 0 \\ 1 \end{bmatrix} \otimes \begin{bmatrix} 0 \\ 1 \end{bmatrix} = \begin{bmatrix} 0 \\ 0 \\ 0 \\ 1 \end{bmatrix} \qquad (2.26)$$

When treating qubits as registers in a computer scientific sense, it is particularly helpful to remember the following: For $\mathbb{C}^{2^n}$, the computational basis is such that:

- When expressed as a ket, the number inside the ket is an n-bit binary number. Let this number be i.
- When written out as a column vector, it has 2^n elements, and each element is equal to 0 except the ith element, which is equal to 1 (where the elements are indexed from 0 to $2^n - 1$).

The correspondence between the binary number and the location of the non-zero element relies on numbering of elements from 0 to $2^n - 1$, which contravenes the earlier definition of vector element numbering commencing at one and is, therefore, slightly unsatisfactory. To some extent, this is unavoidable; however, in certain contexts, it is useful to use the computational basis, but numbered from $|1\rangle$ to $|2^n\rangle$, and this we shall term the *standard basis* (which, therefore, differs from the computational basis just by the way it is numbered, not by the basis vectors themselves – and is appropriate as the terms 'computational basis' and 'standard basis' are usually taken as synonymous). The use of the symbol 'n' as the length of the binary string within the ket (when a computational basis state) is ubiquitous throughout the book, and as such capital '$N = 2^n$' is usually used as the dimension of the corresponding vector.

When referring to computational basis states, $|0^n\rangle = |0\rangle^{\otimes n}$ refers to the state with the n bit string $00\ldots 0$ inside the ket (and similarly for $|1^n\rangle = |1\rangle^{\otimes n}$).

2.3.4 Subspaces

Quantum algorithms are sometimes analysed in terms of their action not on the entire vector space, but rather a *subspace* thereof. To give some visual intuition for what a subspace is, consider a three-dimensional set of Cartesian coordinates; a two-dimensional subspace is then any plane. It follows that any pair of non-parallel vectors therein suffice to define the subspace, and these vectors are said to *span* the subspace. This principle generalises to any dimension: a subspace is not defined by any *particular* set of vectors, but rather *any* set of vectors that span the subspace. Usually, in quantum computation, spanning vectors are given that are mutually orthogonal.

2.4 Matrix properties

There are a few important things to know about the matrices that occur in the topics studied in the remainder of the book. This section introduces the most important of these matrix properties and definitions.

2.4.1 Transpose and conjugate transpose (adjoint)

Every matrix has a number of important closely-related matrices. Let A be the $n \times m$ matrix:

$$A = \begin{bmatrix} a_{11} & \cdots & a_{1m} \\ \vdots & \ddots & \\ a_{n1} & & a_{nm} \end{bmatrix} \quad (2.27)$$

A can be 'transposed' by swapping its rows and columns, from which the $m \times n$ matrix A^T (read 'A transpose') is obtained:

$$A^T = \begin{bmatrix} a_{11} & \cdots & a_{n1} \\ \vdots & \ddots & \\ a_{1m} & & a_{nm} \end{bmatrix} \quad (2.28)$$

Combining complex conjugation with the transpose gives the *conjugate transpose* or *adjoint* of a matrix:

$$A^\dagger = (A^*)^T = (A^T)^* = \begin{bmatrix} a_{11}^* & \cdots & a_{n1}^* \\ \vdots & \ddots & \\ a_{1m}^* & & a_{nm}^* \end{bmatrix} \qquad (2.29)$$

which is particularly important in quantum mechanics. When manipulating matrix equations, it is useful to note that $(AB)^T = B^T A^T$ and $(AB)^\dagger = B^\dagger A^\dagger$.

2.4.2 Hermitian and unitary matrices

The definition of the adjoint of a square matrix enables two of the most important classes of matrices in quantum mechanics to be introduced. These classes of matrices are both *normal*, which means that they have the property $A^\dagger A = A A^\dagger$,

- A matrix, H, is *Hermitian* if $H = H^\dagger$.
    - The eigenvalues (defined in Section 2.4.5) of Hermitian matrices are all real.
- A matrix, U, is *unitary* if $U^\dagger U = UU^\dagger = I$.
    - By definition, $AA^{-1} = I$, where A^{-1} is the inverse of A, and hence, unitary matrices have the property that their adjoint is equal to their inverse (which is another unitary matrix).
    - The eigenvalues of unitary matrices have modulus one.
    - A matrix is unitary if and only if its columns form an orthonormal basis (also the case for the rows).
    - Unitary operators preserve inner products: if U is unitary, $|u'\rangle = U|u\rangle$ and $|v'\rangle = U|v\rangle$, then:

$$\begin{aligned} \langle u'|v'\rangle &= (U|u\rangle)^\dagger (U|v\rangle) \\ &= (\langle u|U^\dagger)(U|v\rangle) \\ &= \langle u|(U^\dagger U)|v\rangle \\ &= \langle u|I|v\rangle \\ &= \langle u|v\rangle \end{aligned} \qquad (2.30)$$

2.4.3 Sparsity and rank

Two properties of matrices, which are relevant when characterising the potential of certain quantum algorithms to speed-up various linear algebraic tasks, are their rank and row-sparsity:

- The rank of a matrix is the dimension of the vector space spanned by its columns, which is equal to the maximum number of linearly independent columns (and also the number of linearly independent rows). A matrix that has rank equal to the minimum of its number of rows and column is called *full-rank* (that is, it has the greatest possible rank for any matrix of the same size).
- The row-sparsity of a matrix is the largest number of non-zero elements in any row. A matrix is called *s*-sparse if it has at most *s* non-zero elements in any row.

As well as row-sparsity, there is a notion of overall sparsity, which is defined as the number of non-zero elements in the whole matrix (or the ratio of this to the total number of elements in the matrix), and as such it may appear that rank and sparsity are related. Certainly if a particular matrix is filled with zeros to such an extent that many entire columns are all-zero, then it will be both sparse and low-rank. However, beyond this slightly extreme example, sparsity (and in particular row-sparsity) and rank are not particularly connected. For example, the identity (which has the familiar matrix representation of the elements of the leading diagonal all being equal to one, and all other elements being zero) is full-rank; however, it is 1-sparse.

Rank and sparsity differ in another important way: rank is an invariant property of a matrix if it is represented in a different basis, that is, the rank is the same for all *similar* matrices (see Section 2.4.5), whereas sparsity is not.

2.4.4 Trace and determinant

Two additional important properties of square matrices are their trace and determinant:

- The trace of a square matrix is defined as the sum of the elements of its leading diagonal.
- The determinant of a square matrix is a scalar value that is a function of the whole matrix, denoted as $\det(\cdot)$. The calculation of the determinant varies by matrix size, but in the case of a 2×2 matrix $A = \begin{bmatrix} a_{11} & a_{12} \\ a_{21} & a_{22} \end{bmatrix}$, $\det(A) = a_{11}a_{22} - a_{12}a_{21}$.

2.4.5 Eigenvalues, eigenvectors, and the spectral decomposition

Any vector $|u\rangle = \begin{bmatrix} u_1 & u_2 & \cdots & u_n \end{bmatrix}^T$ can be expressed as a weighted sum of standard basis vectors:

$$|u\rangle = a_1|1\rangle + a_2|2\rangle + \cdots + a_n|n\rangle \qquad (2.31)$$

Similarly, any matrix can be expressed as a double sum over the outer products of standard basis vectors:

$$\begin{bmatrix} a_{11} & \cdots & a_{1m} \\ \vdots & \ddots & \\ a_{n1} & & a_{nm} \end{bmatrix} = \sum_{i=1}^{n} \sum_{j=1}^{m} a_{ij} |i\rangle \langle j| \quad (2.32)$$

where each $|i\rangle\langle j|$ is a matrix with all zeros except the (i,j)th element, which is equal to 1, and hence, this simple decomposition can be thought of as picking out each element of A in turn and summing up. Two more sophisticated matrix decompositions are important for quantum mechanics and computation. The first of these is the *spectral decomposition*, which represents certain matrices in terms of their *eigenvalues* and *eigenvectors*; and the second is the *singular value decomposition* (SVD), given in Section 2.4.6.

If an $n \times n$ matrix, A, has the effect of scaling a given (non-zero) vector, $|u\rangle$ by a constant, λ, then that vector is known as an *eigenvector*, with corresponding *eigenvalue* λ:

$$A|u\rangle = \lambda|u\rangle \quad (2.33)$$

The eigenvalues of a matrix are the roots of the *characteristic polynomial*, defined by:

$$\det(A - \lambda I) = 0 \quad (2.34)$$

which can be solved for λ, and substituted into Eqn. (2.33) to obtain the eigenvectors. Each $n \times n$ square matrix has at least one and at most n eigenvalues (i.e., there are at most n unique solutions to Eqn. (2.34)). In the case where distinct eigenvectors have the same eigenvalue (sometimes termed a repeated eigenvalue), then this is known as a degeneracy, and there is no unique set of eigenvector directions. For example, if the vectors $|u_1\rangle$ and $|u_2\rangle$ each have the same eigenvalue, then any linear combination of $|u_1\rangle$ and $|u_2\rangle$ will also be an eigenvector with this eigenvalue.

The set of eigenvalues can also be used to compute the determinant, trace, and rank:

- The determinant of a matrix is the product of its eigenvalues.
- The trace of a matrix is the sum of its eigenvalues.
- The rank of a matrix is the number of non-zero eigenvalues. (For the matrices studied in this book.)

In the case where the eigenvectors of an $n \times n$ complex matrix, A, form an orthonormal set (or can be chosen to form an orthonormal set, in the case of degeneracy), A can be expressed in the following form:

$$A = \sum_{i=1}^{n} \lambda_i |\lambda_i\rangle \langle\lambda_i| \qquad (2.35)$$

where λ_i is the ith eigenvalue of A, corresponding to the ith eigenvector, $|\lambda_i\rangle$. The right-hand side of Eqn. (2.35) is called as the spectral decomposition of A, and matrices that can be decomposed in this manner are said to be diagonalisable. Matrices are diagonalisable if and only if they are normal. (Note that this definition of diagonalisability is standard in quantum mechanics, but is somewhat stronger than elsewhere in mathematics, where a matrix is often described as diagonalisable if its eigenvectors form *any* basis – not necessarily an *orthonormal* basis as here.)

Another (equivalent) form of the spectral decomposition can be obtained by defining a *diagonal matrix*, Λ (that is, a matrix with non-zero elements only on the leading diagonal) containing the eigenvalues

$$\Lambda = \begin{bmatrix} \lambda_1 & 0 & \cdots & 0 \\ 0 & \lambda_2 & \cdots & 0 \\ \vdots & \vdots & \ddots & \\ 0 & 0 & & \lambda_n \end{bmatrix} \qquad (2.36)$$

From this, it is easy to see that $\lambda_i = \langle i|\Lambda|i\rangle$, which can be substituted into Eqn. (2.35) to obtain the following equivalent form:

$$\begin{aligned} A &= \sum_{i=1}^{n} \lambda_i |\lambda_i\rangle \langle\lambda_i| \\ &= \sum_{i=1}^{n} |\lambda_i\rangle \lambda_i \langle\lambda_i| \\ &= \sum_{i=1}^{n} |\lambda_i\rangle \langle i|\Lambda|i\rangle \langle\lambda_i| \\ &= \sum_{i=1}^{n} |\lambda_i\rangle \langle i|\Lambda|i\rangle \langle\lambda_i| + \underbrace{\sum_{i}\sum_{i\neq j} |\lambda_i\rangle \langle i|\Lambda|j\rangle}_{=0} \langle\lambda_j| \\ &= \sum_{i=1}^{n}\sum_{j=1}^{n} |\lambda_i\rangle \langle i|\Lambda|j\rangle \langle\lambda_j| \\ &= \left(\sum_{i=1}^{n} |\lambda_i\rangle \langle i|\right) \Lambda \left(\sum_{j=1}^{n} |j\rangle \langle\lambda_j|\right) \\ &= U\Lambda U^{\dagger} \end{aligned} \qquad (2.37)$$

where $U = \sum_{i=1}^{n} |\lambda_i\rangle \langle i|$, is the matrix such that its columns are $|\lambda_i\rangle$ and, hence, form an orthonormal set – meaning that by the definition in Section 2.4.2, it is unitary.

The spectral decomposition also prompts the notion of matrix similarity. Two matrices A and B are said to be *similar* if

$$A = QBQ^{-1} \qquad (2.38)$$

for some *change of basis matrix* Q and its inverse Q^{-1}. The *spectral theorem* states that every normal matrix is *unitarily* similar (i.e., in Eqn. (2.38), Q is unitary) to a diagonal matrix – a result that is implicit in the above analysis using the fact that for unitary matrices, $U^{-1} = U^\dagger$. Certain matrix properties are necessarily equal for all similar matrices: Section 2.4.3 highlights that this is the case for the matrix rank, and the determinant and trace are also properties that are invariant for similar matrices.

A final important implication of the spectral decomposition is that every orthonormal set, $\{|u_i\rangle\}$, has the following property:

$$\sum_i |u_i\rangle\langle u_i| = I \qquad (2.39)$$

This is because the $n \times n$ identity is a full-rank matrix with all n eigenvalues equal to 1. Therefore, it has degeneracy equal to n, and so the entire space can be thought of as a degenerate subspace. It follows that any orthonormal set can be chosen as the basis and the spectral theorem (2.35) with $\lambda_i = 1$ for all i directly gives Eqn. (2.39).

2.4.6 The singular value decomposition

Not every matrix is square, and not every square matrix has the property that its eigenvectors form an orthonormal set. So it follows that the spectral decomposition only applies to a relatively restricted set of matrices – albeit an extremely important set in quantum mechanics. However, matrix decompositions are important in general, and the most relevant matrix decomposition (for the purposes of studying quantum algorithms) that applies to all matrices is the SVD. Let A be any $m \times n$ matrix; then the SVD of A is:

$$A = \sum_i \sigma_i |u_i\rangle\langle v_i| = U\Sigma V^\dagger \qquad (2.40)$$

where U is an $m \times m$ unitary matrix with columns $|u_i\rangle$; Σ is an $m \times n$ diagonal matrix, with min(m, n) non-negative real values, σ_i, known as the *singular values*, on the leading diagonal; and V is an $n \times n$ unitary matrix with columns $|v_i\rangle$. The sets of vectors $\{|u_i\rangle\}$ and $\{|v_i\rangle\}$ are sometimes termed the left and right singular basis vectors, respectively.

In the case of square matrices that are diagonalisable, the SVD is closely-related, but not necessarily identical to the spectral decomposition. This is because the stipulation that the singular value must be real and non-negative means that any such non-negativity and complexity must be 'absorbed' into one of the left or right basis vectors.

2.4.7 Parameterised matrices and matrix operations

The matrices introduced thus far have either been arrays of numerical values or symbolic terms standing for general numerical values. It is, however, the case that some interesting classes of matrices are defined in terms of parameterised values in the arrays. The most obvious example is that of rotation matrices, such as the R_y operation (introduced in full in Chapter 4), which is defined in a manner that is parameterised by the rotation angle, θ:

$$R_y(\theta) = \begin{bmatrix} \cos\frac{\theta}{2} & -\sin\frac{\theta}{2} \\ \sin\frac{\theta}{2} & \cos\frac{\theta}{2} \end{bmatrix} \quad (2.41)$$

Matrices where certain elements are functions of parameters are not to be confused with matrices obtained by taking the function of some other matrix. In general, taking a function of matrix – where that function is one whose familiar use is as applied to a scalar – may be thought of in various ways. Most formally, if a smooth analytic function is applied to a square matrix, then the Taylor series of the function may be taken as its definition, with matrix multiplication taking the place of multiplication in the same Taylor series for the function applied to a scalar. Let some smooth function, $f : \mathbb{C} \to \mathbb{C}$, have Taylor series $f(x) = \sum_{j=0}^{\infty} a_j x^j$, in which case the corresponding matrix Taylor series of the function applied to the matrix A is:

$$f(A) = \sum_{j=0}^{\infty} a_j A^j \quad (2.42)$$

where A^0 is defined as the identity.

In the case of diagonalisable matrices, it is easy to show that this is equivalent to applying the same function to the eigenvalues, when expressed using the spectral decomposition:

$$f(A) = \sum_{j=0}^{\infty} a_j A^j = \sum_{j=0}^{\infty} a_j \left(\sum_i \lambda_i |\lambda_i\rangle \langle \lambda_i| \right)^j = \sum_{j=0}^{\infty} \sum_i a_j \lambda_i^j |\lambda_i\rangle \langle \lambda_i| = \sum_i f(\lambda_i) |\lambda_i\rangle \langle \lambda_i| \quad (2.43)$$

which follows because, by their orthonormality, terms $\langle \lambda_i | \lambda_{i'} \rangle$ are equal to 1 when $i = i'$ and 0 otherwise.

This, in turn, reveals a connection between unitary and Hermitian matrices, which is important in quantum mechanics. Using the fact that Hermitian matrices are self-adjoint and so normal, and therefore always have a spectral decomposition, let H be any (finite) Hermitian matrix, with spectral decomposition $H = \sum_i E_i |\lambda_i\rangle \langle \lambda_i|$; by its Hermiticity, its eigenvalues are all real. Consider now taking the exponential of H multiplied by the square root of -1,

$$e^{iH} = \sum_i e^{iE_i} |\lambda_i\rangle \langle \lambda_i| \quad (2.44)$$

which is therefore a diagonalisable matrix with the same eigenvectors and eigenvalues e^{iE_i}; however, as E_i are all real, these eigenvalues are all complex numbers with absolute value 1, and so, by definition, this is a unitary matrix. (Note that here $\{E_i\}$ rather than $\{\lambda_i\}$ has been used for the set of eigenvalues: in quantum mechanics, Hamiltonian matrices – which are representations of the energy of quantum systems – are Hermitian, and the eigenvalues correspond to *energy levels*. For this reason, the eigenvalues of Hermitian matrices are usually denoted as E_i in the remainder of the book.)

2.4.8 Matrix inversion and pseudo-inversion

If a square matrix, A, is such that another matrix A^{-1} has the property that $AA^{-1} = A^{-1}A = I$, then A is said to be invertible and A^{-1} is its inverse. This is taken as the definition of the inverse, and for diagonalisable matrices, it is also the case that the same matrix is obtained if $(.)^{-1}$ is treated as a scalar function applied to the eigenvalues in the manner described in Section 2.4.7. This clearly shows why a matrix must be full rank (have all non-zero eigenvalues) – and by extension have non-zero determinant (the determinant is the product of the eigenvalues, as stated in Section 2.4.5) – to be invertible.

For linear systems of equations with non-invertible matrices, that is, those of the form:

$$|u\rangle = A|v\rangle \qquad (2.45)$$

where A is non-invertible (it is not square and / or not full rank), it is still sometimes the case, in practice, that one wishes to find a matrix to apply to $|u\rangle$ to obtain some estimate of $|v\rangle$. The standard approach is to take the *matrix pseudo-inverse*, which can be defined in terms of the SVD of A. If A has SVD $A = \sum_i \sigma_i |u_i\rangle\langle v_i|$, then the pseudo-inverse, A^+, is given by:

$$A^+ = \sum_i f(\sigma_i)|v_i\rangle\langle u_i| \qquad (2.46)$$

where

$$f(x) = \begin{cases} \frac{1}{x} & \text{if } x \neq 0 \\ 0 & \text{otherwise} \end{cases} \qquad (2.47)$$

In the case of invertible matrices, the pseudo-inverse is equal to the inverse. (Note that this is true even though the SVD is not necessarily quite the same as the spectral decomposition for invertible matrices.)

2.4.9 Matrix (operator) norms

As is the case for vectors, it is sometimes useful to measure the 'size' of the matrix. There are various ways to do this, indeed, the determinant provides such a measure in one sense. There are myriad other definitions, of which the *operator norm* is the most widely used in quantum information and computation. The ℓ^2 operator norm of the matrix A is defined as follows:

$$||A|| = \sup_{|u\rangle} \left(\frac{||A|u\rangle||}{|||u\rangle||} \right) \tag{2.48}$$

where $|u\rangle \neq 0$, and sup is the *supremum*, which may be thought of as similar to the maximum. Therefore, the ℓ^2 operator norm is the maximum ℓ^2 (vector) norm that can be obtained when the operator is applied to a unit vector. In the same way that 'norm' is typically used in place of the full 'ℓ^2 norm' when referring to vectors, 'operator norm' without further qualification should be taken as meaning the ℓ^2 operator norm.

2.5 Notational miscellany

Throughout the book, standard notation has been used. Different sub-communities within quantum computing (theoretical computer scientists, quantum computational chemists, etc.) use slightly different notational conventions, which means that there are occasionally slight differences from chapter to chapter; however, each chapter is self-consistent and symbol re-use is avoided as far as possible. Where there is no risk of ambiguity, clarity and brevity have been preferred over strict formality; for example, sets of indexed elements are often referred to as (for example) $\{x_*\}$ rather than $\{x_i\}_{i \in \{0,1,\ldots\}}$, similarly, where the range of summations/products is clear, sometimes just the index is given (e.g., $\sum_i x_i$). In a similar vein, $a_{*,i}$ is the vector formed by taking the ith column of the matrix A (and similarly for matrix rows and row vectors).

Bit strings are treated as numbers, in particular unsigned binary numbers, and so are written in normal (rather than bold) font, although sometimes individual bits of the string are referenced by a subscript, as if they were vectors of the same length. $\oplus$ is used for bitwise modulo-2 addition between two bit strings of the same length.

Standard notation for probability is used throughout: $\Pr(\cdot)$ is probability; $\mathbb{E}(\cdot)$ is expectation; $\Pr(\cdot|\cdot)$ and $\mathbb{E}(\cdot|\cdot)$ denote conditional probability and conditional expectation, respectively; and $\binom{n}{k}$ is the combinatorial term 'n choose k'.

Chapter problems

1. If I is the two-dimensional identity matrix and $H = \frac{1}{\sqrt{2}} \begin{bmatrix} 1 & 1 \\ 1 & -1 \end{bmatrix}$ (i.e., the Hadamard gate), give matrix representations of the operators

(a) $I \otimes H$

(b) $H \otimes I$

2. Let A and B be 2×2 matrices, and $|\psi\rangle$ and $|\phi\rangle$ be single qubit states (i.e., 2×1 vectors). Show that:

$$(A \otimes B)(|\psi\rangle \otimes |\phi\rangle) = (A|\psi\rangle) \otimes (B|\phi\rangle) \tag{2.49}$$

3. Let $|v_1\rangle, ..., |v_n\rangle$ be a basis for a vector space V; also let $|u_1\rangle, ..., |u_{n+1}\rangle$ be any collection of $n + 1$ vectors in V. Show that $|u_1\rangle, ..., |u_{n+1}\rangle$ cannot all be linearly independent, i.e., one of them must be expressible as a linear combination of the others.

4. Find the eigenvalues and associated eigenvectors of

(a) $\frac{1}{\sqrt{2}} \begin{bmatrix} 1 & -1 \\ 1 & 1 \end{bmatrix}$

(b) $\begin{bmatrix} 0 & -i \\ i & 0 \end{bmatrix}$

5. Find the trace and determinant of $\begin{bmatrix} 1 & 2 \\ 3 & 4 \end{bmatrix}$. Confirm that the 'direct' calculation matches the values found by first calculating the eigenvalues.

6. Express each of the two linear operators

(a) $\frac{1}{\sqrt{2}} \begin{bmatrix} 1 & -1 \\ 1 & 1 \end{bmatrix}$

(b) $\begin{bmatrix} 0 & -i \\ i & 0 \end{bmatrix}$

as a linear combination of outer products of computational basis vectors.

7. Show that unitary operations are norm preserving. That is, if U is unitary, then the norm of $U|\psi\rangle$ equals the norm of $|\psi\rangle$, for all $|\psi\rangle$.

8. Let $A = \begin{bmatrix} 1 & 2 & 3 \\ 4 & 5 & 6 \\ 5 & 7 & 9 \end{bmatrix}$.

(a) The third row of A is the sum of the first and second. Does A have an inverse?

(b) A has singular value decomposition:

$$A = \begin{bmatrix} -0.2354 & 0.7818 & -0.5774 \\ -0.5594 & -0.5948 & -0.5774 \\ -0.7948 & 0.1870 & 0.5774 \end{bmatrix} \begin{bmatrix} 15.6633 & 0 & 0 \\ 0 & 0.8126 & 0 \\ 0 & 0 & 0 \end{bmatrix}$$

$$\begin{bmatrix} -0.4116 & -0.5638 & -0.7160 \\ -0.8148 & -0.1243 & 0.5662 \\ -0.4082 & 0.8165 & -0.4082 \end{bmatrix} \tag{2.50}$$

what is the pseudo-inverse of A?

Further reading

Bra-ket notation was invented by Dirac [1939], and the remainder of this chapter consists of long-known linear algebraic prerequisites. *The Matrix Cookbook* [2008] is a useful source of definitions and properties of a wide range of matrices.

3
Quantum mechanics and quantum information

As highlighted at the start of the book, the discovery of physical systems exhibiting quantum mechanical behaviour was unquestionably one of the most astounding events in the history of science. Phenomena like ostensibly mutually exclusive states co-existing in superposition, collapsing wave functions leading to effects that *in principle* cannot be predicted, and distant particles apparently being entangled rocked the very foundations of what was previously understood to be the fundamental nature of reality. Moreover, the discovery suggested information-theoretic capabilities of quantum mechanical systems, such as superluminal signalling, that had hitherto been considered impossible.

Unlike, for example, *Newtonian* (classical) mechanics, in which the laws dictate the precise behaviour of physical entities, quantum mechanics is not a physical theory *per se* but rather an abstract mathematical framework. This framework is defined by a set of rules – the *postulates of quantum mechanics* – and any physical system that adheres to these (or that can be well-modelled as adhering to these) *is*, therefore, a quantum mechanical system. From the postulates, algorithms and communication protocols can be derived, which show that quantum mechanics does indeed provide a very powerful paradigm for computation and information processing – but this power is somewhat subtle. For instance, on closer inspection, superluminal signalling is *not* possible in a quantum mechanical system.

The four postulates of quantum mechanics concern the following:

1. **State space**: how to describe a quantum state.
2. **Evolution**: how a quantum state changes with time.
3. **Measurement**: the effect on a quantum state of interaction with a classical system that extracts classical information therefrom.
4. **Composition**: How to compose multiple quantum systems.

Postulates 1, 2, and 4 describe the behaviour of a *closed* quantum system existing and evolving in isolation; conversely, postulate 3 details how a quantum system is affected when information is extracted by the outside 'classical' world. For this reason, it makes sense to define and discuss the postulates in these two groupings, rather than in numerical order.

3.1 State space, composition, and evolution

Postulate 1: State space. *Associated with any isolated physical system is a complex vector space with an inner product (that is, a Hilbert space) known as the state space of the system. The system is completely described by its state vector, which is a unit vector in the system's state space.*

This book considers only qubits – quantum states in the space $\mathbb{C}^2$ – and compositions thereof, although higher dimensional 'qudit' states are sometimes present in quantum computing literature.

Postulate 4: Composition. *The state space of a composite physical system is the tensor product of the state spaces of the component physical systems.*

That is, if the systems are numbered 1 through n, such that system number i is prepared in the state $|\psi_i\rangle$, then the joint state of the total system is $|\psi_1\rangle \otimes |\psi_2\rangle \otimes \cdots \otimes |\psi_n\rangle$.

Together, postulates 1 and 4 state that an n-qubit quantum state is represented by a 2^n-element complex unit vector.

Postulate 2: Evolution. *The time evolution of the state, $|\psi\rangle$, of a closed quantum system is described by the* Schrödinger equation:

$$i\hbar \frac{d|\psi\rangle}{dt} = H|\psi\rangle \tag{3.1}$$

where $\hbar$ is the physical constant, Planck's constant *(a real number), and H is a fixed Hermitian operator known as the* Hamiltonian *of the closed system.*

Physically, the Hamiltonian represents the energy of a system, and the eigenvectors (or eigen*states*) are the quantised 'energy levels' that the system can take. Moreover, the Hamiltonian is linear, and so if a composite physical system is described by multiple Hamiltonians corresponding to the interactions between various subsets of the component parts, then the overall Hamiltonian is the sum of these. (Note that some care is needed when using this linearity, as each of the component Hamiltonians should extend to the entire composite system, but with identity operators for the parts that are not interacting. This is made explicit in the example in Eqn. (11.3).)

The Schrödinger equation can be solved between some initial time t_0 and a final time t_1 to give

$$|\psi_{t_1}\rangle = \exp\left(\frac{-iH(t_1 - t_0)}{\hbar}\right)|\psi_{t_0}\rangle \tag{3.2}$$

where $U \equiv \exp\left(\frac{-iH(t_1 - t_0)}{\hbar}\right)$ is unitary (as shown in Section 2.4.7, when a Hermitian operator is multiplied by i and then exponentiated, the result is a unitary operator). For the majority of the algorithms and protocols studied in this book, this unitary evolution is all that matters, and so it is helpful to introduce a modified (simplified) form of postulate 2.

Postulate 2': Unitary evolution. *The change in the state of a closed quantum system from t_0 to t_1 is described by the unitary transformation:*

$$|\psi_{t_1}\rangle = U|\psi_{t_0}\rangle \qquad (3.3)$$

Notably, unitary operators are the unique linear maps that preserve the norm:

$$|||\psi_{t_1}\rangle|| = ||U|\psi_{t_0}\rangle|| = |||\psi_{t_0}\rangle|| = 1 \qquad (3.4)$$

which is important, as it means that every unitary operation maps an n-qubit state to another n-qubit state.

3.1.1 Important single-qubit unitary operations

Amongst the most important unitary operations in quantum mechanics are the X, Y, and Z Pauli operators, whose matrix forms are:

$$X = \begin{bmatrix} 0 & 1 \\ 1 & 0 \end{bmatrix} \qquad Y = i\begin{bmatrix} 0 & -1 \\ 1 & 0 \end{bmatrix} \qquad Z = \begin{bmatrix} 1 & 0 \\ 0 & -1 \end{bmatrix} \qquad (3.5)$$

and which have the following effect on the computational basis states:

$$X|0\rangle = \begin{bmatrix} 0 & 1 \\ 1 & 0 \end{bmatrix}\begin{bmatrix} 1 \\ 0 \end{bmatrix} = \begin{bmatrix} 0 \\ 1 \end{bmatrix} = |1\rangle \qquad X|1\rangle = \begin{bmatrix} 0 & 1 \\ 1 & 0 \end{bmatrix}\begin{bmatrix} 0 \\ 1 \end{bmatrix} = \begin{bmatrix} 1 \\ 0 \end{bmatrix} = |0\rangle \qquad (3.6)$$

$$Y|0\rangle = i\begin{bmatrix} 0 & -1 \\ 1 & 0 \end{bmatrix}\begin{bmatrix} 1 \\ 0 \end{bmatrix} = \begin{bmatrix} 0 \\ i \end{bmatrix} = i|1\rangle \qquad Y|1\rangle = i\begin{bmatrix} 0 & -1 \\ 1 & 0 \end{bmatrix}\begin{bmatrix} 0 \\ 1 \end{bmatrix} = \begin{bmatrix} -i \\ 0 \end{bmatrix} = -i|0\rangle \qquad (3.7)$$

$$Z|0\rangle = \begin{bmatrix} 1 & 0 \\ 0 & -1 \end{bmatrix}\begin{bmatrix} 1 \\ 0 \end{bmatrix} = \begin{bmatrix} 1 \\ 0 \end{bmatrix} = |0\rangle \qquad Z|1\rangle = \begin{bmatrix} 1 & 0 \\ 0 & -1 \end{bmatrix}\begin{bmatrix} 0 \\ 1 \end{bmatrix} = \begin{bmatrix} 0 \\ -1 \end{bmatrix} = -|1\rangle \qquad (3.8)$$

The Pauli matrices are of size 2×2 and, therefore, operate on a single qubit. Thus, they are termed 'single-qubit unitaries'. Another important single-qubit unitary is the *Hadamard* operation:

$$H = \frac{1}{\sqrt{2}}\begin{bmatrix} 1 & 1 \\ 1 & -1 \end{bmatrix} \qquad (3.9)$$

which has the following effect on the computational basis states:

$$H|0\rangle = \frac{1}{\sqrt{2}}(|0\rangle + |1\rangle) \qquad H|1\rangle = \frac{1}{\sqrt{2}}(|0\rangle - |1\rangle) \qquad (3.10)$$

That is, the Hadamard matrix puts the computational basis states in superposition. The prepared states are extremely common in quantum information, and so are given their own symbols:

$$|+\rangle = \frac{1}{\sqrt{2}}(|0\rangle + |1\rangle) \tag{3.11}$$

$$|-\rangle = \frac{1}{\sqrt{2}}(|0\rangle - |1\rangle) \tag{3.12}$$

As $|+\rangle$ and $|-\rangle$ are unit vectors and $\langle +|-\rangle = 0$, they form an orthonormal basis for a single qubit.

Another important property of H is that it is self-inverse, i.e., $HH = I$, therefore:

$$H|+\rangle = |0\rangle \qquad H|-\rangle = |1\rangle \tag{3.13}$$

This can be thought of as the Hadamard matrix also giving rise to interference; in the case of Eqn. (3.13), this interference applies to a superposition of computational basis states, such that one or other of the computational basis states is obtained.

3.1.2 Evolution of multi-qubit states

It is very important to appreciate that postulate 2 applies equally to composite systems as it does to single-qubit (or qudit) systems (as does postulate 2'). That is, a multi-qubit system may evolve according to any unitary of the correct size. This gives great physical significance to the equivalence (first introduced in Section 2.2.2, now re-expressed with the matrix operations as unitary operations):

$$(U_1 \otimes U_2)(|\psi_1\rangle \otimes |\psi_2\rangle) = (U_1|\psi_1\rangle) \otimes (U_2|\psi_2\rangle) \tag{3.14}$$

Physically, the right-hand side (RHS) may be thought of as two different quantum systems that evolve according to U_1 and U_2. The final state $(U_1|\psi_1\rangle) \otimes (U_2|\psi_2\rangle)$ is then treated as the state of a composite system. By contrast, the left-hand side (LHS) may be thought of as a single composite system undergoing the 'composite' unitary evolution $U_1 \otimes U_2$. Put simply, the RHS says *evolve first, compose second*, whereas the LHS says *compose first, evolve second*; however, this amounts to nothing more than a distinction in how the two systems are *described*, and not to any physical differentiation, and so any physically meaningful model must have these two as equivalent.

$|\psi\rangle = |\psi_1\rangle \otimes |\psi_2\rangle$ is what is known as a *separable* or *product* state. Let $|\psi'\rangle = (U_1 \otimes U_2)(|\psi_1\rangle \otimes |\psi_2\rangle)$, $|\psi'_1\rangle = U_1|\psi_1\rangle$ and $|\psi'_2\rangle = U_2|\psi_2\rangle$; then, according to Eqn. (3.14):

$$|\psi'\rangle = |\psi'_1\rangle \otimes |\psi'_2\rangle \tag{3.15}$$

which is an example of the important general property of the tensor product: a tensor product of operations applied to a separable state results in a separable state. However, not all unitary operations can be decomposed into a tensor product of single-qubit unitaries. Such operations may, even when applied to a product state, yield states that cannot be separated. Such states are known as *entangled* states.

3.1.3 Entangling operations and entangled states

Consider the two-qubit (4 × 4) unitary

$$\text{CNOT} = \begin{bmatrix} 1 & 0 & 0 & 0 \\ 0 & 1 & 0 & 0 \\ 0 & 0 & 0 & 1 \\ 0 & 0 & 1 & 0 \end{bmatrix} \tag{3.16}$$

applied to the state $|+\rangle \otimes |0\rangle = (1/\sqrt{2}) \begin{bmatrix} 1 & 1 \end{bmatrix}^T \otimes \begin{bmatrix} 1 & 0 \end{bmatrix}^T = \begin{bmatrix} 1/\sqrt{2} & 0 & 1/\sqrt{2} & 0 \end{bmatrix}^T$:

$$\text{CNOT}(|+\rangle \otimes |0\rangle) = \begin{bmatrix} 1 & 0 & 0 & 0 \\ 0 & 1 & 0 & 0 \\ 0 & 0 & 0 & 1 \\ 0 & 0 & 1 & 0 \end{bmatrix} \begin{bmatrix} 1/\sqrt{2} \\ 0 \\ 1/\sqrt{2} \\ 0 \end{bmatrix} = \begin{bmatrix} 1/\sqrt{2} \\ 0 \\ 0 \\ 1/\sqrt{2} \end{bmatrix} = \frac{1}{\sqrt{2}}(|00\rangle + |11\rangle) \tag{3.17}$$

which is an entangled state: it cannot be expressed as a tensor product of two single-qubit states. Four important two-qubit entangled states are the Bell states:

$$|\Phi^+\rangle = \frac{1}{\sqrt{2}}(|00\rangle + |11\rangle) \tag{3.18}$$

$$|\Phi^-\rangle = \frac{1}{\sqrt{2}}(|00\rangle - |11\rangle) \tag{3.19}$$

$$|\Psi^+\rangle = \frac{1}{\sqrt{2}}(|01\rangle + |10\rangle) \tag{3.20}$$

$$|\Psi^-\rangle = \frac{1}{\sqrt{2}}(|01\rangle - |10\rangle) \tag{3.21}$$

which form an orthonormal basis for $\mathbb{C}^4$.

3.2 Measurement

Postulate 3: Measurement. *Quantum measurements are described by a collection $\{M_i\}$ of measurement operators. These are operators acting on the state space of the system being measured. The index (here denoted i) refers to the measurement outcomes that may occur in the experiment. Let $|\psi\rangle$ be the quantum state being measured, and M be a random variable that takes value i when the ith measurement outcome occurs; then:*

$$\Pr(M = i) = \langle \psi | M_i^\dagger M_i | \psi \rangle \tag{3.22}$$

and the state of the system after the measurement (obtaining M = i) is

$$\frac{M_i|\psi\rangle}{\sqrt{\langle\psi|M_i^\dagger M_i|\psi\rangle}} \qquad (3.23)$$

As $\Pr(M=i)$ is obtained by the inner product of the vector $M_i|\psi\rangle$ with itself, it is always a positive real number (see Section 2.3.1), as is required for it to be a probability. It is further necessary that the probabilities of all possible outcomes sum to 1, that is

$$\sum_i \Pr(M=i) = \sum_i \langle\psi|M_i^\dagger M_i|\psi\rangle = 1 \qquad (3.24)$$

which is satisfied if and only if the *completeness equation* holds:

$$\sum_i M_i^\dagger M_i = I \qquad (3.25)$$

The completeness equation is sufficient because:

$$\begin{aligned}
\sum_i \Pr(M=i) &= \sum_i \langle\psi|M_i^\dagger M_i|\psi\rangle \\
&= \langle\psi|\left(\sum_i M_i^\dagger M_i\right)|\psi\rangle \\
&= \langle\psi|I|\psi\rangle \\
&= \langle\psi|\psi\rangle \\
&= 1
\end{aligned} \qquad (3.26)$$

Conversely, suppose $\left(\sum_i M_i^\dagger M_i\right) \neq I$; then, there will be some $|\psi\rangle$ such that Eqn. (3.26) does not equal 1, and so Eqn. (3.25) is also necessary.

In the case where the measurement operators are projectors onto an orthonormal basis, i.e., $M_i = |u_i\rangle\langle u_i|$, where $\{|u_i\rangle\}$ is an orthonormal set – a setting known as *projective measurement* – the results in Section 2.3.2 mean that the completeness equation can be simplified to:

$$\sum_i M_i = I \qquad (3.27)$$

which always holds, as shown in Eqn. (2.39). Note that when the measurement operators are projectors onto an orthonormal basis, the basis vectors are referred to as the *measurement basis vectors* (or states).

3.2.1 Computational basis measurement and the Born rule

In quantum computer science, 'measurement' is often used synonymously with 'computational basis measurement', which means that the measurement operators are projectors onto the computational basis states. For an n-qubit state, the dimension is $N = 2^n$, and so the computational basis measurement operators are

$$M_i = |i\rangle\langle i| \tag{3.28}$$

for $i = 0 \ldots N - 1$. M_i is a matrix of all zeros, except for a single one in the $(i + 1)$th position on the leading diagonal (the plus one is needed as the computational basis vectors are numbered from 0, whereas matrix elements are numbered from 1) and so the completeness equation holds.

Consider a general quantum state

$$|\psi\rangle = \sum_j a_j |j\rangle \tag{3.29}$$

which has been expressed as a superposition of computational basis states; then, the outcome probabilities for a computational basis measurement can be derived:

$$\begin{aligned}\Pr(M = i) &= \left(\sum_j a_j^* \langle j|\right) |i\rangle\langle i| \left(\sum_j a_j |j\rangle\right) \\ &= a_i^* \langle i||i\rangle\langle i|a_i|i\rangle \\ &= a_i^* a_i \\ &= |a_i|^2 \end{aligned} \tag{3.30}$$

This is known as the *Born rule* (after Max Born), and it is easy to verify that if the ith measurement outcome is obtained, the post-measurement state is $|i\rangle$. This means that the measurement outcome is the (classical) bit string i.

The Born rule also applies when just a subset of the qubits is measured in the computational basis. In this case, the probability of a measurement outcome can be found by summing up the squared amplitudes of all of the superposed terms that correspond to that outcome, and the post-measurement state is the superposition of the superposed terms that are consistent with the measurement outcome. This is easiest to show with an example.

Suppose that the first qubit of the two-qubit state

$$|\psi\rangle = a|00\rangle + b|01\rangle + c|10\rangle + d|11\rangle \tag{3.31}$$

is measured; then, the probability of obtaining outcome 0 is $|a|^2 + |b|^2$, and in this case, the post-measurement state is:

$$|\psi'\rangle = \frac{1}{\sqrt{|a|^2 + |b|^2}} |0\rangle(a|0\rangle + b|1\rangle) \tag{3.32}$$

Similarly, the probability of obtaining outcome 1 is $|c|^2 + |d|^2$, and in this case, the post-measurement state is:

$$|\psi'\rangle = \frac{1}{\sqrt{|c|^2 + |d|^2}} |1\rangle(c|0\rangle + d|1\rangle) \tag{3.33}$$

This is consistent with the Born rule for the measurement of the entire state, and to see this, let the second qubit now be measured. If the first measurement outcome was 0, then a second application of the Born rule gives probabilities $|a|^2 / (|a|^2 + |b|^2)$ and $|b|^2 / (|a|^2 + |b|^2)$ for measurement outcomes 0 and 1, respectively. If the first measurement outcome was 1, then a second application of the Born rule gives probabilities $|c|^2 / (|c|^2 + |d|^2)$ and $|d|^2 / (|c|^2 + |d|^2)$ for measurement outcomes 0 and 1, respectively. Putting this together gives the measurement outcome probabilities for the four possible 2-bit strings:

$$\Pr(M = 00) = (|a|^2 + |b|^2) \times \frac{|a|^2}{|a|^2 + |b|^2} = |a|^2 \tag{3.34}$$

$$\Pr(M = 01) = (|a|^2 + |b|^2) \times \frac{|b|^2}{|a|^2 + |b|^2} = |b|^2 \tag{3.35}$$

$$\Pr(M = 10) = (|c|^2 + |d|^2) \times \frac{|c|^2}{|c|^2 + |d|^2} = |c|^2 \tag{3.36}$$

$$\Pr(M = 11) = (|c|^2 + |d|^2) \times \frac{|d|^2}{|c|^2 + |d|^2} = |d|^2 \tag{3.37}$$

(where the values that the random variable, M, can take are given in binary to emphasise the correspondence between the measurement outcome and the post-measurement computational basis state). These measurement outcome probabilities are identical to those obtained by applying the Born rule for the two-qubit state $|\psi\rangle$, as defined in Eqn. (3.31), and this principle holds in general: performing a computational basis measurement on a (multi-qubit) state is the same as performing computational basis measurements on each qubit therein individually.

3.2.2 Global and relative phase

The measurement postulate reveals the insignificance of *global* phase. Consider that any quantum state can be expressed as:

$$|\psi\rangle = e^{i\omega} \sum_i a_i |i\rangle \tag{3.38}$$

where $e^{i\omega}$ has been factored out such that a_0 is a positive real number.

There are no observable differences between the state $|\psi\rangle$ and a state, $|\psi'\rangle = \sum_i a_i |i\rangle$, which is equal to $|\psi\rangle$ *up to a global phase factor*. Suppose that $|\psi\rangle$ undergoes some unitary evolution, U, in which case:

$$U|\psi\rangle = Ue^{i\omega}|\psi'\rangle = e^{i\omega}U|\psi'\rangle \tag{3.39}$$

i.e., the global phase factor, $e^{i\omega}$, is just a factor that remains 'out the front' whatever unitary evolution occurs. Now, suppose that some measurement is taken; then, the probability of the ith outcome (for any set of measurement operators) is given by the measurement postulate

$$\langle\psi|U^\dagger M_i^\dagger M_i U|\psi\rangle = \langle\psi'|U^\dagger e^{-i\omega} M_i^\dagger M_i e^{i\omega} U|\psi'\rangle = \langle\psi'|U^\dagger M_i^\dagger M_i U|\psi'\rangle \quad (3.40)$$

where M_i is the ith measurement operator. Thus, each measurement outcome occurs with the same probability, regardless of whether $|\psi\rangle$ or $|\psi'\rangle$ is measured, and hence, there is no observable difference between the two. For this reason, global phase factors *are* typically neglected in analysis, and quantum states are normally expressed such that the first computational basis state does indeed have a positive real coefficient.

Conversely, *relative* phase most certainly does lead to observable differences. To see an example of this, consider measurements of $|+\rangle$ and $|-\rangle$. The Born rule shows that for each of $|+\rangle$ or $|-\rangle$, if measured in the computational basis, the measurement outcomes 0 and 1 will each be obtained with equal probability. However, $|+\rangle$ and $|-\rangle$ are different states, and in this case, it is pertinent that $|+\rangle$ and $|-\rangle$ differ by a *relative* phase. An observable consequence of this occurs if one measures not in the computational basis, but with measurement operators:

$$M_0 = |+\rangle\langle+|; \quad M_1 = |-\rangle\langle-| \quad (3.41)$$

By the measurement postulate, if the state is $|+\rangle$, then the probability that $M = 0$ is:

$$\Pr(M=0) = \langle+|(|+\rangle\langle+|)^\dagger(|+\rangle\langle+|)|+\rangle = (\langle+|+\rangle)^3 = 1 \quad (3.42)$$

whereas if the state is $|-\rangle$, then $M = 0$ will never be obtained:

$$\Pr(M=0) = \langle-|(|+\rangle\langle+|)^\dagger(|+\rangle\langle+|)|-\rangle = \langle-|+\rangle\langle+|+\rangle\langle+|-\rangle = 0 \quad (3.43)$$

3.2.3 The Bloch sphere

Every single-qubit quantum state is of the form $|\psi\rangle = a|0\rangle + b|1\rangle$ for some complex numbers a and b, and for this to be a unit vector, it is further necessary that $|a|^2 + |b|^2 = 1$. Neglecting global phase factors, this means that every single-qubit state can be written in the form

$$|\psi\rangle = \cos\frac{\theta}{2}|0\rangle + e^{i\varphi}\sin\frac{\theta}{2}|1\rangle \quad (3.44)$$

for some angles $\theta \in [0, \pi)$ and $\varphi \in [0, 2\pi)$.

This, in turn, leads to a nice illustration of any single-qubit state, as a point on the surface of the *Bloch sphere*:

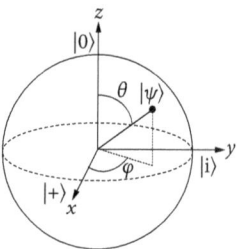

Figure 3.1

Orthogonal states are co-polar points on the Bloch sphere surface (that is, they are not at an angle of $\pi/2$ radians from each other, as one may intuitively expect – this also explains why the term $\theta/2$ appears in Eqn. (3.44), whereas the angle θ appears in Fig. 3.1). As shown in Section 3.2.4, it is always possible to perform a measurement that perfectly distinguishes states that are orthogonal, and so all pairs of co-polar points are distinguishable. Three particularly noteworthy pairs of orthogonal states are those that lie along the x, y, and z axes: these are commonly denoted $|0\rangle, |1\rangle; |i\rangle, |-i\rangle$; and $|+\rangle, |-\rangle$, respectively. The first pair is the computational basis states for a single qubit, and the third pair is defined in Eqns. (3.11) and (3.12). The second pair is:

$$|i\rangle = \frac{1}{\sqrt{2}}(|0\rangle + i|1\rangle) \qquad (3.45)$$

$$|-i\rangle = \frac{1}{\sqrt{2}}(|0\rangle - i|1\rangle) \qquad (3.46)$$

For each of the three cases, the first element of the pair is depicted in Fig. 3.1, and the second element is the co-polar point, i.e., the point '−1' along the corresponding axis.

The Bloch sphere is also useful as a visual aid for picturing single-qubit operations. As a single-qubit unitary maps a single-qubit state to another single-qubit state, a unitary transformation is itself an operation that maps all of the points on the surface of the Bloch sphere to other points on the surface of the Bloch sphere. For instance, the Pauli X, Y, and Z operations are rotations of π radians around the x, y, and z axes, respectively; and the Bloch sphere can also be used to illustrate three other *parameterised* quantum gates: the rotation gates $R_x(\sigma)$, $R_y(\sigma)$, and $R_z(\sigma)$, which are unitaries that rotate the state by σ radians around the x, y, and z axes, respectively (see Chapter 4 for their full definition).

3.2.4 Distinguishing orthogonal and non-orthogonal quantum states

An important property of quantum states is that they can be distinguished *if and only if* they are orthogonal. Let $\{|\psi_i\rangle\}$ be a set of quantum states (of the same size); to distinguish a quantum state in this sense means that any element of $\{|\psi_i\rangle\}$ is prepared and must be identified by a measurement (performed by a party who is not privy to the state preparation, but knows it is an element of the set). If the states of $\{|\psi_i\rangle\}$ are pair-wise orthogonal, then the set can be supplemented with further states to form an orthonormal basis (if the number of elements is not already the dimension of the states therein), and so measurement operators of the form $M_i = |\psi_i\rangle\langle\psi_i|$ can be defined. It is a consequence of the spectral theorem that $\sum_i M_i = I$ (see Eqn. (2.39)). Thus, $\{M_i\}$ forms a complete set of measurement operators, and every state $|\psi\rangle \in \{|\psi_i\rangle\}$ can be distinguished as each measurement operator 'picks out' the corresponding state with probability one (and probability zero for all of the others):

$$\Pr(M = j) = \langle\psi_i|(|\psi_j\rangle\langle\psi_j|)^\dagger(|\psi_j\rangle\langle\psi_j|)|\psi_i\rangle = \begin{cases} 1 & \text{if } i = j \\ 0 & \text{otherwise} \end{cases} \quad (3.47)$$

Turning to the case where the possibilities for an unknown quantum state are not orthogonal, we begin with a motivational example. Let $|\psi\rangle$ be a single-qubit state, which is known to be one of two possibilities, either $|\psi_a\rangle$ or $|\psi_b\rangle$, which are now *not* orthogonal. (*A priori*, each of $|\psi_a\rangle$ and $|\psi_b\rangle$ is treated as equally likely.) There is no measurement that can perfectly (i.e., with 100% accuracy) distinguish these possibilities, but it is possible to perform a measurement that produces *some* information about the likelihood of whether $|\psi\rangle = |\psi_a\rangle$ or $|\psi\rangle = |\psi_b\rangle$. Intuitively, one may expect:

- If one just guesses, then the guess will be correct with probability equal to one half, so any sensible strategy should do at least as well as this.
- The 'closer together' $|\psi_a\rangle$ and $|\psi_b\rangle$ are, the harder they will be to distinguish (i.e., the lower the probability of correctly inferring $|\psi\rangle$)

Such intuition turns out to be correct and is captured by the *Holevo–Helstrom bound*:

The Holevo–Helstrom bound. *If $|\psi\rangle$ is either $|\psi_a\rangle$ or $|\psi_b\rangle$ (with equal probability a priori), then the probability of correctly inferring the state $|\psi\rangle$ is less than or equal to $\frac{1}{2}(1+\sin(\theta/2))$, where $\cos(\theta/2) = |\langle\psi_a|\psi_b\rangle|$.*

Notably, θ is the angle between $|\psi_a\rangle$ and $|\psi_b\rangle$ when represented on the Bloch sphere, which can be pictured (by slicing the Bloch sphere to obtain the great circle containing $|\psi_a\rangle$ and $|\psi_b\rangle$):

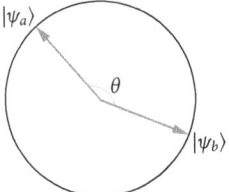

Figure 3.2

Depicting the two possibilities for $|\psi\rangle$ in this way confirms the earlier intuition. If $|\psi_a\rangle$ and $|\psi_b\rangle$ are nearly aligned, so $\theta \approx 0$, then $\sin(\theta/2) \approx 0$ and $\frac{1}{2}(1 + \sin(\theta/2)) \approx \frac{1}{2}$: in which case it is not possible to do much better than just guessing. Conversely, if $|\psi_a\rangle$ and $|\psi_b\rangle$ are orthogonal, that is $\theta = \pi$, then $\frac{1}{2}(1 + \sin(\theta/2)) = 1$. In this case, the Holevo–Helstrom bound is tight, in the sense that an appropriate choice of measurement basis will indeed enable perfect discrimination between the two possibilities. In fact, the Holevo–Helstrom bound is *always* tight: the bounding probability can always be achieved by choosing the measurement operators as projectors onto the eigenvectors of $|\psi_a\rangle\langle\psi_a| - |\psi_b\rangle\langle\psi_b|$.

To see non-orthogonal state discrimination in action, consider the example of distinguishing $|0\rangle$ and $|+\rangle$, which are not orthogonal:

$$\langle 0|+\rangle = \begin{bmatrix} 1 & 0 \end{bmatrix} \begin{bmatrix} \frac{1}{\sqrt{2}} \\ \frac{1}{\sqrt{2}} \end{bmatrix} = \frac{1}{\sqrt{2}} \neq 0 \tag{3.48}$$

Therefore, it is not possible to infer the state of $|\psi\rangle$ with perfect accuracy if all that is known *a priori* is that it is either $|0\rangle$ or $|+\rangle$. If the desire is to obtain the greatest possible probability of inferring the correct state, then the Holevo–Helstrom bound dictates that the measurement basis should be the eigenvectors of

$$|\psi_a\rangle\langle\psi_a| - |\psi_b\rangle\langle\psi_b| = |0\rangle\langle 0| - |+\rangle\langle+| = \begin{bmatrix} \frac{1}{2} & -\frac{1}{2} \\ -\frac{1}{2} & -\frac{1}{2} \end{bmatrix} \tag{3.49}$$

which are $[0.38 \ 0.92]^T$ and $[-0.92 \ 0.38]^T$. In this case, the probability of correctly inferring the state is equal to

$$\frac{1}{2}(1 + \sin(\arccos(1/\sqrt{2}))) = 0.85 \tag{3.50}$$

However, it is not obligatory to measure in the optimal (according to the Holevo–Helstrom bound) basis, and it is interesting to observe what happens if we distinguish $|0\rangle$ and $|+\rangle$ by instead measuring in the computational basis (i.e., with $M_0 = |0\rangle\langle 0|$ and $M_1 = |1\rangle\langle 1|$). Assuming that $|\psi\rangle$ is each of $|0\rangle$ or $|+\rangle$ with equal probability, the measurement outcome probabilities can be tabulated for each possible $|\psi\rangle$ and measurement outcome pair:

	$M = 0$	$M = 1$
$|\psi\rangle = |0\rangle$	$\frac{1}{2}$	0
$|\psi\rangle = |+\rangle$	$\frac{1}{4}$	$\frac{1}{4}$

Therefore

- $\frac{1}{4}$ of the time the measurement outcome is 1, which means $|\psi\rangle = |+\rangle$.

- $\frac{3}{4}$ of the time the measurement outcome is 0, in which case the maximum likelihood estimate is $|\psi\rangle = |0\rangle$, but $\frac{1}{3}$ of the time this will be wrong, and actually $|\psi\rangle = |+\rangle$.

Thus, the overall success probability is $1 - \frac{3}{4} \times \frac{1}{3} = \frac{3}{4}$. This is less than the theoretically achievable 0.85 (from Eqn. (3.50)), but such a choice of measurement basis has the potentially attractive property that on occasion the identity of $|\psi\rangle$ is learned with certainty: if the measurement outcome is 1, then we know for sure that $|\psi\rangle = |+\rangle$.

These two possibilities for the choice of measurement basis can be depicted using a slice of the Bloch sphere (with an orthonormal measurement basis corresponding to an axis of the Bloch sphere).

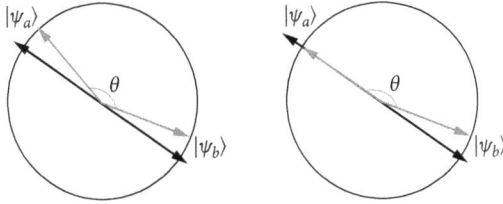

Figure 3.3

Shown on the LHS is the optimal strategy, which amounts to choosing the measurement basis spread equally each side of the states to be distinguished; however, as shown on the RHS, if one of the measurement basis vectors aligns with one the states being distinguished, *sometimes* the measurement outcome determines the state with 100% accuracy. That is, in the right-hand figure, if the measurement outcome corresponding to the downward pointing vector is obtained, then this can only have been because the state is $|\psi_b\rangle$, as $|\psi_a\rangle$ is orthogonal to this measurement basis vector.

3.2.5 Measuring the expectation of an observable

Measurements are often used in the context of measuring the expectation of *observables*. An observable is a Hermitian operator, H, which has spectral decomposition:

$$H = \sum_i E_i |\lambda_i\rangle\langle\lambda_i| \qquad (3.51)$$

Letting $\{M_i\}$ be measurement operators that are projectors onto the eigenstates of H, i.e., $M_i = |\lambda_i\rangle\langle\lambda_i|$, the probability of obtaining a measurement outcome corresponding to the ith measurement operator can be expressed (directly from the measurement postulate):

$$\Pr(M = i) = \langle\psi|M_i^\dagger M_i|\psi\rangle = \langle\psi|M_i|\psi\rangle \qquad (3.52)$$

When measuring the *expectation of an observable*, the measurement outcome corresponding to the ith eigenvector, is interpreted as a numerical value, specifically the ith eigenvalue (as opposed to the arbitrary index value as in the definition of the random variable, M). From this, the expectation when measuring the observable, H applied to the state $|\psi\rangle$ is defined:

$$\mathbb{E}(H \mid |\psi\rangle) = \sum_i E_i \Pr(M = i \mid |\psi\rangle)$$

$$= \sum_i E_i \langle \psi | M_i | \psi \rangle$$

$$= \langle \psi | \left(\sum_i E_i M_i \right) | \psi \rangle$$

$$= \langle \psi | \left(\sum_i E_i |\lambda_i\rangle\langle\lambda_i| \right) | \psi \rangle$$

$$= \langle \psi | H | \psi \rangle \qquad (3.53)$$

Measuring the expectation of an observable is sometimes referred to simply as *observable measurement* or *measurement of an observable* and is most relevant in quantum algorithms that simulate physical and chemical systems (see Chapter 11).

3.3 Bell inequalities

One of the most important foundational results in quantum mechanics, which follows directly from the postulates, is the violation of Bell inequalities. The story of Bell inequalities begins with a trio of august physicists, Einstein, Podolsky, and Rosen (EPR), who were adamant that entanglement must amount to the presence of a *hidden variable*, and that apparently entangled states are in fact nothing more than artefacts of (classical) ignorance about this hidden variable. For example, if the only operation that may be applied to the entangled state $|\Phi^+\rangle = \frac{1}{\sqrt{2}}(|00\rangle + |11\rangle)$ is computational basis measurement, there is no way of confirming that this is indeed an entangled state, rather than an equal *mixture* of the (classical) states $|00\rangle$ and $|11\rangle$. It was another brilliant physicist, John Stewart Bell, who realised that measuring (instances of) the same state in different bases is the key to distinguishing a mixture of classical states (any sample from which could be thought of as being pre-determined by a hidden-variable) from an *entangled quantum state* whose measurement outcomes are not pre-determined.

Bell began by supposing that two spatially separated parties, by convention *Alice* and *Bob*, are each issued by a third party, *Charlie*, one half of a physical system, on which they can thereafter perform a measurement. Additionally, Charlie can prepare multiple instances of the same physical system to distribute to Alice and Bob. Alice and Bob are each capable of performing two different measurements on their system, and each of the four measurements (two each for Alice and Bob) can yield outcomes ±1.

Formally, let Alice's half of the system have two properties A_1 and A_2 (each of which takes value ±1 as specified above); and likewise let Bob's half of the system have two properties B_1 and B_2 (again, each of which takes value ±1). Alice and Bob may choose to measure either property (i.e., Alice A_1 or A_2, Bob B_1 or B_2), and they do so simultaneously and independently each time an instance of the system is issued by Charlie. The Bell inequalities consider the expectation of the term:

$$A_1 B_1 + A_2 B_1 + A_2 B_2 - A_1 B_2 = (A_1 + A_2) B_1 + (A_2 - A_1) B_2 \qquad (3.54)$$

It may appear that because Alice and Bob only choose *one* measurement each time, they are issued an instance of the system that this expression is not well-defined. However, owing to the linearity of the expectation ($\mathbb{E}(A_1 B_1 + A_2 B_1 + A_2 B_2 - A_1 B_2) = \mathbb{E}(A_1 B_1) + \mathbb{E}(A_2 B_1) + \mathbb{E}(A_2 B_2) - \mathbb{E}(A_1 B_2)$) and so if one prefers, this can be thought of as the sum of four expectation values corresponding to the four possibilities for the pair of measurement choices Alice and Bob make. The expectation can be bounded

$$
\begin{aligned}
&\mathbb{E}(A_1 B_1 + A_2 B_1 + A_2 B_2 - A_1 B_2) \\
&= \sum_{a_1, a_2, b_1, b_2} \Pr(A_1 = a_1, A_2 = a_2, B_1 = b_1, B_2 = b_2)(a_1 b_1 + a_2 b_1 + a_2 b_2 - a_1 b_2) \\
&= \sum_{a_1, a_2, b_1, b_2} \Pr(A_1 = a_1, A_2 = a_2, B_1 = b_1, B_2 = b_2)((a_1 + a_2) b_1 + (a_2 - a_1) b_2) \\
&\leq \sum_{a_1, a_2, b_1, b_2} \Pr(A_1 = a_1, A_2 = a_2, B_1 = b_1, B_2 = b_2) \times 2 \\
&= 2 \qquad (3.55)
\end{aligned}
$$

where the bound in the penultimate line occurs because a_1 and a_2 are ±1 so exactly one of $(a_1 + a_2)$ and $(a_2 - a_1)$ is equal to zero and the other equal to ±2; and b_1 and b_2 always take value ±1. This can be re-expressed to give the Bell inequality explicitly:

$$\mathbb{E}(A_1 B_1) + \mathbb{E}(A_2 B_1) + \mathbb{E}(A_2 B_2) - \mathbb{E}(A_1 B_2) \leq 2 \qquad (3.56)$$

It is important to appreciate that the analysis to this point is completely general for the classical world, and that the probability distribution $\Pr(A_1 = a_1, A_2 = a_2, B_1 = b_1, B_2 = b_2)$ wholly captures one's ignorance about the aforementioned hidden variable (i.e., it represents *a priori* unknown correlations between the measurement outcomes).

Now, let the physical system that Alice and Bob share be the Bell state $|\Psi^-\rangle = \frac{1}{\sqrt{2}}(|01\rangle - |10\rangle)$, and let the measurements they perform be measurements of observables. In particular:

$$A_1 = Z = \begin{bmatrix} 1 & 0 \\ 0 & -1 \end{bmatrix} \quad (3.57)$$

$$A_2 = X = \begin{bmatrix} 0 & 1 \\ 1 & 0 \end{bmatrix} \quad (3.58)$$

$$B_1 = \frac{1}{\sqrt{2}}(-Z - X) = \frac{1}{\sqrt{2}}\begin{bmatrix} -1 & -1 \\ -1 & 1 \end{bmatrix} \quad (3.59)$$

$$B_2 = \frac{1}{\sqrt{2}}(Z - X) = \frac{1}{\sqrt{2}}\begin{bmatrix} 1 & -1 \\ -1 & -1 \end{bmatrix} \quad (3.60)$$

Each of these four observables has one eigenvalue equal to 1 and one equal to -1, so these do indeed correspond to measurements of the system that can return ± 1, as specified in the set-up.

As the two (simultaneous) measurements of Alice and Bob together act on the whole quantum system, it is convenient to express Alice and Bob's collective actions as a single observable measurement. For instance, if Alice chooses to measure observable A_1 and Bob chooses to measure observable B_2, then the following observable is measured:

$$\begin{aligned} A_1 \otimes B_2 = Z \otimes \left(\frac{1}{\sqrt{2}}(Z - X) \right) &= \begin{bmatrix} 1 & 0 \\ 0 & -1 \end{bmatrix} \otimes \frac{1}{\sqrt{2}} \begin{bmatrix} 1 & -1 \\ -1 & -1 \end{bmatrix} \\ &= \frac{1}{\sqrt{2}} \begin{bmatrix} 1 & -1 & 0 & 0 \\ -1 & -1 & 0 & 0 \\ 0 & 0 & -1 & 1 \\ 0 & 0 & 1 & 1 \end{bmatrix} \end{aligned} \quad (3.61)$$

and similarly for $A_1 \otimes B_1$, $A_2 \otimes B_1$, and $A_2 \otimes B_2$.

It is important to clarify a few things at this point. First, the fact that Alice and Bob independently choose which observable to measure is not overly significant: we treat this choice as an objective (classical) fact, and so for any instance of the prepared physical system, exactly one of $A_1 \otimes B_1$, $A_1 \otimes B_2$, $A_2 \otimes B_1$, and $A_2 \otimes B_2$ is measured. Nor is it particularly pertinent with which probability Alice and Bob choose each observable – just that each is chosen sometimes (and independently), so each of $A_1 \otimes B_1$, $A_1 \otimes B_2$, $A_2 \otimes B_1$, and $A_2 \otimes B_2$ arises many times in the course of many repeats of the experiment. Additionally, it is not actually crucial that Alice and Bob perform the measurements *exactly* simultaneously: formally, it is necessary that they do so in sufficiently quick succession to preclude the possibility of signalling (that is, the time lapse between the measurements must be less than the time it would take for a ray of light to traverse the intervening distance). Finally, not too much should be read into the fact that we treat Alice and Bob's actions as a single observable measurement. As the observable in question is separable (i.e., it can be decomposed as a tensor product), this is purely a convenient way to perform the analysis and does not imply any entanglement or correlation in the measurement itself.

The benefit of treating the measurements of Alice and Bob as a single observable is that it is relatively easily to evaluate the expectation of Alice's measurement outcome multiplied by Bob's, because $\mathbb{E}(A_k B_l | |\psi\rangle) = \langle \psi | (A_k \otimes B_l) | \psi \rangle$ (for $k \in \{0, 1\}$, $l \in \{0, 1\}$). To see this, consider that if $M = M_1 \otimes M_2$ and M_1 has eigenvectors $\{|m_i\rangle\}$ with associated eigenvalues $\{m_i\}$, and M_2 has eigenvectors $\{|n_j\rangle\}$ with associated eigenvalues $\{n_j\}$, then $\{|m_i\rangle \otimes |n_j\rangle\}$ are the eigenvectors of M, with associated eigenvalues $\{m_i n_j\}$. Thus, by measuring the expectation of the observable $A_k \otimes B_l$, which corresponds to Alice *and* Bob's measurements (together), we automatically obtain $\mathbb{E}(A_k B_l)$. So we proceed to evaluate the expectation for each of the four possible measurements on the state $|\Psi^-\rangle$.

First,

$$\mathbb{E}(A_1 B_2) = \langle \Psi^- | (A_1 \otimes B_2) | \Psi^- \rangle$$

$$= \frac{1}{2\sqrt{2}} \begin{bmatrix} 0 & 1 & -1 & 0 \end{bmatrix} \begin{bmatrix} 1 & -1 & 0 & 0 \\ -1 & -1 & 0 & 0 \\ 0 & 0 & -1 & 1 \\ 0 & 0 & 1 & 1 \end{bmatrix} \begin{bmatrix} 0 \\ 1 \\ -1 \\ 0 \end{bmatrix} = -\frac{1}{\sqrt{2}} \quad (3.62)$$

and similarly:

$$\mathbb{E}(A_1 B_1) = \langle \Psi^- | (A_1 \otimes B_1) | \Psi^- \rangle = \frac{1}{\sqrt{2}} \quad (3.63)$$

$$\mathbb{E}(A_2 B_1) = \langle \Psi^- | (A_2 \otimes B_1) | \Psi^- \rangle = \frac{1}{\sqrt{2}} \quad (3.64)$$

$$\mathbb{E}(A_2 B_2) = \langle \Psi^- | (A_2 \otimes B_2) | \Psi^- \rangle = \frac{1}{\sqrt{2}} \quad (3.65)$$

Together, this gives:

$$\mathbb{E}(A_1 B_1) + \mathbb{E}(A_2 B_1) + \mathbb{E}(A_2 B_2) - \mathbb{E}(A_1 B_2) = 2\sqrt{2} > 2 \quad (3.66)$$

which is a violation of the Bell inequality (3.56).

The violation of the Bell inequality proves that entangled states have stronger than classical correlations: no hidden variable can lead to the observed measurement statistics. Moreover, Bell inequalities have been violated in real physical experiments with overwhelming experimental confidence.

The Bell inequalities are one of the seminal results in the field of *quantum foundations*, which seeks to understand the precise physical nature of the quantum world. From a practical point of view, the experimental violations of Bell inequalities can be taken as evidence that the systems in question are indeed quantum mechanical in nature, and testing Bell inequality violation is frequently used as a way to validate that quantum computational hardware really does exhibit quantum mechanical behaviour.

3.4 No-go theorems

No-go theorems (or principles) rule out certain computational or information-processing tasks that one may otherwise think that quantum mechanics could accomplish. In this section, we introduce three important no-go theorems that follow directly from the postulates of quantum mechanics.

3.4.1 The no-signalling principle

The Bell state $|\Phi^+\rangle = \frac{1}{\sqrt{2}}(|00\rangle + |11\rangle)$ is an entangled state (sometimes termed a *Bell pair* or *EPR pair*) and has the following property: if the first qubit is measured (in the computational basis), then the measurement outcome will either be 0 or 1; in the case that it is 0, this means that the state has collapsed to $|00\rangle$, and so if the second qubit is then measured, the outcome is guaranteed also to be 0 (similarly, if the first measurement outcome is 1, then the collapsed state is $|11\rangle$, and so if the second qubit is then measured, the outcome is guaranteed also to be 1). There is no physical requirement that the two entangled qubits are in close proximity to each other (that they are *local* in physics parlance), and even when they are not local, this collapse happens to both qubits instantaneously upon the measurement of the first qubit. This is what Einstein referred to as 'spooky action at a distance' and ostensibly would appear to suggest that entanglement can be used to transfer information faster than the speed of light (known as superluminal signalling), which would violate the theory of relativity.

In fact, collapsing entanglement *cannot* be used to transfer information, as proven by the *no-signalling principle*, a simplified instance of which we now consider.

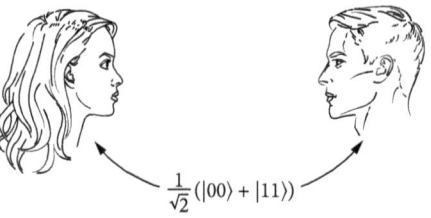

Figure 3.4

- Alice and Bob are at different ends of the universe, but each have one half of a Bell pair: $\frac{1}{\sqrt{2}}(|00\rangle + |11\rangle)$.
- Alice can measure her qubit (in the computational basis) whenever she wants, and this will collapse Bob's to the same state.
- We are interested in whether Bob can infer whether or not Alice has measured her qubit.

- But all that Bob can do to infer whether Alice has measured her qubit is to measure his own qubit (in the computational basis) – therefore, the question reduces to whether the measurement outcome probabilities of Bob's qubit are altered by virtue of Alice having performed her measurement or not.

If Bob can infer from measurement of his qubit whether or not Alice has already measured hers, then this *does* enable superluminal signalling, in the following way:

- Alice and Bob are spatially separated by a distance that takes light Δt seconds to traverse.
- Bob is interested in whether some event that Alice witnesses occurs before time t_B.
- When Alice witnesses the event she will signal to notify Bob.

So there are two alternatives:

1. If Alice uses classical signalling, then if the event occurs less than Δt seconds before t_B, there is no way she can send a signal to Bob that he will receive before t_B.
2. However, if Alice can send a signal by measuring her qubit, then this constitutes instantaneous information transfer and, hence, notifies Bob of the event *any* time up to t_B.

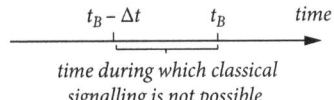

Figure 3.5

However, Alice cannot use shared entanglement to signal to Bob. If Alice *has not* measured her qubit, then the state is $\frac{1}{\sqrt{2}}(|00\rangle + |11\rangle)$, and so Bob has a $\left(\frac{1}{\sqrt{2}}\right)^2 = \frac{1}{2}$ probability of measuring each of 0 and 1. Whereas if Alice *has* measured her qubit, then Bob's qubit has collapsed – it is either in state 0, or state 1 (each with probability 1/2). However, in the absence of any additional information (for example, Alice sending a classical message indicating her measurement outcome), Bob has no knowledge of which of these measurement outcomes Alice observed, and so all he knows is that he will measure each of $|0\rangle$ and $|1\rangle$ with probability 1/2. What this amounts to is that, in the absence of any additional information being transferred from Alice to Bob, Bob's measurement statistics when measuring the uncollapsed quantum state are identical to his ignorance (expressed probabilistically) when measuring the state already collapsed by Alice.

48 Quantum mechanics and quantum information

The simple set-up is useful for explaining why wave function collapse does not trivially allow for superluminal signalling; however, it is hardly a complete refutation of the possibility. Thus, we now look at a more general set-up where:

- Alice and Bob share a general two-qubit (entangled) state;
- Alice can perform any single-qubit projective measurement;
- Bob can perform any single-qubit projective measurement,

and show that the no-signalling principle still holds.

Let Alice measure in the orthonormal basis $|a_1\rangle, |a_2\rangle$, that is, her measurement operators are projectors onto these states, and let Bob measure in the orthonormal basis $|b_1\rangle, |b_2\rangle$. The four states $\{|a_1\rangle|b_1\rangle, |a_1\rangle|b_2\rangle, |a_2\rangle|b_1\rangle, |a_2\rangle|b_2\rangle\}$ form a two-qubit basis, and hence, the entangled state shared between Alice and Bob can be expressed as a superposition of these basis states:

$$|\psi\rangle = \alpha|a_1\rangle|b_1\rangle + \beta|a_1\rangle|b_2\rangle + \gamma|a_2\rangle|b_1\rangle + \delta|a_2\rangle|b_2\rangle \tag{3.67}$$

We show that, as for the first simple example, Bob obtains the measurement outcomes with the same probability regardless of whether Alice has already measured her qubit or not for this generalised set up. If Alice measures her qubit first:

1. With probability $|\alpha|^2 + |\beta|^2$, the measurement outcome is $|a_1\rangle$ and the post-measurement state is $\frac{1}{\sqrt{|\alpha|^2 + |\beta|^2}}|a_1\rangle(\alpha|b_1\rangle + \beta|b_2\rangle)$.
2. With probability $|\gamma|^2 + |\delta|^2$, the measurement outcome is $|a_2\rangle$ and the post-measurement state is $\frac{1}{\sqrt{|\gamma|^2 + |\delta|^2}}|a_2\rangle(\gamma|b_1\rangle + \delta|b_2\rangle)$

Indexing Bob's measurement operators $M_1 = |b_1\rangle\langle b_1|$, $M_2 = |b_2\rangle\langle b_2|$, he obtains the two possible outcomes for his measurement with the following probabilities:

$$\Pr(M = 1) = (|\alpha|^2 + |\beta|^2) \times \frac{|\alpha|^2}{|\alpha|^2 + |\beta|^2} + (|\gamma|^2 + |\delta|^2) \times \frac{|\gamma|^2}{|\gamma|^2 + |\delta|^2}$$

$$= |\alpha|^2 + |\gamma|^2 \tag{3.68}$$

$$\Pr(M = 2) = (|\alpha|^2 + |\beta|^2) \times \frac{|\beta|^2}{|\alpha|^2 + |\beta|^2} + (|\gamma|^2 + |\delta|^2) \times \frac{|\delta|^2}{|\gamma|^2 + |\delta|^2}$$

$$= |\beta|^2 + |\delta|^2 \tag{3.69}$$

which are identical to the measurement probabilities if Bob measures the uncollapsed state, i.e., when the second qubit of the state in Eqn. (3.67) is measured. Thus, the no-signalling principle is upheld in this more general scenario.

Of course, it is possible to go even more general still, and consider any case in which Alice and Bob are spatially separated and share any kind of entangled resource (potentially with third parties involved as well), however even then the no-signalling principle holds – and there is little additional insight to be gained by exhaustively working through all of the possibilities.

3.4.2 The no-cloning theorem

The no-cloning theorem addresses the question of whether there exists a *cloning unitary*, U, such that for any arbitrary quantum state $|\psi\rangle$ and a qubit initially set to $|0\rangle$:

$$U(|\psi\rangle|0\rangle) = |\psi\rangle|\psi\rangle \tag{3.70}$$

That is, a unitary that can copy ('clone') any state, $|\psi\rangle$. In particular, the no-cloning theorem shows that no such unitary exists. Note that the choice of $|0\rangle$ for the initial state of the second qubit is arbitrary – all that is required is that it is in some fixed state – but as any fixed state can be mapped to $|0\rangle$ by a unitary operation, there is no loss of generality in simply addressing the case where it is initialised to $|0\rangle$.

The key to showing the non-existence of a cloning unitary is to focus on the fact that it must clone *any* quantum state, and thus, as well as Eqn. (3.70), the same unitary must also clone some other quantum state, $|\phi\rangle$:

$$U(|\phi\rangle|0\rangle) = |\phi\rangle|\phi\rangle \tag{3.71}$$

The proof proceeds by taking inner products of the LHS and RHS of Eqns. (3.70) and (3.71):

$$\langle\psi|\langle 0|U^\dagger U|\phi\rangle|0\rangle = (\langle\psi|\psi\rangle)(\langle\phi|\phi\rangle)) \tag{3.72}$$

$$\implies \langle\psi|\phi\rangle\langle 0|0\rangle = (\langle\psi|\phi\rangle)^2 \tag{3.73}$$

$$\implies \langle\psi|\phi\rangle = (\langle\psi|\phi\rangle)^2 \tag{3.74}$$

which is only true if $|\psi\rangle = |\phi\rangle$ (so $\langle\psi|\phi\rangle = 1$) or $|\psi\rangle$ and $|\phi\rangle$ are orthogonal (so $\langle\psi|\phi\rangle = 0$). Therefore, there is no unitary that can clone general quantum states.

In theoretical physics, the no-cloning theorem is very important for a whole plethora of reasons. From the point of view of quantum computing, the fact that cloning is not possible makes quantum error correction harder (or at least less similar to classical error correction). Cloning would also enable an infinite amount of classical information to be compressed into a single qubit and then recovered afterwards by the following simple protocol:

1. Map a classical bit-string to a unique qubit state (qubit states are continuously parameterised so some encoding to achieve this for any bit-string length is always possible).
2. Communicate the single qubit.
3. Receive the qubit, make an arbitrary number of copies by cloning, and perform *quantum state tomography* to recover the original classical information.

(Quantum state tomography consists of making several different measurements of copies of a quantum state to obtain the classical description of the state.)

3.4.3 The no-deleting theorem

Inverting the no-cloning theorem yields the no-deleting theorem: there does not exist a unitary that can delete one of two copies of a quantum state. That is, there does not exist a unitary, $\tilde{U}$, that can achieve:

$$\tilde{U}(|\psi\rangle|\psi\rangle) = |\psi\rangle|0\rangle \tag{3.75}$$

for an arbitrary state $|\psi\rangle$. This follows immediately from the proof of the no-cloning theorem, as all unitary matrices have inverses, and so the existence of such a unitary $\tilde{U}$ would imply a cloning unitary, namely its inverse, $\tilde{U}^\dagger$.

Chapter problems

1. Which of the following are possible states of a qubit?
   (a) $\frac{1}{\sqrt{2}}(|0\rangle + |1\rangle)$
   (b) $\frac{\sqrt{3}}{2}|1\rangle - \frac{1}{2}|0\rangle$
   (c) $0.7|0\rangle + 0.3|1\rangle$
   (d) $0.8|0\rangle + 0.6|1\rangle$
   (e) $\cos\theta|0\rangle + i\sin\theta|1\rangle$
   (f) $\cos^2\theta|0\rangle - \sin^2\theta|1\rangle$
   (g) $(\frac{1}{2} + \frac{i}{2})|0\rangle + (\frac{1}{2} - \frac{i}{2})|1\rangle$

   For each valid state among the above, give the probabilities of observing $|0\rangle$ and $|1\rangle$ when the system is measured in the computational basis.

   What are the probabilities of the two outcomes when the state is measured in the basis $\frac{1}{\sqrt{2}}(|0\rangle + |1\rangle), \frac{1}{\sqrt{2}}(|0\rangle - |1\rangle)$?

2. A two-qubit system is in the following state $\frac{1}{\sqrt{30}}(|00\rangle + 2i|01\rangle - 3|10\rangle - 4i|11\rangle)$. The first qubit is measured and observed to be 1. What is the state of the system after the measurement? What is the probability that a subsequent measurement of the second qubit will observe a 1?

3. Show that the Hadamard matrix and the three Pauli matrices are unitary.

4. Express each of the three Pauli matrices X, Y, and Z, as a product of the other two multiplied by i.

5. Show that the four Bell states, $|\Phi^+\rangle, |\Phi^-\rangle, |\Psi^+\rangle$, and $|\Psi^-\rangle$ form an orthonormal basis for $\mathbb{C}^4$.

6. Show that the entangled state $\frac{1}{\sqrt{2}}(|00\rangle + |11\rangle)$ cannot be expressed as a tensor product of two single-qubit states. (**Hint:** start with a general expression of a tensor product of two single-qubit states, $(\alpha|0\rangle + \beta|1\rangle)(\gamma|0\rangle + \delta|1\rangle)$ and multiply out.)

7. Show that the quantum states

$$\frac{1}{\sqrt{2}}(|+\rangle + |-\rangle) \text{ and } \frac{1}{\sqrt{2}}(|+\rangle - |-\rangle)$$

can be perfectly distinguished. Give the measurement basis to achieve this in terms of the computational basis states $|0\rangle$ and $|1\rangle$.

8. Let $|\psi\rangle$ be some unknown quantum state, which is either $|1\rangle$ or $\frac{\sqrt{3}}{2}|0\rangle + \frac{1}{2}|1\rangle$. Furthermore, it is known that there is a 75% probability that $|\psi\rangle$ is $|1\rangle$ and a 25% probability that $|\psi\rangle$ is $\frac{\sqrt{3}}{2}|0\rangle + \frac{1}{2}|1\rangle$.

    A measurement must be performed to help identify which state $|\psi\rangle$ is. Give a measurement basis that guarantees to correctly determine $|\psi\rangle$ for one of the measurement outcomes and maximises the overall probability of correctly identifying $|\psi\rangle$. Give the probability of success.

9. An experimenter has prepared a (known) initial two-qubit state, and plans to measure the first qubit in the computational basis, and then measure the second qubit. For each of the following statements, identify the initial states that make it true.

    (a) If the second qubit is measured in the computational basis, the outcome of the second measurement can always be predicted with certainty, once the outcome of the first measurement is known.

    (b) If the second qubit is measured in the computational basis, the outcome of the second measurement can sometimes be predicted with certainty, depending on the outcome of the first measurement.

    (c) Regardless of the choice of basis for the measurement of the second qubit, the probability of obtaining each possible outcome for the second measurement is independent of the outcome of the first measurement.

    The experimenter plans to adapt the scheme so that the basis for the second measurement can be chosen in a way that depends on the first measurement outcome. Is it always possible to choose a measurement basis such that both outcomes for the second measurement have a 50% probability? Justify your answer.

Further reading

The postulates of quantum mechanics have been presented in the form given by Nielsen and Chuang [2010], the Born rule is due to Max Born [1926], and the Bloch sphere is named after Felix Bloch, although the name was not introduced until later, for example by Arecchi *et al* [1972]. Einstein, Podolsky, and Rosen are responsible for the EPR 'paradox' [1935], and the Bell inequalities were formulated by John Stewart Bell [1964]. The Holevo–Helstrom bound presented here is actually a simple special case (i.e., for pure states) of a general result about quantum state distinguishability proved by Helstrom [1969] and Holevo [1973]. The no-signalling principle is due to Ghirardi *et al* [1988]; the no-cloning principle was independently discovered by Wooters and Zurek [1982] and by Dieks [1982], and it was later observed that the result had actually already been derived by Park [1970]; the no-deleting principle was discovered by Pati and Braunstein [2000].

4
Quantum circuits

In this chapter, we shift our perspective from seeing the unitary evolution of a quantum state as a physical process that 'just happens' to instead seeing it as an executable operation in a programmable computer. To do this, we introduce the *quantum circuit model*, which subsequently allows us to present and analyse quantum algorithms and quantum communication protocols. Before doing so, however, it is worth addressing the fact that the name 'quantum circuit model' may be considered something of a misnomer. Certainly it does not imply an analogy with an analogue electrical circuit, where some components are hooked up to a battery, thus literally forming a closed circuit. In fact, the name really arises from the similarity with *digital* circuits, in which logical operations are applied to the constituent bits. It follows that the gates in the quantum circuits are not material entities as such – all that physically exists are the qubits themselves, to which operations (gates) are applied. This gives rise to the alternative name, the *quantum gate model*.

As we develop quantum circuitry as a model of computation there is a subtle but important point to bear in mind: the postulates of quantum mechanics describe what happens to a *closed* quantum system, however treating quantum phenomena as controllable and executable necessarily implies some opening of the system. We speak of 'applying unitaries' to a quantum state when the postulates tell us that unitary evolution is what happens to a system that does not interact with the outside. In Chapter 15, we plug this gap by considering noisy quantum systems and show that efficient quantum computation is still possible, even if the quantum system is not completely closed.

4.1 The essentials of the quantum circuit model

To motivate the quantum circuit model, we start with a simple example. Consider a two-qubit state $|\psi\rangle = \alpha|00\rangle + \beta|01\rangle + \gamma|10\rangle + \delta|11\rangle$, and let a Pauli-$X$ operation be applied to the first qubit. From the previous chapters, the state, $|\psi'\rangle$, after the operation has been applied can be found by computing $(X \otimes I)|\psi\rangle$. A unitary operation can be represented as a *quantum circuit*, read from left to right where:

- wires are qubits (possibly entangled);
- gates are unitary matrices.

For the example of $(X \otimes I)$, the quantum circuit is:

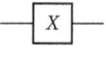

Figure 4.1

As the Pauli-X is a 'not' operation, we immediately get $|\psi'\rangle = \alpha|10\rangle + \beta|11\rangle + \gamma|00\rangle + \delta|01\rangle$ by simply 'flipping' the first bit inside each ket (showing that the same answer is obtained by doing the matrix multiplication is one of the chapter problems). In this way, the quantum circuit model sometimes can provide an easier and more visual way to express the action of unitary matrices on multi-qubit states.

The quantum circuit model is suitable for analysing the quantum algorithms and information processing protocols studied in the later chapters as they invariably do involve building up a multi-qubit operation as a product of elementary single- and two-qubit unitary operations. Of these elementary operations, we have already met the Pauli and Hadamard single-qubit unitary matrices as well as the CNOT two-qubit unitary. The *phase gate*, $S = \begin{bmatrix} 1 & 0 \\ 0 & i \end{bmatrix}$, and *T gate*, $T = \begin{bmatrix} 1 & 0 \\ 0 & e^{i\pi/4} \end{bmatrix}$, are also useful primitives. Thus, *some* of the quantum gates we use throughout the remainder of this book are:

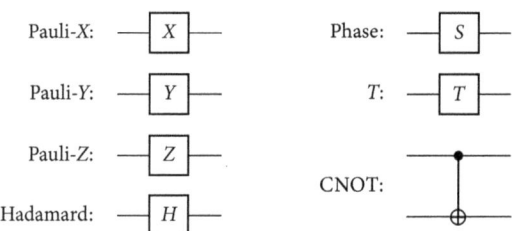

Figure 4.2

With these operations introduced, quantum circuits can be defined: a quantum circuit of n qubits has three stages:

1. Initialisation of all qubits in some n-qubit state. As preparation of the initial state may itself be non-trivial, for the purposes of analysing quantum algorithms, we require that the initial state is $|0\rangle^{\otimes n}$ (or something that can be prepared from $|0\rangle^{\otimes n}$ in a few simple operations) so that we do not undercount the computational complexity by omitting the operations involved in state preparation.
2. Some quantum gates, which represent unitary transformations, applied to one, two, or more of the wires.
3. A final layer of measurements in the computational basis, on some or all of the qubits. In fact, by the *principle of implicit measurement* (see Section 4.2), all

of the qubits can be treated as being measured in the final layer. (Note that a 'layer of operations' is a column of gates/measurements that occur in parallel; for example, in Fig. 4.3, the two X gates are in the same layer.)

Another example of a quantum circuit is thus:

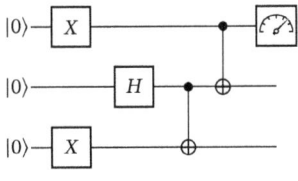

Figure 4.3

where the final symbol on the top qubit indicates computational basis measurement.

The fact that measurements of states prepared by quantum circuits are (at least in this book) necessarily taken in the computational basis may appear to be unduly restrictive. However, this is not actually so, as projective measurement in any basis is always equivalent to some unitary circuit, followed by a computational basis measurement, which can be seen by the following. Letting $\{M_i\} = \{|\lambda_i\rangle\langle\lambda_i|\}$ be a set of measurement operators defined as projectors onto the orthonormal basis $\{\lambda_i\}$, we define:

$$U = \sum_i |\lambda_i\rangle\langle i| \tag{4.1}$$

i.e., a unitary operation whose columns are the vectors $|\lambda_i\rangle$. We now consider the probability of obtaining the ith measurement outcome, when an arbitrary state, $|\psi\rangle$, prepared by some quantum circuit, is measured. From the measurement postulate:

$$\begin{aligned}
\Pr(M = i) &= \langle\psi|M_i|\psi\rangle \\
&= \langle\psi||\lambda_i\rangle\langle\lambda_i||\psi\rangle \\
&= \langle\psi|UU^\dagger|\lambda_i\rangle\langle\lambda_i|UU^\dagger|\psi\rangle \\
&= \langle\psi|U\sum_j |j\rangle\langle\lambda_j||\lambda_i\rangle\langle\lambda_i|\sum_k |\lambda_k\rangle\langle k|U^\dagger|\psi\rangle \\
&= \langle\psi|U|i\rangle\langle\lambda_i||\lambda_i\rangle\langle\lambda_i||\lambda_i\rangle\langle i|U^\dagger|\psi\rangle \\
&= \langle\psi|U|i\rangle\langle i|U^\dagger|\psi\rangle \\
&= \langle\psi'||i\rangle\langle i||\psi'\rangle \tag{4.2}
\end{aligned}$$

where $|\psi'\rangle = U^\dagger|\psi\rangle$. As this holds for each i, the measurement statistics are, therefore, identical to the case where the circuit $U^\dagger$ is first applied to $|\psi\rangle$, and then a computational basis measurement (i.e., with measurement operators $\{|i\rangle\langle i|\}$) is taken.

(Note that $U^\dagger$ is a quantum circuit, as one of the main results towards which this chapter builds is that every unitary operation is some quantum circuit.) Treating measurement as always being taken in the computational basis has the advantage of being notationally simple: when depicting quantum circuits, there is no need to decorate the measurement symbols to indicate a basis. It also provides a fair comparison from the perspective of computational complexity, as there is no possibility of hiding (a possibly large number of) quantum operations by choosing a particularly exotic basis.

Returning to the definition of a quantum circuit in the enumeration above, sometimes when analysing the operation of quantum circuits, it is the quantum state prepared which is of interest, and so measurement is omitted. For example, consider a circuit that prepares the Bell state $|\Phi^+\rangle$:

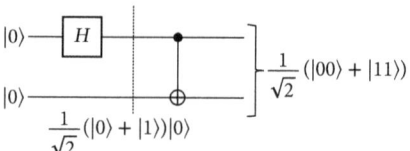

Figure 4.4

As measurement is clearly always required at some point to extract classical information from the state prepared by a quantum circuit, such circuits are usually actually 'sub-circuits'/'circuit fragments' that are connected with other circuit blocks to form a complete circuit (i.e., including measurement).

As well as sometimes omitting measurement altogether, the stipulation that measurement should only occur in a final layer may also be relaxed on occasion, thus allowing *mid-circuit measurement*. This is necessary to incorporate *classical control*; however, in many cases, the *principle of deferred measurement* means that a circuit with mid-circuit measurement can be rewritten as a circuit in the above standard form (that is, with measurement only in the final layer). This is detailed in Section 4.2.

One reason for insisting on a standard definition of a quantum circuit with measurement only allowed to occur in the final layer is that in this case, the quantum circuit, with the initialisation and measurement stages omitted, simply represents a unitary operation. This is because composition of unitary operators by the matrix and tensor product yields a matrix that is itself unitary, and in particular, to find the matrix of a quantum circuit, it is necessary to adhere to the following two rules (which come directly from the postulates of quantum mechanics), *applied in the following order*:

1. Composition of operations acting in parallel (i.e., in the same circuit layer) on different (sets of) wires (qubits) is achieved by the tensor product.
2. Composition of operations acting on the same wire (qubit) is achieved by the normal matrix product, but *right to left*.

For example:

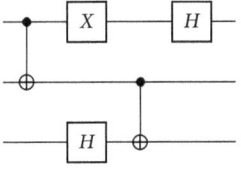

Figure 4.5

corresponds to:

$$(H \otimes I_4) \times (I_2 \otimes \text{CNOT}) \times (X \otimes I_2 \otimes H) \times (\text{CNOT} \otimes I_2) \qquad (4.3)$$

where I_2 is the 2 × 2 identity, and $I_4 = I_2 \otimes I_2$ is the 4 × 4 identity. The composition of matrix products happens from right to left because when an operation is applied to some state, the initial state is written to the right-hand side of the matrix when expressed algebraically, but at the left-hand side of the circuit.

By representing the overall unitary evolution as a product of quantum gates, the quantum circuit model exactly captures what is and is not important about the order of the quantum operations. Naturally, operations on any given qubit occur in the order shown in the circuit, but what about operations on different qubits?

Consider the example in Fig. 4.5: viewed as a physical (unitary) evolution of a quantum state, it may appear important that the final Hadamard gate occurs *after* the CNOT. However, when viewed as a quantum circuit, it appears as if the order of these two is unimportant, as they concern different wires (qubits). That is, we could 'slide' the CNOT along its wire to be in a layer with the Hadamard, or even to come after it. It turns out that sliding operations along their wires does not change the overall unitary operation performed by the quantum circuit. To see this, consider the following example, where U_1 and U_2 each may act on multiple qubits – as indicated by the slash '/' through the wires:

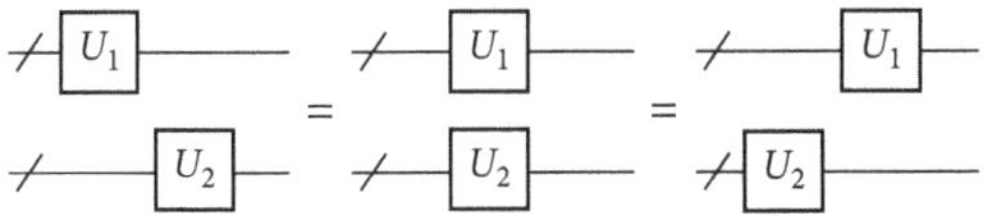

Figure 4.6

because:

$$(I \otimes U_2)(U_1 \otimes I) = (IU_1) \otimes (U_2 I) = U_1 \otimes U_2 = (U_1 I) \otimes (IU_2) = (U_1 \otimes I)(I \otimes U_2) \qquad (4.4)$$

In this way, the quantum circuit model captures unitary evolution in precisely the way one would expect it to from viewing a quantum circuit from an 'electrical engineering' perspective.

It is important to appreciate that this process of 'sliding' gates on different (sets of) qubits past each other is valid for all operations and should not be confused with commutation, where the order of certain unitary operations on *the same* qubit can be swapped. In particular, if two unitary operations (necessarily of the same size) U and V commute, then, by definition, $UV = VU$, and so if U and V appear as sequential gates applied to the same qubit (or set of qubits) in some quantum circuit, then their order can be swapped. Z and S gates provide an example of commuting operators:

$$ZS = \begin{bmatrix} 1 & 0 \\ 0 & -1 \end{bmatrix} \begin{bmatrix} 1 & 0 \\ 0 & i \end{bmatrix} = \begin{bmatrix} 1 & 0 \\ 0 & -i \end{bmatrix} = \begin{bmatrix} 1 & 0 \\ 0 & i \end{bmatrix} \begin{bmatrix} 1 & 0 \\ 0 & -1 \end{bmatrix} = SZ \qquad (4.5)$$

Therefore, in a quantum circuit:

Figure 4.7

A final 'basic' property of quantum circuits is that the inverse (which is equal to the adjoint) of any unitary matrix is itself a unitary matrix. This means that for every quantum gate, its inverse is also a quantum gate. Some gates, such as the Hadamard and Pauli X, Y, and Z operators are self-inverse, and for other operators, it is usual to write the inverse gate with a superscript †. The fact that every quantum *gate* is invertible means that every quantum *circuit* is invertible in the sense that the unitary matrix of the entire circuit itself has an inverse that is a unitary matrix. This property means that quantum circuits are a *reversible* model of computation. Later in the book, Fig. 9.3 gives an explicit example of how to construct the inverse of a quantum circuit from elementary gates.

4.2 Classical control and deferred and implicit measurement

The reason that some qubits may be measured before all of the unitary operations on the other qubits are completed is to allow for the possibility that some of those unitaries are only to be executed if a certain measurement outcome is obtained. This is termed *classical control*. An example of a quantum circuit exhibiting classical control is shown in Fig. 4.8.

(Note that as the two-qubit state is separable throughout, the state of the first qubit after the various gates is shown above the circuit, whereas that of the second qubit is shown below.) By convention, classical information (that is, a *bit* rather than a *qubit*) is represented by a double line. So we should read this classical control as: *if the measurement outcome on the first qubit is zero, then do nothing; if the measurement outcome on the first qubit is one, then perform the Pauli-X operation on the second*

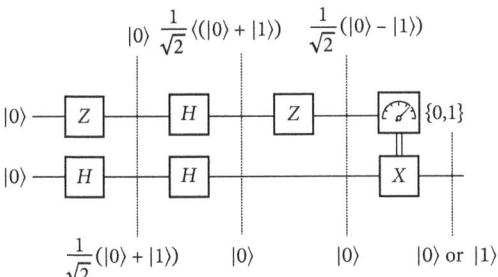

Figure 4.8

qubit. However, any time a circuit involves mid-circuit measurement and classical control, the principle of deferred measurement can be used to write an equivalent circuit where all measurements occur in the final layer.

Principle of deferred measurement. *Any mid-circuit computational basis measurement can be deferred to the end of the circuit, and classical-control conditioned on the measurement outcome replaced by quantum control.*

For example:

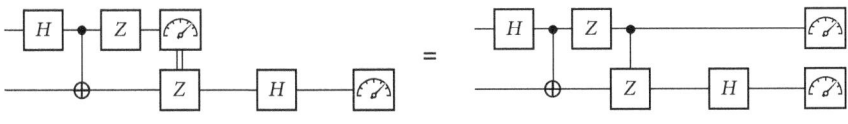

Figure 4.9

The principle of deferred measurement can be reasoned along the lines: *whatever happens in the remainder of the circuit must be consistent with the measurement outcome observed*, and thus, the observed measurement statistics are insensitive to whether that control is classical or quantum with the measurement outcome deferred.

Some care is needed with the principle of deferred measurement: it only works because we have restricted measurement to computational basis measurement, and in particular, it is important to remember that it applies to the equivalence of measurement statistics only – intermediate quantum states will not be the preserved. For this reason, it is sometimes a little unsatisfactory from a physical perspective; for example, if the principle of deferred measurement is applied to the quantum teleportation circuit (Chapter 5), then the physical interpretation is lost.

On a side note (albeit an important one), Fig. 4.9 also introduces general quantum control. Thus far, the only controlled gate we have met is the CNOT, which amounts to quantum control of a Pauli-X gate (sometimes, the CNOT is referred to as the 'CX' gate), such that it could be equivalently written:

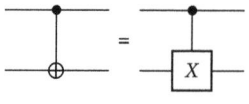

Figure 4.10

Any single-qubit gate could replace the X here, as does Z in the example in Fig. 4.9, and general controlled gates are introduced in full detail in Section 4.3.

Another useful principle concerning measurement is the principle of implicit measurement.

Principle of implicit measurement. *For a complete quantum circuit (i.e., not one where the final state is used as the initial state for further quantum operations in another quantum circuit), any unmeasured qubits can be treated as being measured.*

For example, if for some circuit the only important information concerns the outcome of measuring the first qubit, e.g.:

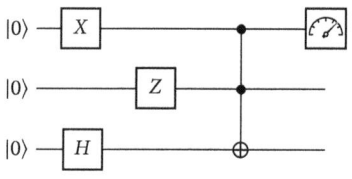

Figure 4.11

then, by causality, it is irrelevant whether or not the other qubits are *later* measured. So there is no difference if the circuit of Fig. 4.11 is replaced by:

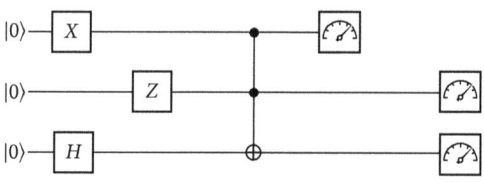

Figure 4.12

but in the quantum circuit model, the measurement operations may be 'slid along' their respective wires, such that the circuit of Fig. 4.12 is equivalent to a circuit where the three measurements occur in the same layer:

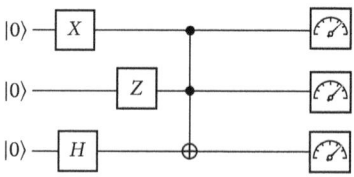

Figure 4.13

The principle of implicit measurement is closely related to the no-signalling principle: if some qubits of an entangled state are measured, then the statistics of those measurements are unchanged by whether or not the other qubits are measured.

4.3 Rotations and controlled gates

The quantum gates that we have met so far largely suffice to define and analyse the quantum algorithms and communication protocols included in this book. Indeed, Section 4.4 shows that the gates already introduced are universal for quantum computation. However, occasionally, it is convenient to use some other gates. First, it is useful to introduce the notion of a general single-qubit gate:

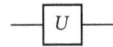

Figure 4.14

where U is an arbitrary 2×2 unitary matrix. Moreover, sometimes, it is also convenient to draw an arbitrary n-qubit unitary operation as a circuit block covering the n wires in question, which we may also typically mark 'U'.

Along with fully defined unitary operations like the Paulis, Hadamard, S, T, and CNOT, and arbitrary unitary operations represented by circuit blocks marked 'U', a third category is the *parameterised* quantum gates, the most notable examples of which are the *rotation gates*. To introduce the rotation gates, first note that for any self-inverse matrix A (i.e., which is such that $A^2 = I$), the following holds for real x:

$$\begin{aligned}
e^{iAx} &= \sum_{k=0}^{\infty} \frac{(iAx)^k}{k!} \\
&= \sum_{\text{even } k} (-1)^{k/2} \frac{x^k}{k!} I + i \sum_{\text{odd } k} (-1)^{(k-1)/2} \frac{x^k}{k!} A \\
&= \cos(x) I + i \sin(x) A
\end{aligned} \quad (4.6)$$

(where $k = 0$ is included in even k). As the Pauli matrices are all self-inverse ($X^2 = Y^2 = Z^2 = I$), the following gates, each parameterised by the angle θ, are defined:

$$R_x(\theta) = \exp(-i\theta X/2) = \cos(\theta/2)I - i\sin(\theta/2)X = \begin{bmatrix} \cos\theta/2 & -i\sin\theta/2 \\ -i\sin\theta/2 & \cos\theta/2 \end{bmatrix} \quad (4.7)$$

$$R_y(\theta) = \exp(-i\theta Y/2) = \cos(\theta/2)I - i\sin(\theta/2)Y = \begin{bmatrix} \cos\theta/2 & -\sin\theta/2 \\ \sin\theta/2 & \cos\theta/2 \end{bmatrix} \quad (4.8)$$

$$R_z(\theta) = \exp(-i\theta Z/2) = \cos(\theta/2)I - i\sin(\theta/2)Z = \begin{bmatrix} e^{-i\theta/2} & 0 \\ 0 & e^{i\theta/2} \end{bmatrix} \quad (4.9)$$

These are known as the rotation gates, as they are unitaries that have the effect of rotating any single-qubit state by the amount θ around the x, y, and z axes, respectively, when represented on the Bloch sphere.

One aspect of the quantum circuit model that is important both for establishing the theoretical capability of computing with quantum circuits and for practical quantum compilation is the idea of decomposing some unitary into a product of other unitaries. An important decomposition of an arbitrary single-qubit unitary, U, is:

$$U = e^{ia} R_z(b) R_y(c) R_z(d) = e^{ia} \begin{bmatrix} e^{-i(b+d)/2} \cos(c/2) & -e^{-i(b-d)/2} \sin(c/2) \\ e^{i(b-d)/2} \sin(c/2) & e^{i(b+d)/2} \cos(c/2) \end{bmatrix} \quad (4.10)$$

where the real numbers a, b, c, and d can be chosen such that U corresponds to any desired single-qubit unitary.

The global phase factor, e^{ia}, has not been omitted here, because we now consider *controlled* unitaries of this form. In this case, the global phase factor of the unitary being controlled is not unobservable – that is, the global phase factor of the single-qubit unitary being controlled is *not* the global phase factor of the two-qubit controlled gate. A two-qubit controlled unitary is represented in quantum circuit form as:

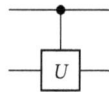

Figure 4.15

This can be read as saying *when the first qubit equals 1, then the unitary on the second qubit is applied* and has block matrix representation

$$\begin{bmatrix} I & 0 \\ 0 & U \end{bmatrix} \quad (4.11)$$

where each element is a 2×2 matrix. It is also easy to verify that this block-defined matrix is unitary for any unitary, U, and thus, a controlled unitary is itself a two-qubit unitary operation and, therefore, a valid quantum gate.

It is often convenient (for both practical and theoretical purposes) to allow the only two-qubit gate used in a quantum circuit to be the CNOT, which is possible as any two-qubit controlled gate can be decomposed into a circuit with only CNOTs and single-qubit unitaries. To see how this is so, we first show that any unitary, U, can be represented in the form:

$$U = e^{ia} AXBXC \quad (4.12)$$

where:

$$A = R_z(b) R_y(c/2) \quad (4.13)$$
$$B = R_y(-c/2) R_z(-(d+b)/2) \quad (4.14)$$
$$C = R_z((d-b)/2) \quad (4.15)$$

which has the properties that $ABC = I$, and $AXBXC = R_z(b)R_y(c)R_z(d)$, i.e., $U = e^{i\alpha}R_z(b)R_y(c)R_z(d)$, which, therefore, suffices to represent any single-qubit unitary, by Eqn. (4.10). Letting $U_\alpha = \begin{bmatrix} e^{i\alpha} & 0 \\ 0 & e^{i\alpha} \end{bmatrix}$ and $U'_\alpha = \begin{bmatrix} 1 & 0 \\ 0 & e^{i\alpha} \end{bmatrix}$, this general unitary decomposition gives:

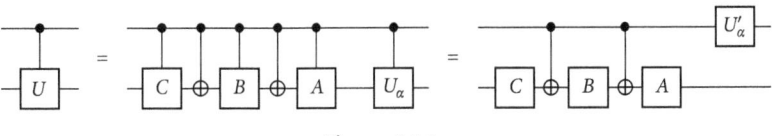

Figure 4.16

The equals signs here is shorthand for 'functional equivalence'. The first equivalence comes from the fact that any controlled unitary can be decomposed into a composition of gates, with each individually controlled (this is a direct consequence of the block matrix form of controlled unitaries, i.e., as in Eqn. (4.11)). In particular, this is applied to the decomposition in Eqn. (4.12), with each of the gates controlled individually. To see why the control for the three matrices A, B, and C can be dropped in the second equivalence, consider the following argument: when the top qubit is equal to one, then A, B, and C would be executed anyway and so dropping the control does not change this, and when the top qubit is 0, then neither of the CNOTs does anything, and so the second qubit is multiplied by $ABC = I$; in this case, omitting the control such that these operations are *all* applied is equivalent to the controlled versions where *none* of them are applied. So, it follows that the same operation is applied even when the control is dropped such that A, B, and C are unconditionally executed on the second qubit. Finally, the controlled U_α (i.e., CU_α) can be rewritten as an unconditional U'_α on the first qubit because:

$$CU_\alpha = \begin{bmatrix} 1 & 0 & 0 & 0 \\ 0 & 1 & 0 & 0 \\ 0 & 0 & e^{i\alpha} & 0 \\ 0 & 0 & 0 & e^{i\alpha} \end{bmatrix} = U'_\alpha \otimes I_2 \tag{4.16}$$

Thus, any controlled single-qubit unitary can be executed using a circuit consisting only of CNOT gates and single-qubit unitaries.

Finally, on the subject of controlled unitaries, we turn our attention to *multi-controlled unitaries*. These are represented in circuit form as:

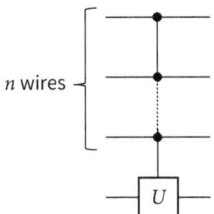

Figure 4.17

which can be interpreted as *if and only if all of the first n qubits equal 1, then execute U on the (n + 1)th qubit*, and is represented by the matrix:

$$C^n U = \begin{bmatrix} I_{2^{n+1}-2} & 0 \\ 0 & U \end{bmatrix} \tag{4.17}$$

To decompose a general multi-controlled gate into a circuit with only CNOTs and single-qubit unitaries, *ancilla qubits* (or simply *ancillas*) are needed. 'Ancilla' refers to an auxiliary qubit that aids the computation, but is not part of the computational state itself *per se*. As a first step, an ancilla, set in the state $|0\rangle$, can be used to replace the multi-controlled unitary with a (functionally equivalent) circuit containing only single-qubit gates, multi-controlled NOT gates, and one single-controlled general unitary:

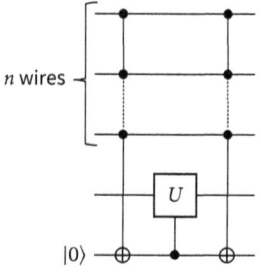

Figure 4.18

The behaviour of this circuit can be described as follows: if all of the first n *control* qubits are in the 1 state, then the NOT gate on the ancilla (bottom-most) *target* qubit is executed. As the ancilla is initialised as $|0\rangle$, this means it is flipped to $|1\rangle$ in which case U is performed on the $(n + 1)$th qubit as required. The second multi-controlled NOT gate affects only the ancilla and has the effect of putting it back in the $|0\rangle$ state (regardless of whether the multi-controlled condition was met or not). This is known as *uncomputing*, and it is conventional to do so such that the ancilla is returned to its original state and does not remain entangled with the computational state itself. Fig. 4.18 also raises another subtle but important point about the quantum circuit model, namely that it is possible to draw gates bridging over wires. Observe that here there is no black circle on the $(n + 1)$th wire indicating control in the multi-controlled NOT gates. This makes writing down the matrix representation of a circuit a little trickier, but Section 4.5 shows how to resolve this.

The second step is to use further ancillas to break down the multi-controlled NOT gate into a cascade of NOT gates conditioned only on the states of two other qubits, which are known as Toffoli gates. For example, a C^4NOT gate can be implemented using five Toffoli gates and two ancillas initialised in the $|0\rangle$ state:

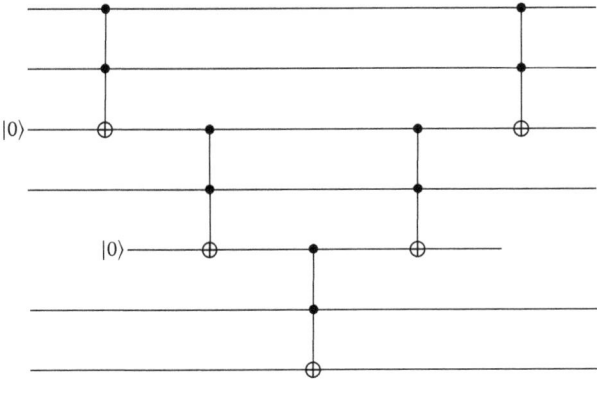

Figure 4.19

Here, the final two Toffoli gates perform the necessary uncomputation. The Toffoli gate is of more general interest than merely being of use in decompositions of this sort, and may be thought of as one of the standard gates that we may use from time-to-time when analysing quantum algorithms. The Toffoli gate has matrix representation (when the first two qubits are the controls, and the third is the target):

$$\text{Toffoli} = \begin{bmatrix} 1 & 0 & 0 & 0 & 0 & 0 & 0 & 0 \\ 0 & 1 & 0 & 0 & 0 & 0 & 0 & 0 \\ 0 & 0 & 1 & 0 & 0 & 0 & 0 & 0 \\ 0 & 0 & 0 & 1 & 0 & 0 & 0 & 0 \\ 0 & 0 & 0 & 0 & 1 & 0 & 0 & 0 \\ 0 & 0 & 0 & 0 & 0 & 1 & 0 & 0 \\ 0 & 0 & 0 & 0 & 0 & 0 & 0 & 1 \\ 0 & 0 & 0 & 0 & 0 & 0 & 1 & 0 \end{bmatrix} \quad (4.18)$$

and can be executed by the following circuit, containing only CNOTs and single-qubit unitaries:

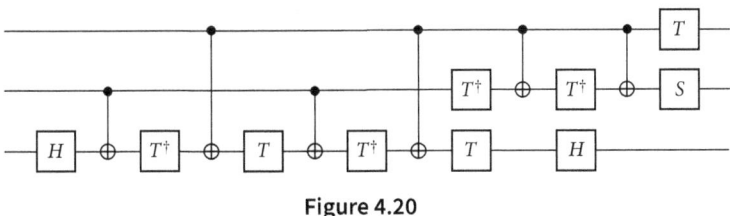

Figure 4.20

The correctness of which can be verified by an easy (if tedious) multiplication of matrices.

Putting together: (i) the decomposition of any multi-controlled single-qubit unitary into a circuit with two multi-controlled NOT gates and a single-controlled single-qubit unitary, in Fig. 4.18; (ii) the decomposition of any single-controlled unitary into a circuit with only CNOTs and single-qubit unitaries, in Fig. 4.16; (iii) the decomposition of a multi-controlled NOT into a circuit with Toffolis and ancillas, as exemplified in Fig. 4.19; and (iv) the decomposition of a Toffoli into a circuit with only single-qubit unitaries and CNOTs, as in Fig. 4.20, suffices to show that *any* multi-controlled single-qubit unitary can be decomposed into a circuit containing only single-qubit unitaries and CNOTs.

In this section, we have met the following gates, which may be thought of as additional primitives:

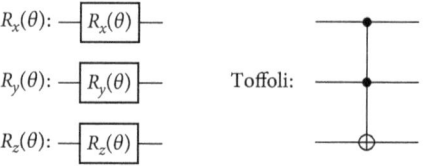

Figure 4.21

4.4 Universality of the quantum circuit model

There are two important notions of universality that we must address for the quantum circuit model. First, there is *computational universality*: that any function that can be computed classically can also be computed using the quantum circuit model. Second, there is *quantum universality*: that the evolution and measurement of any quantum mechanical system, which is a composition of qubits, can be represented using the quantum circuit model. We show that the quantum circuit model indeed satisfies both of these.

4.4.1 Computational universality

We first develop an informal demonstration of the computational universality of the quantum circuit model, showing that any function computable by a classical Boolean circuit can be computed (in a certain sense) in a quantum circuit using gates we have already met. By design, the quantum circuit model expresses unitary evolutions of quantum states in a manner that can be recognised as similar to classical logic circuits. For example, a simple logic circuit to compute $a \cdot b \cdot \bar{c}$ (where $a, b, c \in \{0, 1\}$ are Boolean literals) is

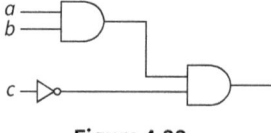

Figure 4.22

There are, however, two important distinctions between quantum and classical circuits: quantum gates have exactly the same number of outputs as they have inputs;

moreover, as the gates represent unitary matrices, they are invertible, for example (for any multi-qubit unitary, U, and quantum state $|\psi\rangle$):

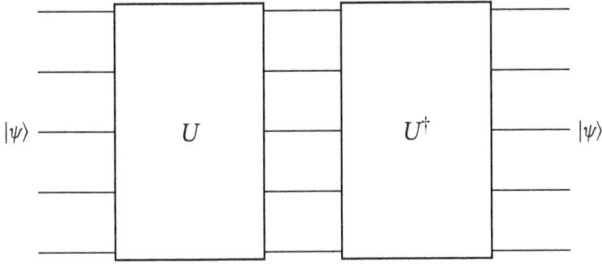

Figure 4.23

So if we consider, for example, the 'AND' gate, then a naive implementation of the AND gate is not possible using quantum gates, as the AND gate takes two inputs to a single output and, therefore, clearly cannot be represented as a unitary matrix. A simple 'patch' would be to give the AND gate a second output, thus constructing a gate of the form:

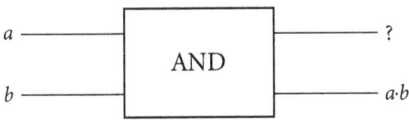

Figure 4.24

Such a gate cannot be reversible and, therefore, cannot be unitary. This is most easily seen by noticing that there are three occasions when the second output is zero ($a = 0, b = 0$); ($a = 0, b = 1$); ($a = 1, b = 0$), and only one bit (the first output) with which to distinguish them. Therefore, the inputs a and b cannot be determined from two outputs of which one is $a \cdot b$, and hence, there is not a reversible circuit of this form.

However, we have already met the solution to the problem of implementing AND in a quantum circuit: use the Toffoli gate. Acting as an AND gate is the very role that the Toffoli gate plays in the multi-controlled gate decompositions in Section 4.3. The Toffoli gate may be thought of as a reversible form of an AND gate, where the first two bits are AND-ed and then added (modulo 2) to the third bit; the first two bits are also returned, which ensures reversibility:

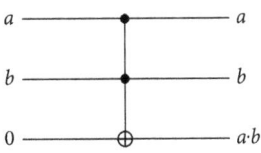

Figure 4.25

The idea of a computationally universal set of reversible gates follows directly from this insight. From classical logic, together, the NAND and FANOUT gates constitute a universal gate-set. That is, any Boolean function on a fixed number, n, of input bits can be implemented using just these gates. Each of these can be implemented using Toffoli and X gates, and some ancilla qubits. For NAND:

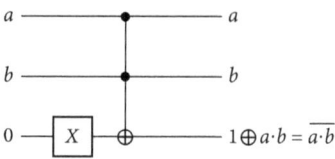

Figure 4.26

An alternative would be to prepare the third qubit in the 1 state directly, which would allow the X gate to be omitted; however, we assume that all ancilla qubits are provided in the $|0\rangle$ state. Similarly, the FANOUT gate can be constructed:

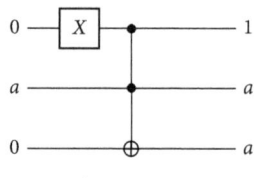

Figure 4.27

Note that this is really a controlled FANOUT; however, the first (control) qubit has been fixed as the constant 1, such that the second and third qubits give the FANOUT operation.

This means that any Boolean function can be computed on a quantum computer, if a supply of ancillas is available. In particular, a function, $f(x)$, can be computed on a quantum computer in the following way:

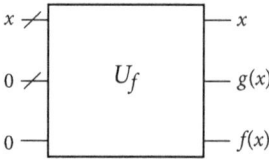

Figure 4.28

where the block 'U_f' is a quantum circuit consisting only of Toffoli and X gates that reversibly computes the function in question. Specifically, U_f takes the bit string x as an input, which is achieved by preparing the top bundle (*register*) of qubits as the corresponding computational basis state; there is also an ancillary register (the middle bundle of qubits); and finally, an output qubit into which the result of the computation is placed. Following the convention established when treating the Toffoli gate as a reversible AND gate, x is itself also returned as an output, as is $g(x)$, the transformed state of the ancillary register (which therefore contains some intermediate results from the computation). For example, consider a simple function that is the AND of each bit in a three-bit input register, i.e., $f(x) = x_1 \cdot x_2 \cdot x_3$ (where x is the three-bit string $x = x_1 x_2 x_3$), which can be implemented using two Tofolli gates. In this case, the additional output could be $g(x) = x_1 \cdot x_2$.

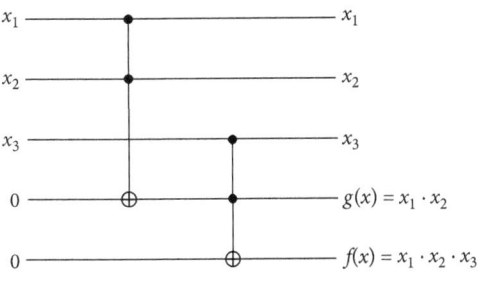

Figure 4.29

Even though this construction is sufficient to show that any (fixed size) Boolean function can be computed as a reversible circuit consisting of Toffoli and X gates, in practice, having $g(x)$ as an output may cause a problem when U_f is applied to a superposition of computational basis states, as in this case the output of the ancillary register would end up entangled with the output of the computation itself. This problem was also present when using a number of Toffoli gates to execute a multi-controlled NOT gate in Section 4.3, and the solution is again to uncompute the ancillary register, which can be achieved in general by constructing the circuit:

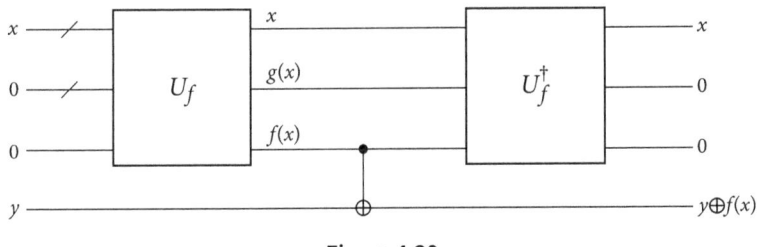

Figure 4.30

Note that here, to represent the general case, rather than forcing the final qubit to be initialised as the value 0, instead the Boolean literal y is taken as an input.

When illustrating reversible functions in quantum circuits, the ancillary register is usually omitted, so *the reversible form of $f(x)$ is typically represented by the circuit block* (now showing the inputs and outputs as quantum states rather than binary strings, to emphasise that this is a block for use in a *quantum* circuit):

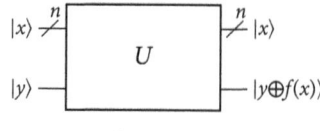

Figure 4.31

Throughout the preceding analysis, the Boolean function in question has been of the form $f: \{0,1\}^n \to \{0,1\}$; however, the same principles apply if there are multiple output bits, such that the function is instead $f: \{0,1\}^n \to \{0,1\}^m$. In this case, the output consists of an m-qubit register, and in Fig. 4.30, some m CNOT gates would be needed to 'copy' the m output values thereto. This is the sense in which any Boolean function can be represented as a quantum circuit: for any Boolean function $f: \{0,1\}^n \to \{0,1\}^m$, there is a unitary, U, which can be implemented using just Toffoli and Pauli-X gates, on $m + n$ qubits (plus possibly some ancillas – which can always be uncomputed back to their original states) that takes as an input $|x\rangle|0^m\rangle$, where $|x\rangle$ is an n-qubit computational basis state, and prepares (returns) $|x\rangle|f(x)\rangle$.

4.4.2 Quantum universality

The quantum circuit model completely captures the postulates of quantum mechanics for any quantum system that is a composition of qubits:

1. **Postulates 1 and 4**: the wires represent the state-space of a composition of qubits, which can be entangled.
2. **Postulates 2**: the gates are just a convenient way of writing down the unitary evolution.
3. **Postulate 3**: Measurement may occur, and in Section 4.1, it was shown that an equivalent circuit with computational basis measurement can be found for any projective measurement; moreover, in Section 4.2, that this measurement can be deferred to the end of the circuit.

Furthermore, there is no loss in generality in assuming that the initial state is always $|0\rangle^{\otimes n}$, as any other state can be treated as a unitary evolution of $|0\rangle^{\otimes n}$ that is then absorbed into the quantum circuit itself.

This justification of quantum universality, however, implicitly allows for arbitrary circuit blocks – so we may simply write a single block 'U' covering all of the wires to represent the desired unitary evolution. A more useful question about quantum universality, then, is whether we can use the quantum circuit model to represent any unitary evolution, when that evolution must broken down into a product of elementary gates.

In fact, it turns out that any n-qubit unitary *can* be approximated to arbitrary accuracy using a product of gates from a finite gate-set. Perhaps surprisingly, only three gates are required, and we now show that $\{CNOT, H, T\}$ is a *universal* gate-set. Before giving a sketch of the proof of the universality of {CNOT, H, T}, it is worth pointing out that all of the elementary non-parameterised single-qubit gates we have met so far can easily be decomposed into a series of H and T gates:

$$S = T^2 \tag{4.19}$$

$$Z = S^2 = T^4 \tag{4.20}$$

$$X = HZH = HT^4H \tag{4.21}$$

$$Y = iXZ = SXSZ = T^2HT^4HT^6 \tag{4.22}$$

We now proceed to show that {CNOT, H, T} is a universal gate-set by first showing that any unitary can be decomposed into a circuit with only CNOTs and single-qubit unitaries, and then that any single-qubit unitary can be constructed from H and T gates.

Any n-qubit unitary can be decomposed into a product of 'two-level' unitaries

In the first step towards showing that {CNOT, H, T} is a universal gate-set, we show that any n-qubit unitary (i.e., represented by a $N \times N$ element matrix, where $N = 2^n$) can be decomposed into a product of 'two-level' unitaries. An N-element two-level unitary, U, is a unitary matrix whose elements are all equal to the corresponding element in the identity matrix of the same dimension, except for in up to four places: u_{ii}, u_{ij}, u_{ji}, and u_{jj} for some $1 \le i, j \le N$. First, consider a general N-element unitary matrix:

$$W = \begin{bmatrix} w_{11} & w_{12} & w_{13} & \cdots \\ w_{21} & w_{22} & w_{23} & \\ w_{31} & w_{32} & w_{33} & \\ \vdots & & & \ddots \end{bmatrix} \tag{4.23}$$

There is a procedure for pre-multiplying W by two-level unitary matrices, such that the second to Nth elements of the first column of the resultant matrix are all zero. The first step in this procedure is to pre-multiply W by the two-level unitary matrix, U_1, defined in the case that $w_{21} \ne 0$ by:

$$U_1 = \frac{1}{\sqrt{|w_{11}|^2 + |w_{21}|^2}} \begin{bmatrix} w_{11}^* & w_{21}^* & 0 & \cdots \\ w_{21} & -w_{11} & 0 & \\ 0 & 0 & 1 & \\ \vdots & & & \ddots \end{bmatrix} \tag{4.24}$$

When w_{21} is zero, then U_1 is instead defined as the identity. In either case, this gives:

$$W^{(1)} = U_1 W = \begin{bmatrix} w_{11}^{(1)} & w_{12}^{(1)} & w_{13}^{(1)} & \cdots \\ 0 & w_{22}^{(1)} & w_{23}^{(1)} & \\ w_{31}^{(1)} & w_{32}^{(1)} & w_{33}^{(1)} & \\ \vdots & & & \ddots \end{bmatrix} \tag{4.25}$$

In a similar vein, if the third element of the first column of $W^{(1)}$ is non-zero, then $W^{(1)}$ is pre-multiplied by U_2, defined when $w_{31}^{(1)} \neq 0$ by:

$$U_2 = \frac{1}{\sqrt{|w_{11}^{(1)}|^2 + |w_{31}^{(1)}|^2}} \begin{bmatrix} (w_{11}^{(1)})^* & 0 & (w_{31}^{(1)})^* & \cdots \\ 0 & 1 & 0 & \\ w_{31}^{(1)} & 0 & -w_{11}^{(1)} & \\ \vdots & & & \ddots \end{bmatrix} \quad (4.26)$$

(Again, if $w_{31}^{(1)}$ is zero, then U_2 is instead the identity.) This gives:

$$W^{(2)} = U_2 W^{(1)} = U_2 U_1 W = \begin{bmatrix} w_{11}^{(2)} & w_{12}^{(2)} & w_{13}^{(2)} & w_{14}^{(2)} & \cdots \\ 0 & w_{22}^{(2)} & w_{23}^{(2)} & w_{24}^{(2)} & \\ 0 & w_{32}^{(2)} & w_{33}^{(2)} & w_{34}^{(2)} & \\ w_{41}^{(2)} & w_{42}^{(2)} & w_{43}^{(2)} & w_{44}^{(2)} & \\ \vdots & & & & \ddots \end{bmatrix} \quad (4.27)$$

Continuing in this manner for the entire first column produces the matrix:

$$W^{(N-1)} = U_{N-1} \ldots U_2 U_1 W = \begin{bmatrix} w_{11}^{(N-1)} & w_{12}^{(N-1)} & w_{13}^{(N-1)} & w_{14}^{(N-1)} & \cdots \\ 0 & w_{22}^{(N-1)} & w_{23}^{(N-1)} & w_{24}^{(N-1)} & \\ 0 & w_{32}^{(N-1)} & w_{33}^{(N-1)} & w_{34}^{(N-1)} & \\ 0 & w_{42}^{(N-1)} & w_{43}^{(N-1)} & w_{44}^{(N-1)} & \\ \vdots & & & & \ddots \end{bmatrix} \quad (4.28)$$

However, $W^{(N-1)}$ has been constructed by multiplying together unitary matrices, and so is itself unitary. This can only be so if $w_{11}^{(N-1)}$ has modulus 1, and so the remainder of the first row of $W^{(N-1)}$ must be equal to zero. The next step is to pre-multiply $W^{(N-1)}$ by:

$$U_N = \begin{bmatrix} (w_{11}^{(N-1)})^* & 0 & 0 & \cdots \\ 0 & 1 & 0 & \\ 0 & 0 & 1 & \\ \vdots & & & \ddots \end{bmatrix} \quad (4.29)$$

(which, by the given definition, is a two-level unitary) to give:

$$\widetilde{W} = U_N W^{(N-1)} = U_N \times \cdots \times U_2 U_1 W = \begin{bmatrix} 1 & 0 & 0 & 0 & \cdots \\ 0 & \tilde{w}_{22} & \tilde{w}_{23} & \tilde{w}_{24} & \\ 0 & \tilde{w}_{32} & \tilde{w}_{33} & \tilde{w}_{34} & \\ 0 & \tilde{w}_{42} & \tilde{w}_{43} & \tilde{w}_{44} & \\ \vdots & & & & \ddots \end{bmatrix} \quad (4.30)$$

which can be rearranged:

$$W = U_1^\dagger \times \cdots \times U_N^\dagger \widetilde{W} \quad (4.31)$$

The inverse of any two-level unitary is itself a two-level unitarity, meaning that W has now been expressed as a product of two-level unitaries and $\widetilde{W}$. As $\widetilde{W}$ can be expressed as a block matrix

$$\widetilde{W} = \begin{bmatrix} 1 & 0 \\ 0 & \widetilde{W'} \end{bmatrix} \qquad (4.32)$$

the same procedure for 'clearing out' the first column can be applied recursively (i.e., starting with $\widetilde{W'}$) to obtain a decomposition of W into a product of two-level unitary matrices.

Any two-level unitary can be expressed in terms of single-qubit unitaries and CNOTs

A two-level matrix is such that it operates non-trivially on (at most) two of the computational basis states. In some cases, a two-level unitary may already constitute a single-qubit unitary controlled by the other qubits. For instance, in the case of the (three qubit) two-level unitary

$$U_a = \begin{bmatrix} 1 & 0 & 0 & 0 & 0 & 0 & 0 & 0 \\ 0 & 1 & 0 & 0 & 0 & 0 & 0 & 0 \\ 0 & 0 & 1 & 0 & 0 & 0 & 0 & 0 \\ 0 & 0 & 0 & 1 & 0 & 0 & 0 & 0 \\ 0 & 0 & 0 & 0 & 1 & 0 & 0 & 0 \\ 0 & 0 & 0 & 0 & 0 & 1 & 0 & 0 \\ 0 & 0 & 0 & 0 & 0 & 0 & u_{11} & u_{12} \\ 0 & 0 & 0 & 0 & 0 & 0 & u_{21} & u_{22} \end{bmatrix} \qquad (4.33)$$

the corresponding circuit is:

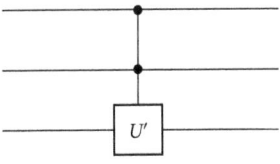

Figure 4.32

where

$$U' = \begin{bmatrix} u_{11} & u_{12} \\ u_{21} & u_{22} \end{bmatrix} \qquad (4.34)$$

As shown in Section 4.3, such a controlled unitary can be decomposed into a circuit that is composed only of CNOT and single-qubit gates.

However, it will not always be the case that the two-level unitary will be in such a convenient form. Consider another two-level unitary:

$$U_b = \begin{bmatrix} u_{11} & 0 & 0 & 0 & 0 & u_{12} & 0 & 0 \\ 0 & 1 & 0 & 0 & 0 & 0 & 0 & 0 \\ 0 & 0 & 1 & 0 & 0 & 0 & 0 & 0 \\ 0 & 0 & 0 & 1 & 0 & 0 & 0 & 0 \\ 0 & 0 & 0 & 0 & 1 & 0 & 0 & 0 \\ u_{21} & 0 & 0 & 0 & 0 & u_{22} & 0 & 0 \\ 0 & 0 & 0 & 0 & 0 & 0 & 1 & 0 \\ 0 & 0 & 0 & 0 & 0 & 0 & 0 & 1 \end{bmatrix} \quad (4.35)$$

which sends $|000\rangle \mapsto u_{11}|000\rangle + u_{21}|101\rangle$ and $|101\rangle \mapsto u_{12}|000\rangle + u_{22}|101\rangle$, with all other computational basis states unchanged.

One way to implement U_b is to decompose it into a product of unitary operations, of which U_a is one and the others are circuits that transpose computational basis states (not to be confused with SWAP gates, as in Section 4.5). That is:

$$U_b = \text{TR}(000, 110)\text{TR}(101, 111)U_a\text{TR}(101, 111)\text{TR}(000, 110) \quad (4.36)$$

where $\text{TR}(x, y)$ sends computational basis states $|x\rangle \mapsto |y\rangle$ and $|y\rangle \mapsto |x\rangle$ and has no effect on the other computational basis states (note that, as x and y are computational basis states, $0 \leq x, y < 2^n$ when the state has n qubits – and are most naturally represented in binary).

To see how the circuit $\text{TR}(\cdot, \cdot)$ can be implemented, let x and y be an arbitrary pair of n-qubit computational basis states that are to be transposed, from which the following unitary operations can be defined:

$$\Pi_x X = |x\rangle\langle x| \otimes X + (I - |x\rangle\langle x|) \otimes I \quad (4.37)$$

$$\Pi_y X = |y\rangle\langle y| \otimes X + (I - |y\rangle\langle y|) \otimes I \quad (4.38)$$

The effect of the circuit in Eqn. (4.37) is to perform a Pauli X operation on the $(n+1)$th qubit, when the first n qubits correspond to the computational basis state $|x\rangle$. For this reason, the operation is sometimes depicted as a circuit block Π_x covering n-qubits, connected to a $\oplus$ on the $(n+1)$th qubit. Similarly, in the case of Eqn. (4.38), a Pauli-X is performed when the first n qubits correspond to the state $|y\rangle$ (again with analogous circuit depiction). An example of these is in Fig. 4.33.

We also define the n-qubit circuit $U_{x,y}$ as a single layer of single-qubit gates such that the gate acting on the ith qubit is X if x and y differ in the ith bit, and the identity otherwise (note that i indexes a qubit not a computational basis state, and so $1 \leq i \leq n$).

Using these definitions, the circuit

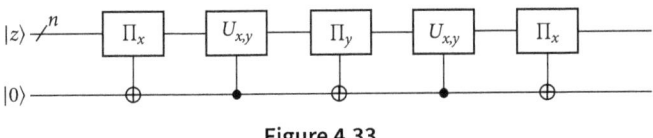

Figure 4.33

transposes the computational basis states x and y, as seen by the following argument for an input $|z\rangle|0\rangle$, where $|z\rangle$ is a computational basis state:

- For $z \neq x, y$, it can easily be seen that the none of the conditions for the controlled operations are met, and so the final state remains $|z\rangle|0\rangle$.
- Turning to the case where $z = x$, then the first circuit block sends $|x\rangle|0\rangle \mapsto |x\rangle|1\rangle$; the controlled-$U_{x,y}$ sends $|x\rangle|1\rangle \mapsto |y\rangle|1\rangle$; the third circuit block sends $|y\rangle|1\rangle \mapsto |y\rangle|0\rangle$; and the remaining operations have no effect. Thus, when the input is $|x\rangle$, the overall effect is $|x\rangle|0\rangle \mapsto |y\rangle|0\rangle$
- Finally, when $z = y$, the first two circuit blocks have no effect, and so the state remains $|y\rangle|0\rangle$; the third circuit block then sends $|y\rangle|0\rangle \mapsto |y\rangle|1\rangle$; the condition for the second controlled-$U_{x,y}$ is therefore met and thus sends $|y\rangle|1\rangle \mapsto |x\rangle|1\rangle$; the final circuit block then sends $|x\rangle|1\rangle \mapsto |x\rangle|0\rangle$. Thus, the overall operation is to send $|y\rangle|0\rangle \mapsto |x\rangle|0\rangle$

This case analysis has, therefore, shown that the circuit indeed performs the desired transposition of $|x\rangle$ and $|y\rangle$. The circuit blocks in question consist only of X gates, and controlled versions thereof: $U_{x,y}$ is nothing more than a tensor product of identities and X gates, whilst $\Pi_x X$ and $\Pi_y X$ are circuits that 'pick out' the computational basis states x and y (respectively) to control the X, and so amount to nothing more than a multi-controlled X gate, with X gates used to 'flip' (and flip back) qubits where the operation is conditional on 0 rather than 1.

It is helpful to see an explicit example of how transposition circuits are constructed, and for this we return to the earlier example in Eqn. (4.36). One of the required transpositions in Eqn. (4.36) is TR(000, 110), and according to construction in Fig. 4.33, this has circuit:

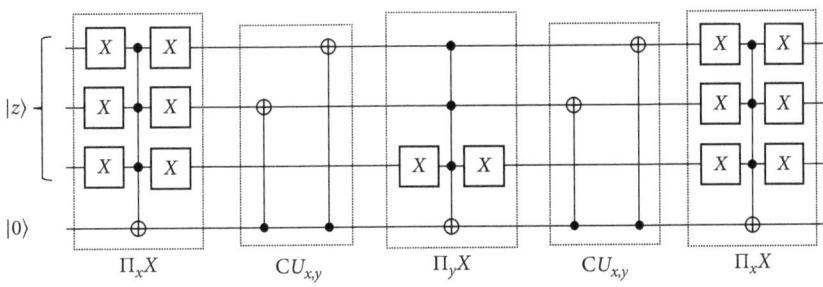

Figure 4.34

As multi-controlled X gates can be decomposed into circuits consisting only of single-qubit unitary and CNOT operations (see Section 4.3), we have, therefore, shown that any two-level unitary can be implemented using only single-qubit unitary and CNOT operations. Using the previous decomposition of any unitary into a product of two-level unitaries, it follows that every unitary can be decomposed into a circuit with only CNOT gates and single-qubit unitaries.

CNOT, H, T is a universal gate-set

As it has been shown that any n-qubit unitary can be decomposed into a product of single-qubit unitaries and CNOTs, all that remains to prove that {CNOT, H, T} is a universal gate-set is to show that H and T can generate any single-qubit unitary. To do this, we need to introduce one further result about rotation gates, namely that an arbitrary single-qubit unitary can be decomposed into rotations around two fixed non-parallel axes. An axis of rotation is defined as a real unit vector, $\hat{n} = (n_x, n_y, n_z)$, which has been expressed in terms of components in the directions of the x, y, and z axes of the Bloch sphere. A rotation of any single-qubit state by an angle θ around $\hat{n}$ is achieved by the following unitary operation:

$$R_{\hat{n}}(\theta) = \exp(-i\theta(n_x X + n_y Y + n_z Z)/2) = \cos(\theta/2)I - i\sin(\theta/2)(n_x X + n_y Y + n_z Z) \quad (4.39)$$

Any single-qubit unitary can be decomposed into three rotations around any two fixed non-parallel axes $\hat{n}$ and $\hat{m}$, that is for any U, there exist real numbers a', b', c', d' such that

$$U = e^{ia'} R_{\hat{n}}(b') R_{\hat{m}}(c') R_{\hat{n}}(d') \quad (4.40)$$

for any (non-parallel) $\hat{n}$ and $\hat{m}$.

This result means that the problem of synthesising arbitrary single-qubit unitaries using H and T reduces to merely using H and T to synthesise rotations around two non-parallel axes. As rotations are operations defined by a continuously varying parameter (i.e., the angle of rotation), we cannot hope to *exactly* synthesise arbitrary single-qubit unitaries from a finite gate-set, but we instead show we can get 'epsilon' close.

The full proof is rather involved and requires a number of mathematical preliminaries, which are beyond the scope of this book. It is, however, possible to gain a solid appreciation of how H and T can be combined to give an epsilon close approximation of any single-qubit unitary by considering a sketch of the proof. First, consider the operator THTH, which is best thought of as HTH followed by T. T is a rotation of $\pi/4$ radians around the z axis of the Bloch sphere, and HTH is a rotation of $\pi/4$ radians around the x axis of the Bloch sphere. So it follows that (up to a global phase factor):

$$\begin{aligned} T(HTH) &= \exp(-i(\pi/8)Z)\exp(-i(\pi/8)X) \\ &= (\cos(\pi/8)I - i\sin(\pi/8)Z)(\cos(\pi/8)I - i\sin(\pi/8)X) \\ &= \cos^2(\pi/8)I - i[\cos(\pi/8)X - i\sin(\pi/8)ZX + \cos(\pi/8)Z]\sin(\pi/8) \\ &= \cos^2(\pi/8)I - i[\cos(\pi/8)X + \sin(\pi/8)Y + \cos(\pi/8)Z]\sin(\pi/8) \quad (4.41) \end{aligned}$$

By the definition in Eqn. (4.39), this amounts to a rotation around an axis in the direction $n_x = \cos \pi/8$, $n_y = \sin \pi/8$, $n_z = \cos \pi/8$, through an angle θ such that $\cos \theta/2 = \cos^2 \pi/8$ – which is an irrational fraction of 2π. Because the angle of rotation is an irrational fraction of 2π, and rotation is an operation that is effectively modulo 2π, a sufficiently large number of applications of $THTH$ can be used to obtain a rotation that is satisfactorily close to any target rotation angle.

We apply the same argument for the operator, $HTHT$, which is best thought of the above operator, $THTH$ pre- and post-multiplied by H. It can be shown that this operation performs a rotation that is an irrational fraction of 2π around an axis in direction $n_x = \cos \pi/8$, $n_y = -\sin \pi/8$, $n_z = \cos \pi/8$. As the axes that $HTHT$ and $THTH$ rotate around are not parallel, it follows from Eqn. (4.40) that *any* unitary can be decomposed into three rotations around these two axes up to a global phase factor. (Note that if the unitary being synthesised is uncontrolled, then this global phase factor is unimportant, whereas if it is controlled, then the global phase can be corrected by an adjustment to U'_α, as in Fig. 4.16.) Moreover, repeated applications of $HTHT$ (respectively, $THTH$) serves to achieve any desired angle of rotation to a specified accuracy around the first (respectively, second) axis. This means that any single-qubit unitary can be decomposed as:

$$U = (THTH)^q (HTHT)^p (THTH)^r \qquad (4.42)$$

for some integers q, p and r.

Perhaps, the most significant thing that this proof sketch has been vague about is exactly what is meant by being 'epsilon close'. The above analysis shows that it is possible to rotate to within any amount, say δ, of the target angle. That is, if $\hat{n}$ is a unit vector in the direction of the rotation axis, then $R_{\hat{n}}(\theta)$ can certainly be repeated a number, n_r, of times such that the difference between some target rotation angle α and $n_r \theta \mod 2\pi$ is such that $|\alpha - n_r \theta \mod 2\pi| \leq \delta$. However, it is important to relate this notion of closeness to a standard measure for closeness of two quantum operators, U, V, such as the discrepancy of operator norm (see Chapter 2):

$$\sup_{|\psi\rangle} ||(U - V)|\psi\rangle|| \qquad (4.43)$$

and it is the case that δ, and hence n_r, can always be chosen such that

$$\sup_{|\psi\rangle} ||(R_{\hat{n}}(\theta))^{n_r} - R_{\hat{n}}(\alpha))|\psi\rangle|| \leq \frac{\epsilon}{3} \qquad (4.44)$$

It can be shown that when approximate unitaries are multiplied, the error – defined as the discrepancy in operator norm – at worst adds, and so for the product of three rotations in Eqn. (4.42), the error is at most ϵ. It is further worth noting that what one often cares about in practice is the discrepancy in measurement outcome probabilities, and it is possible to show that this is at most twice the discrepancy in operator norm for any outcome for any choice of measurement.

The result that any single-qubit unitary can be arbitrarily well approximated using only H and T gates can be put together with the preceding results to complete the proof of the universality of the gate-set {CNOT H, T}. The constructions given have not, however, addressed the question of how efficient this synthesis is (and can in principle be), which is answered in part by the Solovay–Kitaev theorem:

The Solovay–Kitaev theorem. *Any circuit containing m CNOTs and arbitrary single-qubit unitaries can be approximated to an accuracy ϵ by a circuit using $\mathcal{O}(m \log^c(m/\epsilon))$ gates from a finite universal gate-set, where $c \approx 4$.*

The proof of the Solovay–Kitaev theorem is rather intricate and beyond the scope of this book and, although constructively proven, does not usually provide a practical way to synthesise a quantum circuit from a certain universal gate-set. How to do this in practice is the problem of *quantum compilation*.

4.5 Quantum compilation

The fact that any unitary operation can be decomposed into a product of single-qubit unitaries and CNOTs proves that there exist finite universal gate-sets, and any universal quantum computer must, therefore, have one such universal gate-set as its 'native gate-set'. However, quantum computational hardware may be encumbered with additional constraints, for example the qubits therein are physical entities and can often only interact (undergo two-qubit gates) with other *local* qubits. For example, the qubits may be laid out in a linear array:

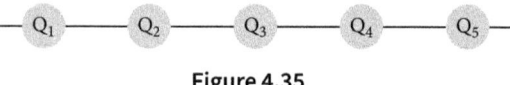

Figure 4.35

If only nearest-neighbour interactions are possible, and a gate is to be executed on qubits 1 and 3, it is necessary to *swap* qubits 1 and 2 (or qubits 2 and 3) such that qubits 1 and 3 are adjacent:

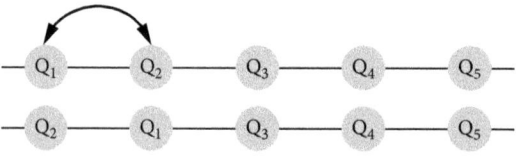

Figure 4.36

Here, 'swap' does not (usually) mean physically moving the qubit, but rather performing a unitary operation on the two qubits to swap their states. This is achieved using the SWAP gate:

$$\text{SWAP} = \begin{bmatrix} 1 & 0 & 0 & 0 \\ 0 & 0 & 1 & 0 \\ 0 & 1 & 0 & 0 \\ 0 & 0 & 0 & 1 \end{bmatrix} \tag{4.45}$$

which is depicted:

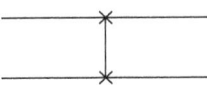

Figure 4.37

and can be constructed from three CNOT gates (showing this is one of the chapter problems). Applying a SWAP operation to a general two-qubit state gives:

$$\alpha|00\rangle + \beta|01\rangle + \gamma|10\rangle + \delta|11\rangle = \begin{bmatrix} \alpha \\ \beta \\ \gamma \\ \delta \end{bmatrix} \xrightarrow{\text{SWAP}} \begin{bmatrix} \alpha \\ \gamma \\ \beta \\ \delta \end{bmatrix} = \alpha|00\rangle + \gamma|01\rangle + \beta|10\rangle + \delta|11\rangle$$

(4.46)

The SWAP gate also resolves one of the problems that came up previously with the quantum circuit model, namely that quantum circuits are allowed to have gates bridging over qubits, which means that it is not obvious how to write down the matrix form of some circuits. For example, consider the following two circuits, each consisting of three qubits, two of which undergo a CNOT gate:

Figure 4.38

The left-hand circuit has matrix form $\text{CNOT} \otimes I_2$, but it is not obvious how to write down the matrix of the right-hand circuit. However, by writing an equivalent circuit including swap gates,

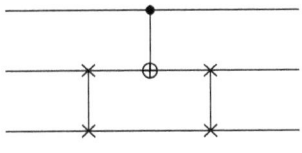

Figure 4.39

the matrix form can now be obtained:

$$(I_2 \otimes \text{SWAP}) \times (\text{CNOT} \otimes I_2) \times (I_2 \otimes \text{SWAP}) \tag{4.47}$$

Returning to the role of the SWAP gate in executing quantum circuits on real hardware, quantum computers do typically have restricted qubit connectivity (maybe more connected than a linear array, but not usually all-to-all) and so swapping qubit states to be physically adjacent is one of the key tasks in quantum compilation. Quantum compilation, broadly speaking, concerns the process of constructing an efficient quantum circuit that has the desired operation and can be executed on a certain target hardware (i.e., with a certain gate-set, qubit connectivity and other properties). This entails, amongst other things, synthesising a circuit that achieves the desired function but uses only gates from the specified gate-set; performing optimisations on the circuit by combining gates to find a functionally equivalent, but simpler circuit (similar in spirit to Boolean algebra simplification in classical logic); and *qubit routing* – the process of inserting swaps into the circuit to enable it to run on the target hardware.

4.6 Post-selection

A piece of quantum jargon that it is useful to know is 'post-selection'. Post selection is an unphysical concept where the user gets to 'choose' the measurement outcome for one or more qubits. So, for example, if a quantum circuit is such that a single qubit is measured in the computational basis, with non-zero probability of getting each of $|0\rangle$ and $|1\rangle$, then one may 'post select' on the outcome $|0\rangle$, thus omitting from further consideration the possibility of getting the outcome $|1\rangle$. The 'unphysicality' of post-selection of course comes from side-stepping the probabilistic rules that actually dictate how frequently certain outcomes arise, and so the closest thing to executing a circuit with post-selection is deploying a 'repeat until success' strategy (that is, repeating until the post-selected outcome occurs).

At first glance, post-selection seems a slightly odd idea – however, when analysing circuits that include measurements (especially mid-circuit measurements), it can often be convenient to post-select the desired outcome and then account for the (expected) number of repeats to obtain that outcome later on. Moreover, using post-selected complexity classes can sometimes provide a surprisingly powerful analytical tool in computational complexity theory – even when deriving results for non post-selected classes.

Chapter problems

1. Let $|\psi\rangle = \alpha|00\rangle + \beta|01\rangle + \gamma|10\rangle + \delta|11\rangle$. Show by explicitly expressing the tensor product and performing the subsequent matrix operation that:

$$(X \otimes I)|\psi\rangle = \alpha|10\rangle + \beta|11\rangle + \gamma|00\rangle + \delta|01\rangle \tag{4.48}$$

2. What is the state that results from the circuit shown below? Is this state entangled?

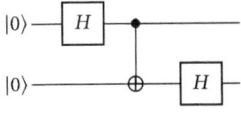

Figure 4.40

3. Show that a swap gate can be constructed from three CNOT gates.
4. Show how the CNOT gate can be constructed from Hadamard gates and the controlled-Z. Demonstrate that the construction is correct by multiplying the corresponding matrices.
5. Obtain the 8×8 matrix for the circuit

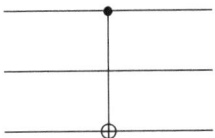

Figure 4.41

using the strategy outlined in the text or otherwise.
6. Let $A = R_z(b)R_y(c/2)$, $B = R_y(-c/2)R_z(-(d+b)/2)$ and $C = R_z((d-b)/2)$, show:
   (a) $ABC = I$;
   (b) $AXBXC = R_z(b)R_y(c)R_z(d)$
7. Show that the circuit in Fig. 4.34 is TR(000, 110), i.e., it transposes the states 000 and 110 and has no effect on the other computational basis states.
8. Show that:
   (a) $\exp(-i\theta(n_x X + n_y Y + n_z Z)/2) = \cos(\theta/2)I - i\sin(\theta/2)(n_x X + n_y Y + n_z Z)$, where $\hat{n} = (n_x, n_y, n_z)$ is a real unit vector;
   (b) $\cos^2(\pi/8)I - i[\cos(\pi/8)X + \sin(\pi/8)Y + \cos(\pi/8)Z]\sin(\pi/8)$ is a rotation through an angle θ such that $\cos\theta/2 = \cos^2 \pi/8$.

Further reading

The quantum circuit model was proposed by Deutsch [1989] and developed further by Yao [1993]. The presentation of the proof of the universality of two-level unitary operators is based on that in Nielsen and Chuang [2010] and originally due to Reck et al [1994]; the transposition circuit construction is due to Herbert et al [2024]; and the proof of the universality of $\{H, T, CNOT\}$ is due to Boykin et al [1999]. The Solovay–Kitaev theorem was proven by Solovay in 1995 in an unpublished manuscript and then independently by Kitaev [1997].

5
Quantum communication protocols

Before getting into quantum computing and quantum algorithms 'proper', it is instructive to look at some quantum communication protocols, which have remarkable results that cannot be achieved classically, even in principle. We study the use of *entanglement as a resource* in teleportation and superdense coding and the use of the non-determinism of quantum measurements to achieve information theoretically (rather than computationally) secure communications in quantum key distribution (QKD). The reason to study these protocols is threefold. First, and most importantly, they are intrinsically important and worthy of study – even though they concern information processing (communication) rather than computation in the purest sense, they certainly belong in a quantum computation textbook. Second, the protocols are relatively simple and so provide a way for readers to familiarise themselves with quantum circuits whilst further building their understanding of quantum information. Finally, even if one is narrowly concerned only with algorithms and computation, it is worth noting that protocols, such as teleportation, are expected to be used in real-world quantum computing, for example, to achieve distributed quantum computation over a network of quantum computers.

5.1 An entangled pair as a resource for information transfer

Let two parties, *Alice* and *Bob*, share an entangled pair, the Bell state $|\Phi^+\rangle = (1/\sqrt{2})(|00\rangle + |11\rangle)$. In Chapter 3, it was shown that, owing to the no-signalling principle, such a resource alone in insufficient for information transfer, and so this is supplemented with an additional communication channel.

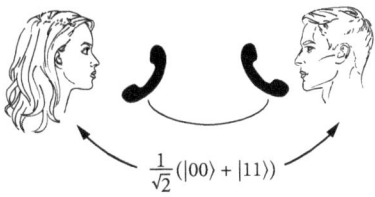

Figure 5.1

This set-up encompasses the two quantum communication protocols:

1. In quantum teleportation, the additional channel is a *classical* communication channel, and the shared entanglement along with two bits of classical information is used to transfer one qubit.
2. In superdense coding, the additional channel is a *quantum* communication channel, and the shared entanglement along with one qubit of quantum information is used to transfer two classical bits.

5.1.1 Teleportation

The quantum teleportation protocol is given by the following circuit, where the zigzag denotes an entangled pair (i.e., the two qubits are prepared in the state $|\Phi^+\rangle$), and the dashed horizontal line indicates that the two parties are spatially separated. The two vertical double lines represent classical control, with each amounting to a bit (corresponding to the measurement outcome) being sent over the classical communication channel.

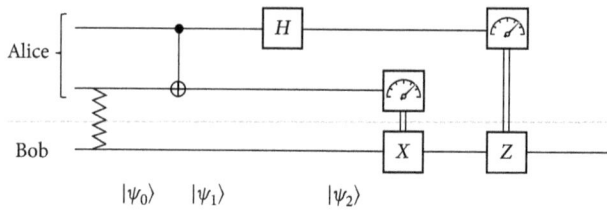

Figure 5.2

The goal of teleportation is for Bob to receive the state of Alice's first qubit, without this qubit being physically transported. In general, the protocol works even if Alice's first qubit is part of a larger entangled state, but, for simplicity, we consider the case where the first qubit is unentangled. In particular, any unentangled single-qubit state can be expressed $\alpha|0\rangle + \beta|1\rangle$ (for some α and β), which is, therefore, the initial state of the first qubit in the circuit. To see that the quantum teleportation protocol achieves its stated aim, the circuit can be simulated 'by hand' up to the state immediately prior to the measurements. This part of the circuit consists of quantum operations performed by Alice.

$$|\psi_0\rangle = \frac{1}{\sqrt{2}}(\alpha|0\rangle + \beta|1\rangle)(|00\rangle + |11\rangle)$$

$$= \frac{1}{\sqrt{2}}(\alpha|0\rangle(|00\rangle + |11\rangle) + \beta|1\rangle(|00\rangle + |11\rangle))$$

$$|\psi_1\rangle = \frac{1}{\sqrt{2}}(\alpha|0\rangle(|00\rangle + |11\rangle) + \beta|1\rangle(|10\rangle + |01\rangle))$$

$$|\psi_2\rangle = \frac{1}{2}(\alpha(|0\rangle + |1\rangle)(|00\rangle + |11\rangle) + \beta(|0\rangle - |1\rangle)(|10\rangle + |01\rangle))$$

$$= \frac{1}{2}(|00\rangle(\alpha|0\rangle + \beta|1\rangle) + |01\rangle(\alpha|1\rangle + \beta|0\rangle) + |10\rangle(\alpha|0\rangle - \beta|1\rangle) + |11\rangle(\alpha|1\rangle - \beta|0\rangle))$$
(5.1)

Notably, in $|\psi_2\rangle$, each of the four possibilities for the first two qubits is entangled with a different state of the third qubit. Alice now measures her two qubits and sends the results to Bob. This measurement outcome is a two-bit string (with the first bit representing the measurement outcome of Alice's first qubit), and the four possibilities can be considered in turn, as shown in Table 5.1. The classical control is then termed the 'correction' that Bob makes to recover the Alice's original state (i.e., the state of the first qubit at the start of the protocol).

Table 5.1 Measurement outcomes and correction operations for quantum teleportation

Measurement	Qubit 3 state	Correction	Qubit 3 after
00	$\alpha|0\rangle + \beta|1\rangle$	I	$\alpha|0\rangle + \beta|1\rangle$
01	$\alpha|1\rangle + \beta|0\rangle$	X	$\alpha|0\rangle + \beta|1\rangle$
10	$\alpha|0\rangle - \beta|1\rangle$	Z	$\alpha|0\rangle + \beta|1\rangle$
11	$\alpha|1\rangle - \beta|0\rangle$	ZX	$\alpha|0\rangle + \beta|1\rangle$

The final column of Table 5.1 shows that, regardless of the measurement outcomes, Alice's original qubit state has now been realised on qubit 3, which is in Bob's possession. It is important to note that teleportation does not violate the no-cloning principle, as Alice's original qubit state has been destroyed (collapsed by measurement) in the process.

5.1.2 Superdense coding

In superdense coding, Alice and Bob share an entangled pair, which Alice uses to send two (classical) bits. That is, she sends one of 00, 01, 10, or 11, and for each of these possibilities, she applies a (different) single-qubit unitary to her qubit (i.e., to the first qubit of the entangled state), as shown in Table 5.2.

Alice then sends her qubit to Bob. Once Bob receives Alice's qubit, he applies the following circuit, treating the first qubit as the qubit received from Alice and the second qubit as his original half of the entangled pair:

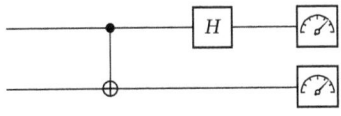

Figure 5.3

Table 5.2 Alice's operations in superdense coding

Initial state	Alice's bit string	Operation	State prior to transmission
$\frac{1}{\sqrt{2}}(\lvert 00\rangle + \lvert 11\rangle)$	00	I	$\frac{1}{\sqrt{2}}(\lvert 00\rangle + \lvert 11\rangle)$
$\frac{1}{\sqrt{2}}(\lvert 00\rangle + \lvert 11\rangle)$	01	X	$\frac{1}{\sqrt{2}}(\lvert 10\rangle + \lvert 01\rangle)$
$\frac{1}{\sqrt{2}}(\lvert 00\rangle + \lvert 11\rangle)$	10	Z	$\frac{1}{\sqrt{2}}(\lvert 00\rangle - \lvert 11\rangle)$
$\frac{1}{\sqrt{2}}(\lvert 00\rangle + \lvert 11\rangle)$	11	ZX	$\frac{1}{\sqrt{2}}(\lvert 01\rangle - \lvert 10\rangle)$

Which yields, the four possible states received after Alice's transmission, shown in Table 5.3. As the state prior to the measurement is, in each case, a computational basis state, the measurement outcome is always deterministic, and the final column shows that Bob always obtains Alice's bit string, and therefore that superdense coding has achieved its stated aim.

Table 5.3 Bob's operations in superdense coding

State received	After CNOT	After H	After measurement
$\frac{1}{\sqrt{2}}(\lvert 00\rangle + \lvert 11\rangle)$	$\frac{1}{\sqrt{2}}(\lvert 00\rangle + \lvert 10\rangle) = \frac{1}{\sqrt{2}}(\lvert 0\rangle + \lvert 1\rangle)\lvert 0\rangle$	$\lvert 00\rangle$	00
$\frac{1}{\sqrt{2}}(\lvert 10\rangle + \lvert 01\rangle)$	$\frac{1}{\sqrt{2}}(\lvert 11\rangle + \lvert 01\rangle) = \frac{1}{\sqrt{2}}(\lvert 0\rangle + \lvert 1\rangle)\lvert 1\rangle$	$\lvert 01\rangle$	01
$\frac{1}{\sqrt{2}}(\lvert 00\rangle - \lvert 11\rangle)$	$\frac{1}{\sqrt{2}}(\lvert 00\rangle - \lvert 10\rangle) = \frac{1}{\sqrt{2}}(\lvert 0\rangle - \lvert 1\rangle)\lvert 0\rangle$	$\lvert 10\rangle$	10
$\frac{1}{\sqrt{2}}(\lvert 01\rangle - \lvert 10\rangle)$	$\frac{1}{\sqrt{2}}(\lvert 01\rangle - \lvert 11\rangle) = \frac{1}{\sqrt{2}}(\lvert 0\rangle - \lvert 1\rangle)\lvert 1\rangle$	$\lvert 11\rangle$	11

Whilst the exposition above gives an explicit circuit such that Bob can obtain Alice's bit string by computational basis measurement, superdense coding can also be explained by the fact that the four possible states that Bob receives are the four Bell basis states, defined in Eqns. (3.18)–(3.21). It is, therefore, equally correct to treat Bob's action as a (two-qubit) Bell basis measurement, that is, with measurement operators $\lvert\Phi^+\rangle\langle\Phi^+\rvert$, $\lvert\Phi^-\rangle\langle\Phi^-\rvert$, $\lvert\Psi^+\rangle\langle\Psi^+\rvert$ and $\lvert\Psi^-\rangle\langle\Psi^-\rvert$, in which case the four basis states can be perfectly distinguished and interpreted as the four corresponding bit strings.

5.1.3 How to think about entanglement as a resource

One way to quantify the resource value of entanglement is to introduce a third unit of information an 'ebit', which is a qubit that is (for the present purposes) one half of an entangled pair prepared as a standard resource state. Thus, the goal is to write down relationships between ebits, qubits, and (classical) bits. For superdense coding and teleportation, these take the following form:

Teleportation: 1ebit + 2bits → 1qubit

Superdense coding: 1ebit + 1qubit → 2bits

On the left-hand side is what Bob ends up with: in each case, his ebit (his half of the original entangled resource pair) and the further information transmitted by Alice; and on the right-hand side is the classical or quantum information he obtains from these.

In the case of superdense coding, the relationship is relatively uncontroversial. Using the Bell basis measurement explanation, in essence, superdense coding is just a statement of the fact that any Bell basis state can be obtained from any other by operating on just one of the two qubits. So, after preparing an initial resource state of two ebits, one ebit remains an ebit, whilst the other is modified – thus must be treated as a qubit – such that the final state encodes any one of four two-bit strings.

The case of teleportation is a little more complicated. Whilst only a single bit of classical information can be obtained by the measurement of a single qubit, as quantum states are continuous, it is appropriate to think of a qubit itself as containing an infinite amount of classical information (in the sense that it would take an infinite number of classical bits to perfectly describe an arbitrary single-qubit state). Yet teleportation shows that any single-qubit state can be obtained by just two bits of classical information, and an ebit: that is, a generic quantum resource that is not 'tuned' to encode anything specific about the qubit state that Alice wishes to teleport.

To this apparent paradox, there are two approaches one may adopt: one pragmatic and one philosophical. The former is to accept that the quantum teleportation circuit undeniably achieves its claimed effect, and not to overly concern oneself with any deeper implications. This is very much in the spirit of the 'shut up and calculate' disposition towards quantum mechanics. The latter is to see that in the quantum teleportation, there is a continuous 'flow' of the quantum state being teleported but that, by virtue of the Bell basis measurement and entangled resource, this flow is both forwards and backwards in time. It is the classical communication of the two bits, and resultant correction that imposes normal causal ordering on the protocol *as a whole*. This is, in some ways, analogous to the situation where the no-signalling principle guarantees that the non-local collapse of spatially separated particles cannot achieve faster than light information transfer, as signalling also requires ordinary classical information transfer. See *further reading* at the end of this chapter for more on the philosophical approach.

5.2 Quantum key distribution

Unlike superdense coding and teleportation, the third quantum communication protocol, QKD, does not use entanglement at all (at least in its original form, as described here). The goal of QKD is to use quantum mechanical phenomena to generate (in a distributed manner) a 'key' shared between two parties that can be used for secure information transfer, and it achieves this by carefully ordering the actions of each party and leveraging the fact that quantum measurements are non-deterministic. The no-cloning principle then precludes any successful attack by an adversary.

5.2.1 The one-time pad

These days most cryptosystems use public-key cryptography, which relies on the one-way nature of some mathematical function (i.e., factoring numbers in the case of the Rivest–Shamir–Adleman cryptosystem (RSA) – which is actually only hard classically, as shown in Chapter 10) to *computationally* guarantee security. A stronger requirement is to *information theoretically* guarantee security, that is to provide a guarantee that holds regardless of the computational resources of the adversary. In the case of the *one-time pad*, an adversary cannot launch an attack that is better than making a random guess:

- At some date in the future, Alice will send Bob an n-bit message.
- Before that, Alice and Bob meet-up and share a *one-time pad* (or key), r, which is a list of n uniformly random bits.
- When the time comes to send the message m, Alice encodes the message by using her copy of r to send $m \oplus r$.
- Bob receives the message and decodes it by using his copy of r: $(m \oplus r) \oplus r = m$.
- Alice and Bob then discard r.

This achieves the desired information-theoretic guarantee of security, as each bit in $m \oplus r$ is (to an eavesdropper without access to m) either 0 or 1, each with 50% probability, and so the eavesdropper can, therefore, do no better than making a random guess as to the encoded value.

To this end, the final step is critical – and gives name the *one time* pad – as multiple uses of the same 'one-time' pad could introduce correlations, which would enable some information to be gleaned by an eavesdropper. To see this, consider two messages sent with the same 'one-time' pad, m_1 and m_2. If some eavesdropper were to intercept two such encoded messages, $m_1 \oplus r$ and $m_2 \oplus r$, then they could obtain $m_1 \oplus r \oplus m_2 \oplus r = m_1 \oplus m_2$, which is an n-bit string consisting only of two (supposedly secure) messages and not the random one-time pad bits themselves. Using the correlations resulting from the structure of the messages being encoded, some information content could, in principle, be discovered in this way. For example, if the message were some text using a standard binary encoding of the 26-letter alphabet, then the (large) non-uniformity of frequency of use of the various letters would introduce such structure, from which a full or partial decryption may be possible.

Whilst helpful for demonstrating what an information-theoretic security guarantee is, the one-time pad is not a particularly practical means of achieving secure communication, owing to the fact that Alice and Bob must first meet in person to share the pad. An alternative would be to share the pad on a communication channel with guaranteed security, but if such a channel were available, then it may as well be used to transfer the message itself.

It is this lack of practicality that has led to the dominance of public-key cryptography; however, QKD offers a way of generating a key (i.e., a one-time pad) in a distributed manner, i.e., without Alice and Bob needing to physically meet ahead of time to share the one-time pad. For QKD, the communication channel requirements are:

1. An authenticated public classical channel.
2. A quantum channel, which could possibly be eavesdropped.

An authenticated public channel means that any party using the channel has absolute confidence in who the message comes from. Momentarily departing from the usual convention of naming the communicating parties 'Alice' and 'Bob' in favour of Alice and *Spartacus*, if all of the other users declare 'I am Spartacus', then as long as Alice is using an authenticated channel, she will be able to discern which one really is Spartacus. It is worth noting that requiring an authenticated public channel is not insignificant, but it is a distinctly weaker requirement than full guaranteed security on a channel between two parties. As well as these communication channels, there are a small number of further requirements:

- Alice and Bob each have a private source of uniformly random classical bits.
- Alice can produce qubits in states $|0\rangle$, $|1\rangle = X|0\rangle$, $|+\rangle = H|0\rangle$, and $|-\rangle = H|1\rangle$.
- Bob can measure qubits in either the computational ($|0\rangle, |1\rangle$) basis, or the $|+\rangle, |-\rangle$ basis.

5.2.2 The BB84 protocol

QKD was discovered in 1984, and the original protocol is known as BB84 after the year of discovery and its discoverers, Charles Bennett and Gilles Brassard. The goal of BB84 is to use the resources given above to generate an n-bit one-time pad that is known by Alice and Bob, but that cannot be known by any other parties.

1. Alice generates a random bit string of length $2n + 2\lceil \frac{1}{2-\log_2 3} n \rceil + \mathcal{O}(\sqrt{n})$, for each bit she either encodes $\{0, 1\}$ as $\{|0\rangle, |1\rangle\}$ or $\{|+\rangle, |-\rangle\}$ (the basis is chosen at random with equal probability). Alice then sends the qubits to Bob.
2. Bob receives the qubits and for each either measures in the $\{|0\rangle, |1\rangle\}$ basis or the $\{|+\rangle, |-\rangle\}$ basis (with the measurement basis chosen uniformly at random and independently each time).
3. For each qubit in turn:
   (a) Bob announces over a public channel in which basis he measured the qubit.

(b) Alice replies over the public channel whether that was the basis in which the qubit was prepared.
(c) If the same basis was indeed used for the preparation and the measurement, then Bob's measurement outcome will equal Alice's bit, and they both retain this bit; otherwise, they discard.
4. With high probability, Alice and Bob now have at least $n + \lceil \frac{1}{2-\log_2 3} n \rceil$ bits that have not been discarded. If the number is greater than $n + \lceil \frac{1}{2-\log_2 3} n \rceil$, they keep the first $n + \lceil \frac{1}{2-\log_2 3} n \rceil$ bits.
5. Alice announces $\lceil \frac{1}{2-\log_2 3} n \rceil$ of the retained bits (chosen at random) that are to be compared by each of Alice and Bob announcing the value on the authenticated public channel. These $\lceil \frac{1}{2-\log_2 3} n \rceil$ bits are then discarded.
6. Alice and Bob now have n identical bits in their possession. If the result of the previous step was that the comparisons were all agreed (Alice and Bob had the same value of bit in each case), they conclude that these n bits constitute a one-time pad that is secret from any eavesdropper.

The $+\mathcal{O}(\sqrt{n})$ in step 1 is necessary to ensure that the basis is agreed in at least $n + \lceil \frac{1}{2-\log_2 3} n \rceil$ cases with high probability (this follows because each basis choice is a Bernoulli random process with probability 1/2 of Alice and Bob agreeing, so the total number of agreements follows a Binomial distribution, and elementary statistics proves the sufficiency of $\mathcal{O}(\sqrt{n})$ additional bits). As this term relates to statistics rather than the quantum protocol itself *per se*, for simplicity, the following worked examples use exactly $2n + 2\lceil \frac{1}{2-\log_2 3} n \rceil$ bits (and the example is then constructed such that Alice and Bob agree the basis for exactly half). In step 5, there is no particular need for it to be Alice who declares which bits are to be compared: it is immaterial whether it is Alice or Bob or some combination of both; however, it is important that the identities of the bits to be compared are not publicly known ahead of the protocol being run.

BB84: worked example without eavesdropping

Let $n = 2$, in which case the protocol requires Alice to use $2n + 2\lceil \frac{1}{2-\log_2 3} n \rceil = 14$ random bits. Table 5.4 shows an instance of the protocol up to the end of Step 3, with the qubits for which the bases do not agree greyed out.

Table 5.4 Example of BB84 (no eavesdropping)

A bit	A basis	Qubit	B basis	B bit
0	$\lvert 0 \rangle, \lvert 1 \rangle$	$\lvert 0 \rangle$	$\lvert 0 \rangle, \lvert 1 \rangle$	0
1	$\lvert + \rangle, \lvert - \rangle$	$\lvert - \rangle$	$\lvert + \rangle, \lvert - \rangle$	1
0	$\lvert 0 \rangle, \lvert 1 \rangle$	$\lvert 0 \rangle$	$\lvert + \rangle, \lvert - \rangle$	{0, 1}
0	$\lvert + \rangle, \lvert - \rangle$	$\lvert + \rangle$	$\lvert 0 \rangle, \lvert 1 \rangle$	{0, 1}
1	$\lvert 0 \rangle, \lvert 1 \rangle$	$\lvert 1 \rangle$	$\lvert 0 \rangle, \lvert 1 \rangle$	1

A bit	A basis	Qubit	B basis	B bit
0	$\vert+\rangle, \vert-\rangle$	$\vert+\rangle$	$\vert+\rangle, \vert-\rangle$	0
1	$\vert 0\rangle, \vert 1\rangle$	$\vert 1\rangle$	$\vert+\rangle, \vert-\rangle$	$\{0, 1\}$
1	$\vert+\rangle, \vert-\rangle$	$\vert-\rangle$	$\vert 0\rangle, \vert 1\rangle$	$\{0, 1\}$
0	$\vert 0\rangle, \vert 1\rangle$	$\vert 0\rangle$	$\vert+\rangle, \vert-\rangle$	$\{0, 1\}$
0	$\vert+\rangle, \vert-\rangle$	$\vert+\rangle$	$\vert 0\rangle, \vert 1\rangle$	$\{0, 1\}$
1	$\vert 0\rangle, \vert 1\rangle$	$\vert 1\rangle$	$\vert 0\rangle, \vert 1\rangle$	1
1	$\vert+\rangle, \vert-\rangle$	$\vert-\rangle$	$\vert+\rangle, \vert-\rangle$	1
0	$\vert 0\rangle, \vert 1\rangle$	$\vert 0\rangle$	$\vert 0\rangle, \vert 1\rangle$	0
0	$\vert 0\rangle, \vert 1\rangle$	$\vert 0\rangle$	$\vert+\rangle, \vert-\rangle$	$\{0, 1\}$

Notice that, in the absence of an eavesdropper, agreeing preparation and measurement bases guarantees that the bit value agrees – which holds in general. The final steps involve the comparison (and subsequent discarding of) $\lceil \frac{1}{2-\log_2 3} n \rceil = 5$ of the bits. As all seven of the non-greyed out bits agree, this will lead to consensus whichever five bits are chosen – leaving two bits that are the one-time pad. To see why it is reasonable to conclude that the remaining two bits constitute a secret shared only between Alice and Bob, it is necessary to further introduce an eavesdropper, Eve.

BB84: worked example with eavesdropping

Eve has access to the public channel, although owing to the fact that it is authenticated she cannot masquerade as Alice and/or Bob. As an eavesdropper, she is also assumed to have tapped the quantum communication channel connecting Alice and Bob, and as such has access to qubit states in the central column 'Qubit' in the above table. At this point, the order of actions in the BB84 is extremely important – Bob *first* receives and measures a qubit, and then announces his (randomly selected) choice of basis *before* Alice replies with the basis she used for the preparation. As such, Bob has to receive a qubit before either the preparation or measurement basis is revealed.

The most obvious attack is for Eve to intercept the qubit transmitted by Alice, make a copy, and re-transmit the original qubit on to Bob, such that she can then obtain the preparation basis when Alice makes the announcement. By choosing to measure in the same basis in which Alice prepares the qubit, Eve can obtain the qubit state without disturbing it. In fact, even obtaining Bob's measurement basis would suffice, as the protocol dictates that the qubit is discarded if Bob's measurement basis does not agree with Alice's preparation basis. However, this attack is precluded by the no-cloning theorem. The options open to Eve, therefore, amount to (i) saving the qubit and sending a different qubit on to Bob; or (ii) measuring the qubit and then re-transmitting on to Bob.

In the first case, whilst this does indeed guarantee that Eve will be able to obtain the correct measurement bases from Bob's announcements, the qubits she transmits to Bob will bear no resemblance to the ones he was intended to receive from Alice. The upshot of this will be that Bob has a probability of 1/2 of agreeing with Alice on the measurement outcome for each of the $\lceil \frac{1}{2-\log_2 3} n \rceil$ qubits that they compare in step 5. If they fail to agree every one of the $\lceil \frac{1}{2-\log_2 3} n \rceil$ measurement outcomes, then this will alert them to the presence of the eavesdropper, so the attack will succeed only with probability $2^{-\lceil \frac{1}{2-\log_2 3} n \rceil}$. Notably, Eve can *guess* the string (or equivalently the one-time pad) with probability $2^{-n} > 2^{-\lceil \frac{1}{2-\log_2 3} n \rceil}$. It is worth remarking on an important subtlety here: in the unlikely event that Eve's attack succeeds, she will knowingly be in possession of the 'secure' one-time pad, whereas a random guess will always be just that. Therefore, it is not quite right to say that this attack is no better than making a random guess, but nevertheless 2^{-n} is a convenient value to suppress the success probability below, in order to show how ineffective any attack can be.

Turning to the second attack, there are various strategies that Eve could employ, but it suffices to consider the case where she randomly chooses a measurement basis (out of $|0\rangle, |1\rangle$ and $|+\rangle, |-\rangle$), obtains the measurement outcome, and then re-transmits the qubit. The previous worked example can be extended to include an eavesdropper attempting this attack, as shown in Table 5.5.

Table 5.5 Example of BB84 with eavesdropping

A bit	A basis	Qubit	E basis	E bit	Qubit	B basis	B bit
0	$\|0\rangle, \|1\rangle$	$\|0\rangle$	$\|0\rangle, \|1\rangle$	0	$\|0\rangle$	$\|0\rangle, \|1\rangle$	0
1	$\|+\rangle, \|-\rangle$	$\|-\rangle$	$\|0\rangle, \|1\rangle$	{0, 1}	$\{\|0\rangle, \|1\rangle\}$	$\|+\rangle, \|-\rangle$	{0, 1}
0	$\|0\rangle, \|1\rangle$	$\|0\rangle$	$\|+\rangle, \|-\rangle$	{0, 1}	$\{\|+\rangle, \|-\rangle\}$	$\|+\rangle, \|-\rangle$	{0, 1}
0	$\|+\rangle, \|-\rangle$	$\|+\rangle$	$\|+\rangle, \|-\rangle$	0	$\|+\rangle$	$\|0\rangle, \|1\rangle$	{0, 1}
1	$\|0\rangle, \|1\rangle$	$\|1\rangle$	$\|+\rangle, \|-\rangle$	{0, 1}	$\{\|+\rangle, \|-\rangle\}$	$\|0\rangle, \|1\rangle$	{0, 1}
0	$\|+\rangle, \|-\rangle$	$\|+\rangle$	$\|+\rangle, \|-\rangle$	0	$\|+\rangle$	$\|+\rangle, \|-\rangle$	0
1	$\|0\rangle, \|1\rangle$	$\|1\rangle$	$\|0\rangle, \|1\rangle$	1	$\|1\rangle$	$\|+\rangle, \|-\rangle$	{0, 1}
1	$\|+\rangle, \|-\rangle$	$\|-\rangle$	$\|0\rangle, \|1\rangle$	{0, 1}	$\{\|0\rangle, \|1\rangle\}$	$\|0\rangle, \|1\rangle$	{0, 1}
0	$\|0\rangle, \|1\rangle$	$\|0\rangle$	$\|+\rangle, \|-\rangle$	{0, 1}	$\{\|+\rangle, \|-\rangle\}$	$\|+\rangle, \|-\rangle$	{0, 1}
0	$\|+\rangle, \|-\rangle$	$\|+\rangle$	$\|+\rangle, \|-\rangle$	0	$\|+\rangle$	$\|0\rangle, \|1\rangle$	{0, 1}
1	$\|0\rangle, \|1\rangle$	$\|1\rangle$	$\|0\rangle, \|1\rangle$	1	$\|1\rangle$	$\|0\rangle, \|1\rangle$	1
1	$\|+\rangle, \|-\rangle$	$\|-\rangle$	$\|0\rangle, \|1\rangle$	{0, 1}	$\{\|0\rangle, \|1\rangle\}$	$\|+\rangle, \|-\rangle$	{0, 1}
0	$\|0\rangle, \|1\rangle$	$\|0\rangle$	$\|0\rangle, \|1\rangle$	0	$\|0\rangle$	$\|0\rangle, \|1\rangle$	0
0	$\|0\rangle, \|1\rangle$	$\|0\rangle$	$\|0\rangle, \|1\rangle$	0	$\|0\rangle$	$\|+\rangle, \|-\rangle$	{0, 1}

Box 5.1 Mutually unbiased bases and the uncertainty principle

For those who have some familiarity with quantum mechanics, the fact that measurement in the 'wrong' basis in BB84 yields no information may be vaguely reminiscent of Werner Heisenberg's celebrated uncertainty principle, and it turns out that there is a deep connection. BB84 relies on the property that the bases $\{|0\rangle, |1\rangle\}$ and $\{|+\rangle, |-\rangle\}$ are such that when a basis state of one basis is measured in the other basis, then each outcome is obtained with 50% probability. The same would have been true had either basis been swapped for the basis $\{|i\rangle, |-i\rangle\}$. Sets of bases with this property are said to be *mutually unbiased*, and mutually unbiased bases play an important role in quantum information theory in general. Mutually unbiased bases can be defined in terms of entropic uncertainty relations, which are closely related to Heisenberg's uncertainty principle. To define entropic uncertainty relations, it is necessary to first introduce information entropy, and for our purposes, we can restrict our attention to a measurement in some basis, $\{|m_*\rangle\}$, with associated measurement outcomes $\{m_*\}$. In this case, the information entropy is:

$$\mathcal{H}_m = -\sum_i \Pr(m_i) \log(\Pr(m_i)) \quad (5.2)$$

If we have two measurement bases, $\{|b_*\rangle\}$ and $\{|\tilde{b}_*\rangle\}$, and define $c = \max_{i,j} \langle b_i | \tilde{b}_j \rangle$, then the following entropic uncertainty relation always holds:

$$\mathcal{H}_b + \mathcal{H}_{\tilde{b}} \geq -2 \log c \quad (5.3)$$

In the case where the space is N-dimensional and $\langle b_i | \tilde{b}_j \rangle = \frac{1}{\sqrt{N}}$ for all i, j, then

$$\mathcal{H}_b + \mathcal{H}_{\tilde{b}} \geq \log N \quad (5.4)$$

and the bases are mutually unbiased. For values of N that are prime powers (as are all qubit systems where $N = 2^n$), one can always obtain a set of exactly $N + 1$ mutually unbiased bases. For example, a single-qubit system has $N + 1 = 2^1 + 1 = 3$ unbiased bases, e.g., $\{|0\rangle, |1\rangle\}, \{|+\rangle, |-\rangle\}$ and $\{|i\rangle, |-i\rangle\}$ as identified above.

Again, the greyed out lines show the occasions where Alice and Bob have already identified a lack of agreement in basis, and so these are discarded and as such are unaffected by the eavesdropping. However, in the case of the second, fifth, and twelfth rows, Alice and Bob have agreed on the basis; however, Eve has happened to guess wrong. The effect of this wrong guess by Eve is to randomise the qubit measurement outcome that Bob obtains. Thus, in steps 5 and 6, when Alice and Bob randomly choose $\lceil \frac{1}{2-\log_2 3} n \rceil = 5$ of the non-discarded qubits to compare, Eve's attack is likely to be detected by there being a disagreement in bit value for one or more of these five comparisons. Such disagreement immediately alerts Alice and Bob to the presence of the eavesdropper, as in the absence of such an eavesdropper, agreeing the measurement basis guarantees agreement of the bit value. Having detected the presence of the eavesdropper, they discard the shared one-time pad.

The remaining question is: how likely is this detection? In principle, Eve could still luck out by (i) always guessing the correct basis; and/or (ii) Bob's value happening to match Alice's for the compared bits, even when Eve guesses wrong. For every intercepted qubit, Eve has a probability of 3/4 that Alice and Bob will agree the bit value: 1/2 of the time Eve will guess the correct basis, and when she does not, there is still a 50–50 chance that Bob will obtain the same bit value as Alice anyway. This means that the probability of the eavesdropping going undetected is:

$$\left(\frac{3}{4}\right)^{\lceil \frac{1}{2-\log_2 3} n \rceil} \leq \left(\left(\frac{3}{4}\right)^{\frac{1}{2-\log_2 3}}\right)^n = \left(\frac{1}{2}\right)^n \tag{5.5}$$

Thus, the probability of evading detection when deploying this attack is upper-bounded by 2^{-n}, which is the same as the probability of guessing the one-time pad itself.

Chapter problems

1. Let Alice and Bob each have one qubit of a Bell pair ($\frac{1}{\sqrt{2}}(|00\rangle + |11\rangle)$); let Bob and Charlie also each have one half of another Bell pair. If Bob uses the Bell pair he shares with Charlie, to teleport his qubit from the Bell pair he (Bob) shares with Alice, show that the result is that Alice and Charlie now share a Bell pair.
2. Consider the three-qubit state $\frac{1}{\sqrt{2}}(|000\rangle + |111\rangle)$.
   (a) If Alice holds one of the three qubits, and Bob holds the other two, can Alice use this state to teleport one qubit to Bob? Explain your answer.
   (b) If instead Alice and Bob each hold one qubit, and now some third party 'Charlie' holds the third qubit, then give a protocol to enable teleportation between Alice and Bob whereby Charlie first applies a Hadamard gate to his qubit, then measures in the computational basis, and sends the measurement outcome to Alice and Bob.
3. Suppose that an eavesdropper, Eve, intercepts the qubit transmitted by Alice in the superdense coding protocol. Can Eve infer which of the four pairs of bits 00, 01, 10, or 11 Alice was trying to transmit? If so, how? If not, why not?
4. Suppose that Alice transmits the two-bit string '00' using the superdense coding protocol, and an evesdropper, Eve, intercepts the qubit transmitted by Alice, measures it in the computational basis, and then re-transmits to Bob. Find the probability that Bob correctly receives 00.
5. Consider the circuit:

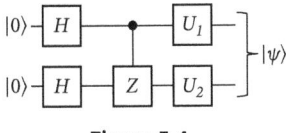

Figure 5.4

(a) What should U_1 and U_2 be to prepare the entangled state $|\psi\rangle = \frac{1}{\sqrt{2}}(|00\rangle + |11\rangle)$?

(b) Is state $|\psi\rangle$ always entangled, or does its entanglement depend on U_1 and U_2? Give reasons for your answer.

(c) If U_1 is a Pauli-X gate and U_2 is a Pauli-Z gate, and the first qubit is sent to 'Alice', and the second to 'Bob', then how can Alice use the shared entangled state to send two (classical) bits of information to Bob, by transmitting only a single qubit?

(d) If Alice knows only that U_1 is either a Pauli-X or Pauli-Z gate, but Bob knows which of these it is (and U_2 is still a Pauli-Z gate which is known to both Alice and Bob), can Alice still send two bits by transmitting only a single qubit? Explain your answer.

6. Alice and Bob wish to use QKD to prepare a private shared one-time pad. Can they do so if they have the ability to prepare and measure quantum states in the following bases?

(a) $\{|+\rangle, |-\rangle\}$ and $\{|i\rangle, |-i\rangle\}$;

(b) $\{|0\rangle, |1\rangle\}$ and $\{\frac{1}{2}|0\rangle - \frac{\sqrt{3}}{2}|1\rangle, \frac{\sqrt{3}}{2}|0\rangle + \frac{1}{2}|1\rangle\}$.

Further reading

Teleportation was discovered by Bennett et al [1993], superdense coding by Bennett and Wiesner [1992], and BB84 by Bennett and Brassard [2014]. For the reader who wants to stretch themselves, a categorical formalism of quantum information, such as Duncan and Coecke's ZX calculus [2011], gives a more intuitive picture of the temporal and spatial ordering of the flow of quantum information in various algorithms and protocols, and Coecke and Kissinger [2017] is a lengthy but accessible resource for this. In particular, Coecke and Kissinger [2017, (4.67)] illustrates teleportation in a way that shows the flow of quantum information 'forwards and backwards in time'. The exclamation 'shut up and calculate' is often misattributed, but N. David Mermin seems to be the origin [1989]. Entropic uncertainty relations are due to Maassen and Uffink [1988].

6
Quantum advantage

One question, more than any other, pervades the field of quantum computing: *Is there a quantum advantage?* This question may be asked either in general, that is, of quantum mechanics as a model of computation, or for individual algorithms. In both cases, to even pose this question in a rigorous and meaningful way, a number of non-trivial prerequisites are required. These prerequisites mainly belong in the domain of computational complexity theory. Once the formalities have been attended to, it is reassuring to see that for the most part, an intuitive and informal assessment of quantum computational complexity generally corresponds to the result obtained by the full analysis.

6.1 How big is Hilbert space?

To get a sense of why one may hope that there will be spectacular improvements in computational efficiency by using a quantum rather than the classical model of computation, it is helpful to examine the shear scale of the theatre, namely Hilbert space, in which quantum computations take place. This section highlights two important features: first, in order to reach every part of Hilbert space, a very large number of quantum operations are required; and second to (naively) simulate even a *single* quantum operation, a comparably large number of classical operations are required.

6.1.1 The gate complexity of approximately preparing every quantum state

To answer the question *how many operations are required to (approximately) prepare every n-qubit quantum state?* we first ask *how many different n-qubit states are there?* An n-qubit quantum state is represented by a 2^n-element complex unit vector, and the value of each element varies continuously, so there are infinitely many quantum states. However, beyond some finite precision, states may be treated as the same for all practical purposes, and so in practice only a finite set of states is of interest. (Indeed, even rigorous theoretical analysis is based on approximately preparing states.) That is to say, a finitely-spaced 'grid' can be overlaid on the Hilbert space such that any state can be 'snapped' to its nearest grid point, which is then treated as an acceptable approximation of the original state. As such, it is only necessary to consider the preparation of a finite number of states corresponding to those at the grid points. Such an argument is necessary, as otherwise it is completely nonsensical to even contemplate counting the number of operations required to prepare each of an uncountably infinite number of quantum states!

We can lower-bound the number of states by considering just those states where each element is either 0 or $1/C$ for some constant C ensuring correct normalisation: by elementary combinatorics, there are $2^{2^n} - 1$ such states (the minus one as the case where every element is equal to zero cannot be normalised to give a valid quantum state). As these are distinct states, the circuits required to prepare each, when applied to a fixed initial state (say $|0^n\rangle$), must also be distinct. The quantum universality result of Section 4.4.2 shows that these circuits do indeed exist, as long as a universal gateset is used. Thus, there is a set of 2^{2^n} different circuits that prepare these (distinct) states, and the question of how many operations is required to prepare any state, therefore, amounts to the question of how many gates (from some fixed, finite gateset) the largest of these circuits has. This can be answered by viewing a quantum circuit as a branching process, where the circuit is an ordered list of gates (in cases where disjoint gates appear in the same 'layer' of the quantum circuit, the order can be arbitrarily chosen), and upper-bounding the number of quantum states that can be prepared by a list of at most some k gates.

As a first step towards this, we bound the number of different states that can be prepared when any one of d different gates (from a fixed gateset) is applied to a fixed n-qubit state. If each of the gates is a single-qubit unitary, then there are dn different ways to apply a single gate, and if instead each gate acts on at most c different qubits, then the number of states, s, is at most:

$$s \leq d\left(\frac{n!}{(n-c)!}\right) \leq dn^c \qquad (6.1)$$

(Note that this expression accounts for the fact that qubit order matters for multi-qubit gates, e.g., if two qubits are acted upon by a CNOT it matters which is the target and which is the control.) In general, c must be at least two, to enable interactions between qubits, and in practice, it is likely that c would equal 2 or 3 – the latter, for example, if the Toffoli gate were included in the gateset. Notwithstanding this practical insight, for the following analysis, all that is required is that c is a constant that does not grow with n.

Considering some standard initial state (conventionally, $|0^n\rangle$), if one considers a set of circuits with *exactly* k gates, at most $s^k \leq (dn^c)^k$ distinct states can be prepared. Thus, by summing a geometric series, the number of different states that can be prepared by a set of circuits, each of which has *at most* k gates, is upper-bounded by $(dn^c)^{k+1}$. This upper bound is not a tight, as, for example, lists of gates which differ only by the order of two successive gates acting on disjoint sets of qubits are counted as distinct circuits preparing distinct states, even though they actually prepare the same state.

Nevertheless, even a loose upper bound suffices to derive the desired final result: the minimum k that could possibly reach all 2^{2^n} states (itself a loose lower-bound on the total number of states) can be found by solving

$$2^{2^n} \leq (dn^c)^{k+1} \tag{6.2}$$

$$\implies k \geq \frac{2^n}{\log_2(d) + c\log_2 n} - 1 \tag{6.3}$$

This shows that the number of gates needed to prepare all n-qubit states is at least (very nearly) exponential in n.

Asymptotic notation

It is worth making a brief tangential comment to explain exactly what 'exponential in n' means, and why it is significant. Computational problems are defined as consisting of differently-sized problem instances, and computational complexity is then given in terms of how the number of operations needed to solve the problem grows with the size of the input. For instance, the problem of adding two vectors of size n is *linear* in n – i.e., the number of operations required is proportional to n. ('Number of operations' is, of course, open to different interpretations, and so it is important that the fields of computation theory and computational complexity theory have been founded on a fundamental model of computation, as detailed in Section 6.2.)

One pertinent point about the simple example of adding two vectors is that in declaring the complexity as 'linear' – i.e., incurring a number of operations proportional to n – the constant of proportionality itself is conspicuous by its absence. Neglecting such a 'pre-factor' (as the constant is often called) is normal in computational complexity, as such constants will, at some problem size, fail to dictate anything about the relative complexities of two algorithms for which the term after the pre-factor differs. For instance, if the standard approach to solving some problem requires a number of operations that grows as $10n$, and a different approach is proposed that requires $100000000n^{0.999}$ operations, then even though $100000000 \gg 10$ and $0.999 \approx 1$, it is still the case that for a large enough n, $100000000n^{0.999} < 10n$.

This motivates the common asymptotic computational complexity notations:

- $\mathcal{O}(f(n))$: there is a function $f: \mathbb{N} \to \mathbb{R}$ (i.e., a function from the integers to the real numbers), such that the number of operations required to solve the problem of size n is upper-bounded by $cf(n)$ for some real constant c. Note that $\mathcal{O}(1)$ means upper-bounded by a constant for all n.
- $\Omega(f(n))$: there is a function $f: \mathbb{N} \to \mathbb{R}$, such that the number of operations required to solve the problem of size n is lower-bounded by $cf(n)$ for some real constant c.
- $\Theta(f(n))$: there is a function $f: \mathbb{N} \to \mathbb{R}$, such that the number of operations required to solve the problem of size n is lower-bounded by $c_1 f(n)$ and upper-bounded by $c_2 f(n)$ for some real constants c_1 and c_2.

It is further worth noting that Poly(n) is the class of problems for which $\mathcal{O}(n^k)$ holds for *some* k. Similarly, Poly($\log n$) (also sometimes written Poly-log(n)) is the class of problems for which $\mathcal{O}(\log^k n)$ holds for some k. A function that is $\mathcal{O}(n \log^k n)$ for some k is called 'quasi-linear'. Alternatively, the function f may be an exponential, and it is important to know that *any* function that is exponential in n grows faster than *every* function that is a polynomial of n – and for this reason 'exponential in n' is generally taken as indicative of being computationally expensive without further qualification.

6.1.2 Exploring Hilbert space classically: state vector simulation

A classical computer programme that takes as an input a quantum circuit and returns the result had the same circuit been executed on an actual quantum computer is said to classically *simulate* the quantum circuit. The simulation overhead is the uplift in computational complexity incurred by the simulation and *efficient* classical simulation is usually defined as the case where the overhead is only a polynomial in the number of qubits.

The 'naive' means of classical simulation is *state vector simulation*, in which the n-qubit state is saved as a 2^n-element vector, and the quantum circuit is 'directly' simulated by performing the corresponding $2^n \times 2^n$ matrix operations. The state-of-the-art means of matrix multiplication for some $N \times N$ matrix requires a number of operations proportional to $N^{2.371}$; however, in our case $N = 2^n$ and so state vector simulation is exponential in n – and hence is deemed inefficient. When addressing the question of *is there a quantum advantage?* one important line of inquiry is to identify classes of quantum circuits where some alternative method can be deployed to achieve efficient classical simulation. This is the subject of Section 6.3, but before coming to that, it is necessary to put the entire subject of computation and complexity on a firmer footing.

6.2 Computation theory and computational complexity theory

Computation theory addresses the seminal question: *Is some mathematical problem computable?* To the uninitiated, it may appear that the ability to write down a problem in a formal manner implies its computability – in the same way that being able to write down a Boolean expression implies that there certainly exists a Boolean circuit to implement it. However, the great power of computation is also its Achilles' heel, as it is possible to encode logical paradoxes into computer programmes in such a way that uncomputable problems can, in fact, be defined. Whilst computability is a fascinating topic, it does not directly impact quantum computing, except for the important result that the class of problems that are computable is the same for classical and quantum models of computations. (That is, quantum computing is *Turing complete*, as defined in Section 6.2.2.) A much more pertinent question is whether there are problems that can be computed much more efficiently using a quantum rather than classical computer, and this brings us into the domain of computational complexity theory.

Computational complexity theory is the study of the resources – most notably the time and space – required to compute the solution to mathematical problems. Consider, as a simple example, the problem of factoring large semi-prime numbers (a semi-prime is the product of exactly two primes), which is generally believed to be a *hard* problem (one that requires considerable resources) at least with any classical computer. An algorithm for factoring takes as an input a large semi-prime and returns either of its (prime) factors. Consider an algorithm that always returns the output 1000000007 (which is prime); then, most of the time, the algorithm returns an incorrect solution. However, on occasions, when the input is a semi-prime obtained by multiplying 1000000007 by another prime, then the returned answer is correct.

As the same constant value is returned regardless of the input, the algorithm clearly has low resource requirements and for *some* inputs returns the correct answer – but, by any reasonable assessment, it cannot be said to have computed the answer. Such a vacuous example is perhaps too simplistic, but consider the generalisation where the algorithm is allowed to make random decisions, and with high probability returns 1000000007 and otherwise deploys some standard factoring algorithm to compute the correct answer for the given input, which it then returns. If the probability of returning 1000000007 immediately is set to be sufficiently large (and possibly such that it grows with the value of the semi-prime that is inputted), then the algorithm still has modest resource requirements *on average* but now also has the attractive property that for *all* possible inputs the correct answer is *sometimes* returned (before the correct answer was never returned for most possible inputs), and for some inputs the correct answer is *always* returned. Clearly, the picture is now more complicated, but fundamentally the algorithm is no better than the standard factoring algorithm that is deployed with low probability. Fortunately computational complexity theory formalises and (sometimes) answers questions such as: *should the algorithm be efficient on average or in the worse case?* and *if the correct answer is only returned with a certain probability, what probability is sufficient to declare the algorithm effective?* in a manner that is coherent and cuts through such daft proposals as the one above.

To dive deeper into computational complexity theory, it is necessary to specify the following:

1. a standard model of computation (which is also needed to properly define *computability*);
2. the relevant types of mathematical problems;
3. some criterion to decide if an algorithm is efficient or inefficient.

6.2.1 Classical and quantum models of computation

In common parlance, it is usual to speak of computational load in terms of the number of operations, which begs the question: what constitutes an operation? In the case of a quantum computation executed as a quantum circuit, the answer may appear obvious: an operation is a gate; however, this still requires some standard gateset to be defined. (Otherwise, every circuit is one operation: the unitary operation of the circuit itself!) In the case of classical computation, in some sense, Boolean circuits play the same role, with elementary Boolean functions constituting the operations, and there are indeed some computational complexity classes defined in terms of Boolean circuits. Moreover, counting gates is usually the most practical way to quantify the computational cost of a quantum algorithm, and in many cases, this provides a valid proxy for the computational complexity (as shown in Section 6.2.3).

However, Boolean and quantum circuits are not considered as fundamental models of computation; instead, the standard fundamental model of computation is usually taken as the *Turing machine*. Turing machines (in some sense) generalise finite automata, and the relative simplicity of finite automata provides a useful analogy for the relationship between *deterministic* Turing machines, *non-deterministic* Turing machines, *probabilistic* Turing machines, and *quantum* Turing machines.

Finite automata
A finite automaton consists of:

- A set of n_s states.
- An input alphabet of size n_a.
- A set of state transitions: usually represented in the form of a $n_s \times n_s$ matrix for each of the n_a letters (the form of the matrix is restricted for each type of automaton).
- An initial 'start' state.
- One or more 'accept' state(s) (marked in the examples below by a black circle).

The *accepted language* is the set of strings of letters (or *symbols*) from the alphabet, such that the final state is the (or an) accept state. The simplest class of finite automata is *deterministic* finite automata. Deterministic automata have the property that for each state–letter pair, there is exactly one transition. Indexing the states and associating each with a vector $|k\rangle$ – i.e., a vector of zeros, except a single one in the kth element – if $|i\rangle$ is the start state, and $s_1 s_2 \cdots s_n$ is input the string, then the final state is $|f\rangle = M_{s_n} M_{s_{n-1}} \cdots M_{s_1} |i\rangle$ (where M_{s_j} is the transition matrix for the jth symbol, s_j). For $|f\rangle$ to indicate a unique final state, as deterministic finite automata require by definition, it is necessary that the transition matrices have exactly one non-zero element equal to 1 in each column.

A simple example of a two-state deterministic automaton can be depicted:

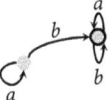

Figure 6.1

with the left-hand state, $|0\rangle$, as the starting state; the right-hand state, $|1\rangle$, as the accept state; and the transitions for each letter in the alphabet, $\{a, b\}$, shown by the arrows. The transition matrices in this example are:

$$M_a = \begin{bmatrix} 1 & 0 \\ 0 & 1 \end{bmatrix}; \quad M_b = \begin{bmatrix} 0 & 0 \\ 1 & 1 \end{bmatrix} \quad (6.4)$$

In this case, the accepted language is the set of all strings containing at least one 'b', e.g., if the input string is abb, then the transition is $M_b M_b M_a$ operating on the initial state:

$$\begin{bmatrix} 0 & 0 \\ 1 & 1 \end{bmatrix} \begin{bmatrix} 0 & 0 \\ 1 & 1 \end{bmatrix} \begin{bmatrix} 1 & 0 \\ 0 & 1 \end{bmatrix} \begin{bmatrix} 1 \\ 0 \end{bmatrix} = \begin{bmatrix} 0 \\ 1 \end{bmatrix} \quad (6.5)$$

Another important class of finite automata is *non-deterministic* finite automata, which can have any number of outgoing arrows for each state–letter pair, so the transition matrices are binary matrices with at least one 1 in each column. An input string is part of the accepted language if there is *some* path finishing in the accepted state. Clearly deterministic finite automata are special cases of non-deterministic finite automata, and the previous example can be extended to a case which is no longer deterministic:

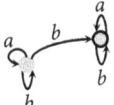

Figure 6.2

with the initial and accepting the states same as before. The transition matrices are now:

$$M_a = \begin{bmatrix} 1 & 0 \\ 0 & 1 \end{bmatrix}; \quad M_b = \begin{bmatrix} 1 & 0 \\ 1 & 1 \end{bmatrix} \quad (6.6)$$

And it is the still the case that the accepted language is the set of all strings containing at least one 'b'.

The third class of finite automata of interest is *probabilistic* automata. Like for non-deterministic automata, in probabilistic automata, each state may transition to more than one other for any given symbol – except now the state transitions are given (positive) weightings, such that the sum of the weights of the transitions for each state–symbol pair equals 1. This means that the transition matrices are *stochastic matrices* (as are the transition matrices of *Markov chains*) and are such that each column sums to 1. The accepted language can be defined either as the set of all strings that end in the final state with certainty or with probability above some threshold.

The example can now be further modified to

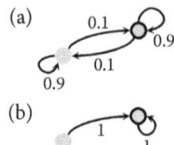

Figure 6.3

which has transition matrices

$$M_a = \begin{bmatrix} 0.9 & 0.1 \\ 0.1 & 0.9 \end{bmatrix}; \quad M_b = \begin{bmatrix} 0 & 0 \\ 1 & 1 \end{bmatrix} \quad (6.7)$$

So all strings (with at least one symbol) are accepted with some probability, and all strings ending with 'b' are accepted with certainty.

The final class of finite automata of interest is the quantum automata. For quantum automata, the transition matrices are unitary matrices consisting of positive and negative complex numbers. A special case of quantum automata is reversible automata, where the transition matrices are binary permutation matrices (exactly one 1 in each column and row – and so these are also deterministic automata). In general, after the transitions corresponding to the input string have been executed, the automaton will be in a superposition of states, and the Born rule is used to determine the probability distribution over the states. As for probabilistic automata, this probability distribution can be used to define the accepted language.

An example quantum automaton is:

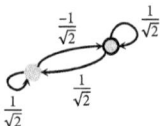

Figure 6.4

with the left-hand state, $|0\rangle$, as the starting state (and $|1\rangle$ is the accept state). This single-letter alphabet automaton has transition matrix:

$$M_a = \frac{1}{\sqrt{2}} \begin{bmatrix} 1 & 1 \\ -1 & 1 \end{bmatrix} \quad (6.8)$$

and it is interesting to observe that

$$M_a M_a = \begin{bmatrix} 0 & 1 \\ -1 & 0 \end{bmatrix} \qquad (6.9)$$

which, when applied to the initial state, gives $M_a M_a |0\rangle = -|1\rangle$. Thus, the probability distribution after the input string aa has been executed has all of the probability mass concentrated at the accept state $|1\rangle$, in spite of the fact that there are two paths of length 2 from $|0\rangle$ back to itself ($|0\rangle \to |0\rangle \to |0\rangle$ and $|0\rangle \to |1\rangle \to |0\rangle$). This is because these two paths interfere in such a way that there is actually zero chance of ending up in $|0\rangle$ and is the same familiar phenomenon seen in quantum circuits, for example when the Hadamard gate is applied to $|+\rangle$.

Turing machines

A Turing machine combines the states of a finite automaton with an infinitely long read-write tape, upon which the input string is initially written. At any time, a 'head' is over one space on the tape and can read the symbol written there (initially the head is at the left-hand end of the tape).

At a given time, the automaton is in a certain state, and the head is over a symbol which it reads. Given this state–symbol pair, a transition function determines:

- which symbol to overwrite on the current space on the tape;
- whether to move the head left or right (in some models a third option is to stay in the same position – but this does not increase the generality);
- which next state the finite automaton moves to.

Conventionally, the transition function is more accurately a *partial* function, and if the state–symbol pair is such that no transition instruction is specified, then the Turing machine is said to *halt*. (However, some models have explicit *Halt-Accept* and *Halt-Reject* states.)

For finite automata, it is natural to associate computational basis states with the states of the automaton. With Turing machines, the 'state' of the machine is now given by three components: the state of the automaton; the contents of the string; and the position of the head. To avoid ambiguity, we introduce the term 'configuration' for this triple. As the tape is infinite, this constitutes an infinite dimensional space; however, we may still associate a computational basis state with each configuration, and this, in turn, allows the four flavours of Turing machine to be defined analogously with their respective finite automata:

- In a deterministic Turing machine, each configuration transitions deterministically to the next. If the machine halts such that the automaton is in an accept state, then the input is part of the accepted language.
- In a non-deterministic Turing machine, each configuration may transition to a plurality of onwards configurations, hence meaning that the entire computational path can be represented by a tree.

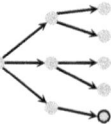

Figure 6.5

If there is some path to an accept state in which the computation halts, then the input is part of the accepted language.
- In a probabilistic Turing machine, a branching process again represents the entire computation, but now with a probability distribution over each branching occurrence.

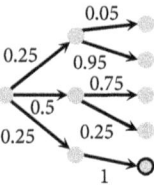

Figure 6.6

The probability of acceptance is given by the probability of halting in an accept state.
- In a quantum Turing machine, the overall transformation on the configuration space must be unitary. The probability of acceptance is given by the probability of halting in an accept state, using the Born rule to collapse the final superposition. (Note that there are a few variations – all equivalent – regarding the exact definition of a quantum Turing machine. For example, the configuration space is expanded to include explicit input, 'workspace', and output tapes in some models.)

For probabilistic and quantum Turing machines, the accepted language is defined in terms of the probability of halting in an accept state, as further detailed in Section 6.2.2.

6.2.2 Computability and the Church–Turing thesis

A Turing machine is not a physical entity, but rather an abstract model that captures completely any reasonable definition of computation (that is, it defines universal computation) – as famously enshrined in the *Church–Turing thesis*.

The Church–Turing thesis. *Anything that can be computed can be computed by a Turing machine.*

Any model of computation that can *simulate* a Turing machine is, therefore, also a universal model of computation and said to be *Turing complete*. The *Lambda Calculus* is the most well-known example of a model of computation that is ostensibly quite distinct from the Turing machine model, but is still Turing complete. The development of the Lambda Calculus was actually undertaken by Alonzo Church (*Church* of the Church–Turing thesis) around the same time that Alan Turing conceptualised the Turing machine, and so it was a remarkable result that two equivalent universal models of computation were proposed almost simultaneously – and one that reinforced the intuition that there is a unique notion of universal computation.

Quantum Turing machines represent a Turing complete model of computation. To give an intuitive explanation of why this is so, it is helpful to build on the argument of Section 4.4.1 that a quantum computer can compute a Boolean function with a fixed number, n, of input bits. By contrast, a *computable* function can be thought of as a function that takes any finite size of input. For instance, a *decision problem* can be seen as a function $F : \{0, 1\}^* \to \{0, 1\}$. The asterisk here can be taken as meaning 'any n', and for each fixed input size, n, this gives a function $F_n : \{0, 1\}^n \to \{0, 1\}$, which can be evaluated by a Boolean circuit and, as shown in Section 4.4.1, also implemented as a quantum circuit. For the function F to be computable (in the classical sense) means that it can be computed by a Turing machine, and so the circuits evaluating F_n can all be generated from a single pattern. Formally, the function that takes n to the circuit evaluating F_n should itself be computable. We can then define a quantum circuit model of computation as one where we have a (classically) computable function that takes each value of n to a quantum circuit on $n + c$ qubits for some value of c. This model is Turing complete by the fact that quantum circuits can simulate any Boolean circuit.

It is worth adding that quantum computing is not a model of computation that contradicts the Church–Turing thesis: there is no problem that is computable on a quantum computer than cannot, in principle at least, be computed on a classical computer. An informal, but helpful, way to appreciate this is by recalling that quantum computers (at least gate-based quantum computers) are nothing more than machines for multiplying very large matrices – and so the state vector simulation of Section 6.1.2 can always be employed.

6.2.3 Decision problems and computational complexity classes

The input to a Turing machine is initially written on the tape, and as such has a well-defined size. This input size is often known as the *problem size*. When the input is an integer, and for simplicity (and without loss of generality) letting the Turing machine have set of symbols {0, 1, BLANK} – thus meaning that the integer is input in binary – the problem size is the number of bits required to express the input integer in binary and not the value of the integer itself. As any mathematical problem is thus defined for varying problem sizes (for example in factoring, the problem size is the number of bits required to express the number being factored), the notion of computational efficiency is thus defined not in terms of the number of operations for a given problem instance, but rather in terms of how the number of operations (or Turing machine transitions) grows with the problem size. As pre-empted in 6.1.1, a widely agreed upon definition of *efficient computation* is when the number of operations grows in a way that is upper-bounded by some polynomial function of the problem size.

Decision problems, briefly mentioned in Section 6.2.2, are the most fundamental type of computational problem – so much so, that their definition (i.e., a more formal definition than that in Section 6.2.2) is effectively implicit in the set-up of finite automata and Turing machines. In a decision problem, for any input string that is part of the *accepted language*, a Turing machine computing the problem should halt in an accept state, and halt in a non-accept state for other input strings. That is, the Turing machine *decides* whether any input is part of the accepted language. From these definitions, four important complexity classes can be defined:

- Any decision problem that can be computed by a deterministic Turing machine, such that its number of transitions before halting (or as it is more usually put, its *running time*) is upper-bounded by a polynomial of the problem size is said to be in the class P.
- Any decision problem that can be computed by a non-deterministic Turing machine, such that the height of the tree is upper-bounded by a polynomial of the problem size is said to be in the class NP. Clearly, $P \subseteq NP$, as each decision could consist of just a single branch.
- Any decision problem that can be computed by a probabilistic Turing machine with running time upper-bounded by a polynomial of the problem size, and such that:

$$\Pr(\text{accept}) \begin{cases} \geq \frac{2}{3} & \text{if the input is in the accepted language} \\ \leq \frac{1}{3} & \text{if the input is not in the accepted language} \end{cases} \quad (6.10)$$

is said to be in the complexity class BPP. $P \subseteq BPP$, as the probability distribution in the latter case could always be set as a certain transition that occurs with certainty.
- Any decision problem that can be computed by a quantum Turing machine with running time upper-bounded by a polynomial of the problem size, and such that:

$$\Pr(\text{accept}) \begin{cases} \geq \frac{2}{3} & \text{if the input is in the accepted language} \\ \leq \frac{1}{3} & \text{if the input is not in the accepted language} \end{cases} \quad (6.11)$$

is said to be in the complexity class BQP. Quantum computation generalises both deterministic and randomised classical computation, in the sense that $P, BPP \subseteq BQP$.

The values $\frac{2}{3}$ and $\frac{1}{3}$ in the above definitions are somewhat arbitrary; all that is required is that some fixed constants above and below $\frac{1}{2}$, respectively, are used. Any such constants define equivalent computational complexity classes, as it is always possible to define an 'outer loop' of the Turing machine to boost the probability to any constant value required, whilst maintaining an overall polynomial run-time. (See chapter problems for a simple example of this.)

For practical purposes, the following informal descriptions of the four complexity classes are usually more useful than their formal counterparts:

- P is the set of problems that can be decided in polynomial time by some deterministic classical algorithm.
- BPP is the set of problems that can be decided in polynomial time by some randomised classical algorithm (i.e., a classical algorithm with access to a source of randomness).
- BQP is the set of problems that can be decided by a family of quantum circuits (one for each problem size), with each consisting of a number of gates from a fixed, standard gateset that is bounded by a polynomial function of the problem size (and therefore the number of qubits is itself a polynomial of the problem size). Furthermore, it is necessary that the circuit for each problem size is *uniformally generated*: a classical algorithm running in time that is a polynomial of the problem size must be able to generate the circuit (using some standard representation thereof). In all of the algorithms studied in this book, the circuits are highly structured and uniform generation is essentially implied by the way that the circuits are presented, and it is therefore safe to take the gate count as a valid proxy for computational complexity.
- NP is the set of problems that can be verified – that is, if some solution is presented it can be checked that it is indeed a solution – by some deterministic classical algorithm running in polynomial time.

There is a subclass of NP that is particularly noteworthy

- NP-complete is the set of the hardest problems in NP: any problem in NP can be mapped to ('reduced to') any NP-complete problem in a way that incurs only a polynomial computational overhead.

It is perhaps not immediately apparent why the class NP-complete is important; however, the existence of a 'complete' subclass for any complexity class is highly significant, as it means that any algorithm that solves a problem therein can solve (with a modest overhead) any problem in the entire class (hence justifying the name 'complete'). On the other hand, in the case of the complexity class NP, there are many known NP-complete problems, and after years of intense effort, the best known algorithms for these do not run in polynomial time. Therefore, it is widely expected (although not proven) that P is a strict subset of NP, and that would automatically mean that there can be no polynomial time algorithm for *any* NP-complete problem.

Whether P is a strict subset of NP or not is the most famous open problem in theoretical computer science, and it also gives further motivation for why polynomial run time is a good choice for a definition of efficiency. Consider any problem in NP; then, it is easy to construct an exponential time algorithm: (1) enumerate all of the possible inputs (which is exponential in the input size: if binary is used as the input, there are 2^n possible input bit strings); (2) check for each input in turn whether the solution is accepted (by virtue of being in NP, this can be achieved in polynomial time). Thus, any algorithm that improves significantly on this 'exhaustive search' strategy should be celebrated as effective. Of course, it would be possible to choose some sub-exponential but superpolynomial function (or even some specific polynomial) to be the threshold of 'efficiency', but that would seem to be rather arbitrary – and choosing polynomial time has an elegant simplicity.

As well as the widely-held belief that P is indeed a proper subset of NP, there are a number of other important and widely-held conjectures about the relationships between the various complexity classes introduced here:

- It is not known whether BPP is a subset of NP or vice versa, but it is conjectured that P = BPP.
- As the decision version of factoring (deciding whether the input has a factor less than the value of a second input, which thus acts as a threshold) is widely believed to be super-polynomial classically (even with a probabilistic Turing machine), the existence of Shor's algorithm (see Chapter 10) is taken as evidence that BPP $\subsetneq$ BQP.
- It is widely believed that NP-complete problems cannot be solved in polynomial time on a quantum computer (unless P = NP), so NP $\not\subseteq$ BQP.
- ... but it is also believed that there are problems outside of NP, which *can* be solved in polynomial time on a quantum computer, so BQP $\not\subseteq$ NP.

These conjectured inclusions can be depicted:

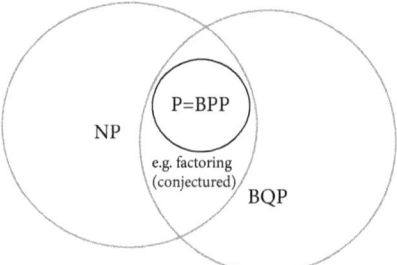

Figure 6.7

Even though in an ideal world one may strive to conclusively prove the above conjectures about complexity class inclusions, these problems are hard long-standing open problems that some of the greatest minds in theoretical computer science have tried and failed to solve. It is, therefore, often the case that results in quantum computing (and sometimes classical computing too) are proven 'up to complexity conjectures', as is the case for the circuits introduced in Section 6.2.5.

6.2.4 Other types of computational problems

Decision problems have been widely studied, and many of the most famous open problems in theoretical computer science concern decision problem complexity classes. Moreover, even for mathematical problems that are not posed as a decision problem, there is often a closely-related decision problem that can be defined and analysed to obtain bounds on the complexity of the original problem. For instance, to obtain the factor of a large semi-prime number, one can deploy the decision version of factoring (where the objective is to *decide* whether an input number has a factor smaller than some, also input, threshold) inside of a binary search over the threshold. Noting that the semi-prime certainly has a factor at most equal to its square root, the largest threshold one need ever consider is square root of the value of the first input, N, (the number being factored). The 'outer loop' that varies the threshold according to the binary search only requires a number of iterations that is logarithmic in $\sqrt{N}$, and hence is linear in the problem size. As factoring is at least polynomial in the problem size, this increase in complexity makes no material difference, and as such the complexity of *deciding* if a large number has a factor less than a certain threshold equates to the complexity of *producing* the factors.

However, it is certainly not the case that decision problems are the only computational problems of interest, and the other classes of problems for which quantum algorithms are presented in the subsequent chapters of this book are:

- **Search problems**: informally, an algorithm tackling a decision problem *decides* whether there is a solution, whereas an algorithm tackling a search problem *produces* a solution. The valid solutions form a set, and any element thereof is an acceptable output. (Or, in certain definitions when the solution set is allowed to be empty, the algorithm should output that no solution exists.)
- **Optimisation problems**: in a search problem, every element of the solution set is deemed equally valid as an output (and only one solution need be obtained), whereas in an optimisation problem, the solutions are equipped with a further 'cost value', which is a real number, and the goal is to output the solution that minimises this. The goal of optimisation may be either to produce the minimal solution itself and / or the cost value or the minimal solution. (Note that there is no loss of generality in defining optimisation as minimisation, as maximisation can be trivially converted to minimisation by negation or inversion.)
- **Sampling problems**: the goal is to sample a given probability distribution (exactly, or more usually, approximately).

The formal definitions of each of these include an input bit string, and the solution set (in the first two cases) varies from input to input, as does the probability distribution to be sampled in the third case.

6.2.5 Complexity-theoretic evidence that certain quantum circuits are hard to simulate classically

As specified in Chapter 4, every quantum circuit consists of three parts: (i) a standard initial state (usually $|0^n\rangle$); (ii) unitary evolution; and (iii) measurement of the final state (without loss of generality, this can be taken as a computational basis measurement). The principles of deferred and implicit measurement (Section 4.2) guarantee that every circuit can be re-factored into such a form – although it is sometimes important to include mid-circuit measurement and classical control to retain the physical significance of the overall operation, as in teleportation (Section 5.1.1). Regardless of the point in the quantum circuit at which various measurements occur, it follows directly from the postulates of quantum mechanics that the only way to extract classical information from a quantum state is by making a measurement. As measurement is a random process, the classical information (usually a bit string) obtained therefrom is a sample from some probability distribution. A foundational question in quantum computing is to identify when the same distribution can be sampled by an efficient classical algorithm, and this provides a more precise definition of efficient classical simulation (i.e., than the informal description given in the opening of the chapter).

As sketched out in Section 6.1.2, the naive approach to simulating even a single quantum operation is exponential in the number of qubits in the circuit, and so it may seem that in *most* or at least *many* cases, the distribution sampled by measuring a quantum state should be hard to replicate by any efficient classical means. It is indeed the case that for many quantum circuits, there is no *known* efficient classical algorithm for sampling the distribution obtained by measuring the prepared quantum state; however, this is not quite the same as finding classes of quantum circuits where it is provably the case that no efficient classical simulation exists. In fact, the term 'provably' here is too strong – very little is known *unconditionally* about the power of quantum computing – and so instead hardness results are proven up to plausible computational complexity conjectures.

One such class of quantum circuits is IQP circuits, which are of the form:

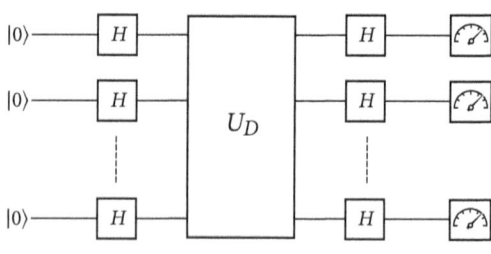

Figure 6.8

where U_D represents the unitary that is diagonal in the computational basis. Any product of gates that are diagonal in the computational basis is itself diagonal in the computational basis, and gates that have this property include T, S, and Z, as well as controlled versions of these.

Efficient classical simulation of (polynomial-sized) IQP circuits would imply the *collapse of polynomial hierarchy to its third level*. The polynomial hierarchy is an infinite series of complexity classes built on top of P and NP, and 'collapse' means that a subset that is generally assumed to be proper is actually equal to its corresponding 'super'set. Whilst the proof of this (and even a proper explanation of it) is beyond the scope of the book, this is generally viewed as extremely unlikely and, hence, provides strong complexity-theoretic evidence that there is indeed an exponential separation in computational power between quantum and classical computation. Another class of quantum circuits that are expected to be hard to simulate classically is *random circuits*, which were used in Google's celebrated *quantum supremacy* experiment.

6.3 Quantum circuits that can be efficiently classically simulated

A theoretician may worry that two things are equal until they are unconditionally proven to be distinct. A practitioner has no such luxury in the case of classical simulation of quantum circuits: the fact that relatively few quantum circuits are proven (up to complexity conjectures) to be hard to efficiently classically simulate does not disguise the fact that for a wide range of quantum circuits, there is no known efficient classical simulation algorithm. This means that those classes of quantum circuits which *are known* to be efficiently classically simulable assume heightened importance. This section introduces three important classes of quantum circuit which have the property that a classical simulation can be conducted with complexity Poly(n), where n is the number of qubits, and the number of gates in the quantum circuit is itself upper-bounded by some polynomial in n.

6.3.1 Efficient classical simulation of permutations

Perhaps the most obvious class of circuits that can be simulated classically is those circuits that already consist of classical gates, namely X, CNOT, and Toffoli. Such circuits can be directly simulated with Boolean algebra, using the direct correspondence between these quantum gates and their (classical) logic gate counterparts:

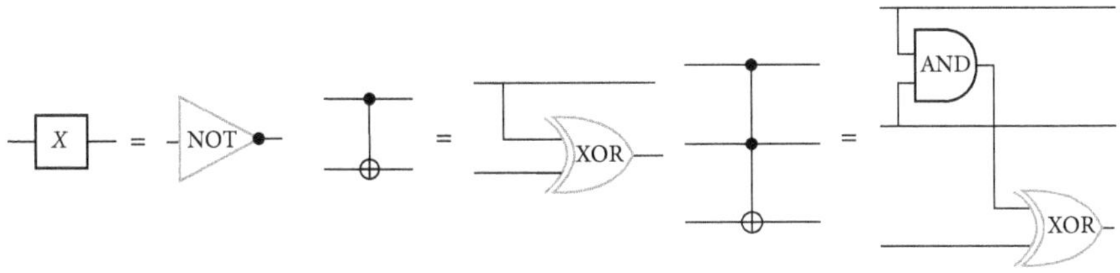

Figure 6.9

For each of these, the matrix of the gate is a permutation:

$$X = \begin{bmatrix} 0 & 1 \\ 1 & 0 \end{bmatrix} \quad \text{CNOT} = \begin{bmatrix} 1 & 0 & 0 & 0 \\ 0 & 1 & 0 & 0 \\ 0 & 0 & 0 & 1 \\ 0 & 0 & 1 & 0 \end{bmatrix} \quad \text{Toffoli} = \begin{bmatrix} 1 & 0 & 0 & 0 & 0 & 0 & 0 & 0 \\ 0 & 1 & 0 & 0 & 0 & 0 & 0 & 0 \\ 0 & 0 & 1 & 0 & 0 & 0 & 0 & 0 \\ 0 & 0 & 0 & 1 & 0 & 0 & 0 & 0 \\ 0 & 0 & 0 & 0 & 1 & 0 & 0 & 0 \\ 0 & 0 & 0 & 0 & 0 & 1 & 0 & 0 \\ 0 & 0 & 0 & 0 & 0 & 0 & 0 & 1 \\ 0 & 0 & 0 & 0 & 0 & 0 & 1 & 0 \end{bmatrix}$$

(6.12)

which is significant as any $2^n \times 2^n$ permutation can be expressed as an n-qubit quantum circuit containing these gates. It, therefore, follows that any permutation can be classically simulated – in about the same number of classical operations as gates in the quantum circuit.

In making this argument, it is necessary to be a little careful about how the quantum state evolves. By definition, the initial state is taken as $|0^n\rangle$, and permutations have the property that each computational basis state is sent to another computational basis state. For this reason, such circuits really amount to nothing more than transformations of classical information.

Interestingly though, even if the permutation is applied to a state that is in a superposition of computational basis states, efficient classically simulation is still sometimes possible. Let $|\psi\rangle$ be any quantum state which has the property that, when measured in the computational basis, the measurement outcomes sample a probability distribution that can alternatively be sampled by an efficient classical algorithm. In this case, a slight generalisation of the principle of deferred measurement (see Section 4.2) can be invoked to show that any permutation applied to $|\psi\rangle$ samples a distribution that can itself be classically simulated using about the same number of operations as are required in the quantum circuit:

Generalisation of the principle of deferred measurement. *Circuits of the form*

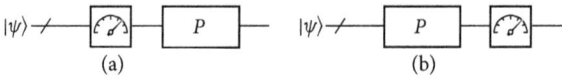

Figure 6.10

sample the same distribution, for any given input quantum state $|\psi\rangle$ and permutation circuit, P.

This equivalence can be shown to hold by considering the probability of measuring some output bit string x. First address circuit (b), in which case the probability is $|\langle x|P|\psi\rangle|^2$. Turning to the circuit (a), the probability of sampling the bit string x is

$|\langle y|\psi\rangle|^2$ where $|y\rangle = P^\dagger|x\rangle$. That is, because P is a permutation, x will be obtained if and only if the measurement outcome, y, is such that $P|y\rangle = |x\rangle$. The probability of this happening is:

$$|\langle y|\psi\rangle|^2 = |(|y\rangle)^\dagger|\psi\rangle|^2 = |(P^\dagger|x\rangle)^\dagger|\psi\rangle|^2 = |\langle x|P|\psi\rangle|^2 \tag{6.13}$$

and thus the probability of measuring any bit string, x, is identical for the two circuits (a) and (b).

As a result, it is possible to classically simulate a quantum circuit if it can be re-factored into the form of a sub-circuit followed by a permutation, where th first sub-circuit is such that when measured in the computationally basis, a distribution that can be efficiently prepared classically is sampled.

One circuit structure that occurs regularly in quantum algorithms is (i) a layer of Hadamard gates to prepare a superposition; (ii) a permutation; and (iii) some final sub-circuit to interfere the superposition in such a way that probability mass concentrates on the answer (this interference is often achieved by the quantum Fourier transform). As a bank of Hadamard gates, when measured, samples the uniform distribution, any quantum circuit consisting just of part (i) followed by part (ii) can be efficiently classically simulated – and so this highlights the importance of the final interference layer to obtain a quantum advantage. Moreover, this observation substantiates the assertion in Chapter 1 that if the coefficients of the superposition of a quantum state contain no more information than the probability distribution sampled if the state were to be measured, then quantum computation would be no more powerful than randomised classical computation.

6.3.2 Efficient classical simulation of circuits without entanglement

Another class of quantum circuits that can be efficiently classically simulated consists of those circuits that involve no entanglement. Consider an n-qubit circuit that contains only single-qubit gates: this can be state vector simulated as n separate single-qubit sub-circuits, and as such each state vector simulation consists of a two-dimensional space. The overall complexity of the simulation is thus only $\Theta(n)$.

More generally, if a non-entangling circuit consists of single- and two-qubit gates, but with the promise that each two qubit gate prepares an unentangled state, then the same essential strategy can be deployed. The state can be represented as a tensor product of n single-qubit spaces, with two such spaces temporarily joined to form a four-dimensional space when a two qubit gate is to be simulated – with the promise that the resultant state is unentangled being deployed to re-factor the resultant state back into the product of n single-qubit spaces. One example of a class of circuits of this form is that consisting of circuits with just single-qubit gates and swap gates.

This simple result shows that entanglement is at the heart of the power of quantum computation, and more advanced results show that it is not merely the no-entanglement setting that is efficiently classically simulable, but also cases where the entanglement is restricted in various ways.

6.3.3 Efficient classical simulation of stabiliser circuits

Stabiliser circuits, also known as Clifford circuits, are those that contain (i) gates from the Clifford group, which can be generated by $\{H, S, \mathrm{CNOT}\}$; and (ii) computational basis measurements. States that can be prepared by applying such circuits to $|0^n\rangle$ are termed 'stabiliser states' and even though they may be highly entangled, the celebrated Gottesman–Knill theorem proves that efficient classical simulation of stabiliser circuits is possible. The original presentation of the Gottesman–Knill theorem took advantage of the *stabiliser formalism* to give a group-theoretic proof. Whilst the stabiliser formalism (from which the common name of these circuits comes) is important in quantum computation in general, notably in quantum error correction, a simpler proof of the Gottesman–Knill theorem is possible using the *quadratic form expansion* of a stabiliser state. In particular, the quadratic form expansion of an n-qubit stabiliser state is:

$$|\psi\rangle = \frac{(\sqrt{i})^g}{\sqrt{2^r}} \sum_{x \in \{0,1\}^r} i^{x^T Q x} |Ax \oplus b\rangle, \qquad (6.14)$$

where $0 \leq r \leq n$, Q is an $r \times r$ symmetric integer matrix, A is an $n \times r$ binary matrix, $b \in \{0,1\}^n$, and g is an integer. Owing to their roles in the exponent of i and $\sqrt{i}$, respectively: (i) the diagonal elements of Q may be reduced modulo-4 and the off-diagonal elements of Q may be reduced modulo-2 (this is taken as implicit in the following updates); and (ii) g may be reduced modolu-8. Operations on A and b are naturally taken modulo-2. Note that, matching the original presentation of this method of stabiliser simulation, bit strings and binary vectors are somewhat conflated – for example the binary vector $Ax \oplus b$ is interpreted as the bit string that then specifies the corresponding computational basis state when contained in a ket. The quadratic form expansion keeps track of the global phase, $(\sqrt{i})^g$, which is necessary when a branching process is used to simulate circuits that are predominantly filled with gates from $\{H, S, \mathrm{CNOT}\}$ along with a smattering of T gates. In this case, the global phase becomes the relative phase of superposed terms.

To give an example of the quadratic form expansion, the state $|\psi_1\rangle = \frac{1}{2}(|00\rangle + i|01\rangle + |10\rangle + i|11\rangle)$ can be represented:

$$|\psi_1\rangle = \frac{1}{2} \sum_{x \in \{0,1\}^2} i^{[x_1\ x_2]\begin{bmatrix}0 & 0\\ 0 & 1\end{bmatrix}[x_1\ x_2]} |x_1 x_2\rangle \qquad (6.15)$$

That is, $g = 0$, $r = n = 2$, $A = I_2$, and $b = 0$, and $Q = \begin{bmatrix} 0 & 0 \\ 0 & 1 \end{bmatrix}$.

One thing to notice about the quadratic form expansion is that, unlike the naive approach to general quantum circuit simulation (i.e., state vector simulation), which incurs exponential (in n) memory usage, the quadratic form expansion requires only a polynomial (specifically a quadratic) amount of memory to store Q, A, and b. This is an essential property of stabiliser states, and is at the heart of this method of efficient stabiliser circuit simulation.

As r may in general be equal to zero, it is necessary to define that, in this case, the state is $(\sqrt{i})^g |b\rangle$. This is formally correct as the $\oplus |b\rangle$ refers to the modulo-2 addition of an n-element vector that does not depend on r, and hence (in a sense that is hard to convey using Dirac notation) properly exists outside of the summation – therefore, even if the summation is over zero terms, $|b\rangle$ is still present. Setting $g = 0$, $r = 0$, and $b = 0$ yields the state $|0^n\rangle$. This provides the base case for an inductive proof of the Gottesman–Knill theorem. We will show that the representation of a stabiliser state as a quadratic form expansion is closed under the action of the gates $\{H, S, \text{CNOT}\}$, in the sense that the state produced by the application of these states can always be represented by another quadratic form expansion. Moreover, in all cases, the transformation of the quadratic form expansion resulting from the application of these gates can be *efficiently* simulated.

To get an idea of how the simulation of Clifford operations works in the quadratic form expansion, it is instructive to consider simulation of the Pauli X, and Z gates. Whilst these can be decomposed as products of H and T (see Eqns. (4.20), and (4.21)), they also have simple direct updates. If a Pauli-X gate being applied to (say) the jth qubit, the state is transformed by simply flipping the jth qubit (in each term in the superposition), which corresponds to:

$$X_j|\psi\rangle = \frac{(\sqrt{i})^g}{\sqrt{2^r}} \sum_{x \in \{0,1\}^r} i^{x^T Q x} |Ax \oplus b \oplus e_j\rangle \tag{6.16}$$

where e_j is a binary vector of all zeros except for a single one in the jth place. (Note that e_j is used at various places in the following analysis, and its length is always implicitly defined by the operation in which it is involved.) This means that the update rule for a Pauli-X is simply:

Algorithm 6.1 Pauli-X applied to jth qubit

1: $b_j \leftarrow b_j \oplus 1$

Simulation of the Pauli-X provides a suitable introduction to the simulation of the controlled Pauli-X gate, that is the CNOT, which operates on *two* qubits, the control and target – denoted h and j, respectively. The action of the CNOT is to flip the jth qubit in superposed terms where the hth qubit is equal to one. This means that the jth qubit is flipped if $e_h^T(Ax \oplus b)$ is equal to one – that is, if the modulo-2 sum of the elements hth row of A and b_h is equal to one. To update the quadratic form expansion accordingly, the effect of this conditional control amounts to adding the hth row of A onto the jth (in place) and likewise the hth element of b onto the jth. Algebraically, this can be expressed by defining $E_{j,h} = I + e_j e_h^T$, which acts on vectors via left-multiplication by adding row h into row j (and similarly for matrices), such that

$$\text{CNOT}_{h,j}|\psi\rangle = \frac{(\sqrt{i})^g}{\sqrt{2^r}} \sum_{x \in \{0,1\}^r} i^{x^T Q x} |E_{j,h} Ax \oplus E_{j,h} b\rangle \tag{6.17}$$

which, therefore, corresponds to the update

Algorithm 6.2 CNOT operation controlled by the hth qubit targeting the jth qubit

1: $b_j \leftarrow b_j \oplus b_h$
2: **for** $k = 1 \ldots r$ **do**
3: $\quad A_{j,k} \leftarrow A_{j,k} \oplus A_{h,k}$
4: **end for**

The update when a Pauli-Z is applied to the jth qubit is slightly more complicated – and itself provides a good basis to then introduce the simulation of the S gate – in that each term in the superposition may accumulate a relative phase equal to $-1 = i^2 = (\sqrt{i})^4$ depending on whether the jth element is equal to one. Using e_j^T to 'pick out' the jth element of a vector, this can be expressed

$$Z_j|\psi\rangle = \frac{(\sqrt{i})^g}{\sqrt{2^r}} \sum_{x \in \{0,1\}^r} (-1)^{e_j^T(Ax \oplus b)} i^{x^T Q x} |Ax \oplus b\rangle$$

$$= \frac{(\sqrt{i})^g (\sqrt{i})^{4b_j}}{\sqrt{2^r}} \sum_{x \in \{0,1\}^r} i^{x^T Q x} i^{2 e_j^T A x} |Ax \oplus b\rangle$$

$$= \frac{(\sqrt{i})^{g+4b_j}}{\sqrt{2^r}} \sum_{x \in \{0,1\}^r} i^{x^T Q x + 2 x^T \text{diag}(a_{j,*}) x} |Ax \oplus b\rangle \tag{6.18}$$

where diag(.) takes as an input a vector and returns a matrix with the vector on the leading diagonal. In the above, the second line follows because an integer exponent of -1 may always be taken as automatically modulo-2; and the third line follows by noticing that for $x_i \in \{0,1\}$, $x_i = x_i^2$. As an integer exponent of $\sqrt{i}$ may always be taken as modulo-8, and similarly an integer exponent of i may always be taken as modulo-4, this gives the following update (where the modulo operations keep the values in the declared ranges):

Algorithm 6.3 Pauli-Z applied to jth qubit

1: $g \leftarrow g + 4b_j \mod 8$
2: $Q \leftarrow Q + 2\text{diag}(a_{j,*})$

This update for the application of a Pauli-Z gate also holds when $r = 0$, i.e., the global phase is correctly updated if the state is a computational basis state.

The operation of the S gate is similar to the Z gate, except that it is now necessary to be a little more careful about the exponent. For this, we define:

$$A \star B = AB \mod 2 \tag{6.19}$$

that is, ordinary matrix multiplication reduced modulo-2. Treating vectors as special cases of matrices, and using the fact that $y^2 = 0 \mod 4$ for even y and $y^2 = 1 \mod 4$ for odd y, this is used for $a_{j,*} \star x \in \{0, 1\}$, to establish the following identity:

$$a_{j,*} \star x = (a_{j,*}x)^2 \mod 4$$
$$= x^T a_{j,*}^T a_{j,*} x \tag{6.20}$$

Another simple identity that is used in the quadratic form expansion update (when transitioning between modulo and non-modulo operations) is, for $u, v \in \{0, 1\}$:

$$u \oplus v = u + v - 2uv \tag{6.21}$$

Together, these can be used to simulate the update for an S operation applied to the jth qubit:

$$S_j|\psi\rangle = \frac{(\sqrt{i})^g}{\sqrt{2^r}} \sum_{x \in \{0,1\}^r} i^{e_j^T A x \oplus b_j} i^{x^T Q x} |Ax \oplus b\rangle$$

$$= \frac{(\sqrt{i})^g}{\sqrt{2^r}} \sum_{x \in \{0,1\}^r} i^{a_{j,*} \star x \oplus b_j} i^{x^T Q x} |Ax \oplus b\rangle$$

$$= \frac{(\sqrt{i})^g}{\sqrt{2^r}} \sum_{x \in \{0,1\}^r} i^{x^T a_{j,*}^T a_{j,*} x \oplus b_j} i^{x^T Q x} |Ax \oplus b\rangle$$

$$= \frac{(\sqrt{i})^g}{\sqrt{2^r}} \sum_{x \in \{0,1\}^r} i^{x^T a_{j,*}^T a_{j,*} x + b_j - 2b_j(x^T a_{j,*}^T a_{j,*} x)} i^{x^T Q x} |Ax \oplus b\rangle$$

$$= \frac{(\sqrt{i})^g}{\sqrt{2^r}} \sum_{x \in \{0,1\}^r} i^{x^T (Q + (1-2b_j)a_{j,*}^T a_{j,*})x + b_j} |Ax \oplus b\rangle$$

$$= \frac{(\sqrt{i})^{g+2b_j}}{\sqrt{2^r}} \sum_{x \in \{0,1\}^r} i^{x^T (Q + (1-2b_j)a_{j,*}^T a_{j,*})x} |Ax \oplus b\rangle \tag{6.22}$$

which in algorithmic form gives:

Algorithm 6.4 S applied to jth qubit

1: $g \leftarrow g + 2b_j \mod 8$
2: $Q \leftarrow Q + (1 - 2b_j)a_{j,*}a_{j,*}^T$

The operations so far have shared the conspicuous property that they do not change the number of terms in the superposition. This is not the case for the Hadamard gate. For example, a Hadamard applied to a single qubit in the $|1\rangle$ state prepares the superposition:

$$|-\rangle = \frac{(\sqrt{i})^0}{\sqrt{2}} \sum_{x \in \{0,1\}^1} i^{x^T[2]x} |[1]x \oplus 0\rangle \tag{6.23}$$

and so r has increased from 0 to 1. To update the quadratic form expansion when a Hadamard gate is applied to the jth qubit, it is helpful to first focus on a single computational basis state (i.e., a single term of the sum in the quadratic form expansion) and further to let $K_j = I \oplus e_j e_j^T$ represent the map which acts on vectors and matrices by left-multiplication, to zero out row j. For some n-qubit computational basis state $|Ax \oplus b\rangle$, the action of a Hadamard is to send the jth qubit to: either $|+\rangle$ if the jth element of $Ax \oplus b$ is 0 or $|-\rangle$ if the jth element of $Ax \oplus b$ is 1, leaving the rest of the computational basis state unchanged. In the quadratic form expansion, this is given by:

$$H_j|Ax \oplus b\rangle = \frac{1}{\sqrt{2}} \sum_{z \in \{0,1\}} (-1)^{(e_j^T(Ax \oplus b))z} |K_j(Ax \oplus b) + ze_j\rangle$$

$$= \frac{1}{\sqrt{2}} \sum_{z \in \{0,1\}} (-1)^{a_{j,*}xz + b_j z} |(A'x + ze_j) \oplus b'\rangle, \tag{6.24}$$

where $A' = K_j A$ and $b' = K_j b$. Applying this representation straightforwardly to a quadratic form expansion yields:

$$H_j|\psi\rangle = \frac{(\sqrt{i})^g}{\sqrt{2^{r+1}}} \sum_{\substack{x \in \{0,1\}^r \\ z \in \{0,1\}}} i^{x^T Qx + 2a_{j,*}xz + 2b_j z} |(A'x + ze_j) \oplus b'\rangle \tag{6.25}$$

Letting z be instead the $(r+1)$th element of x, and updating Q first so that both updates to A and b can be done in one go (so the intermediate matrix A' and vector b' are no longer required), the quadratic form expansion can be updated by the following:

- $Q \leftarrow \begin{bmatrix} Q & a_{j,*}^T \\ a_{j,*} & 2b_j \end{bmatrix}$;
- $b \leftarrow K_j b$;

- $A \leftarrow [K_j A, e_j]$ (the matrix obtained by adjoining the vector e_j as an additional $(r+1)$st column to the matrix $K_j A$);
- $x \leftarrow [x, z]^T$ (which is achieved by $r \leftarrow r + 1$).

which is then in the standard form (6.14).

If we were to impose no constraint on the number of columns of the expansion matrix A, this would suffice to produce a representation of $H_j |\psi\rangle$. However, it may be that A is no longer full rank – indeed, if r is now greater than n, then A is certainly not full rank. If we return to the example of $|-\rangle$ expressed as a quadratic form expansion as in Eqn. (6.23), and simply apply the above updates when a further H is applied, then we get the following:

$$H|-\rangle = \frac{(\sqrt{i})^0}{2} \sum_{x \in \{0,1\}^2} i^{x^T \begin{bmatrix} 2 & 1 \\ 1 & 0 \end{bmatrix} x} |[0,1] x \oplus 0\rangle \qquad (6.26)$$

However, we know that $H|-\rangle = |1\rangle$, and this can be seen by expanding the sum and cancelling terms:

$$H|-\rangle = \frac{1}{2} \sum_{\substack{x_1 \in \{0,1\} \\ x_2 \in \{0,1\}}} i^{2x_1 + 2x_1 x_2} |x_2\rangle = \frac{1}{2} \left(i^0 |0\rangle + i^2 |0\rangle + i^0 |1\rangle + i^4 |1\rangle \right) = |1\rangle \qquad (6.27)$$

The fact that A is not full rank means that the same computational basis state appears multiple times when the sum is written out, which is an undesirable property that holds whenever the rank of A is smaller than r.

However, in general, the rank of a matrix can be efficiently checked by Gaussian elimination, and when A is not full rank, a change of variable can be made to find an equivalent quadratic form expansion, with r reduced such that A now is full rank. The algorithm for updating the quadratic form expansion when a Hadamard gate is applied to the jth qubit is given in Algorithm 6.5. More details on how the Hadamard gate is simulated in practice are given in Appendix. A.

Algorithm 6.5 H applied to jth qubit

1: $Q \leftarrow \begin{bmatrix} Q & a_{j,*}^T \\ a_{j,*} & 2b_j \end{bmatrix}$
2: $b \leftarrow K_j b$
3: $A \leftarrow [K_j A, e_j]$
4: $r \leftarrow r + 1$
5: Compute the rank of A
6: **if** A is not full rank **then**
7:     Efficient further updates to obtain equivalent quadratic form expansion with full-rank A
8: **end if**

With these procedures given to update the quadratic form expansion when a CNOT, S, or Hadamard gate is applied to any qubit(s), we have nearly completed the demonstration of this method of stabiliser simulation. What remains is to show that there exists a means of efficiently simulating measurement (that is, obtaining a random number sampled from the correct probability distribution). For this, we are aided by the fact that every computational basis state that appears in the superposition does so with equal probability, and so to simulate a measurement, it suffices to uniformly sample an r bit string, say y and substitute this in to obtain the measurement outcome $Ay \oplus b$. (This only works for full rank A, where each $Ay \oplus b$ is distinct; if this were not the case then a more complicated – and less efficient – procedure would be needed.)

As well as measuring the entire state, the quadratic form expansion facilitates the simulation of the measurement of individual qubits therein. It is a feature of stabiliser states that a single-qubit computational basis measurement will either be deterministic, or the outcomes 0 and 1 will each occur with 50% probability. If the jth qubit is measured in the computational basis, and the jth row of A is all zero, then the outcome will certainly be equal to b_j. In this case, the measurement can trivially be 'simulated', and there is no alteration to the state. In the case where the Ath row of j is not all zeros, then a coin can be flipped to obtain the outcome, $m \in \{0,1\}$, and A, b, and Q can be updated to be consistent with the (random) measurement outcome, m – which will entail a reduction of r by one (that is, $a_{j,*}x \oplus b_j = m$ can be substituted in and the number of independent equations thus reduced by one).

One of the strengths of simulating stabiliser circuits with the quadratic form expansion is that it can be seen as an automation of how one would simulate circuits with these gates 'by hand', with direct updates to the expansion over computational basis states for X, CNOT, Z, S, and Hadamard operations. As such, it makes it relatively easy to appreciate why this constitutes an efficiently classically simulable class of quantum circuits: all of the information needed to define a stabiliser state is contained in matrices/vectors of size, 1×1, $n \times 1$, $r \times r$, and $n \times r$ (where $r \leq n$) and so the data structures describing the state are of size commensurate with the number of qubits, n, rather than the size of the state vector, 2^n, as in generic state vector simulation. Moreover, all of the updates amount to simple operations on these vectors and matrices, and so even a naive implementation would incur complexity $\mathcal{O}(n^3)$ (bottle necks being Gaussian elimination and matrix multiplication); however, it is actually possible to improve this to $\mathcal{O}(n^2)$ with a little care – and further improvements are available when the matrices are sparse.

6.3.4 Generic techniques for quantum circuit simulation

Keeping track of a branching structure (each branch being a stabiliser circuit) whenever a T gate occurs in a (computationally universal) $\{H, S, T, CNOT\}$ circuit means that stabiliser-simulation-based methods can be used to achieve classical simulation of quantum circuits in general – albeit at the cost of exponential complexity in the

Table 6.1 Means of classically simulating quantum circuits

Simulation technique	Exponential in...
State vector	number of qubits
Tensor network	e.g., bond dimension
Stabiliser-based	T-gate count

T-gate count. It is a general feature of classical simulation methods that they are 'exponential in something', for example, state vector simulation is exponential in the number of qubits. This means that state vector simulation is, by definition, an inefficient means of classical simulation; however, other classical simulation methods may prove to be efficient if the circuit structure is restricted in some way, e.g., if there are very few T gates, then stabiliser-based simulation methods prove effective.

Another common technique for performing general classical simulation of quantum circuits involves representing the quantum circuit as a tensor network, and performing the resultant contraction. This is often the most efficient simulation technique when the goal is to estimate some observable expectation, rather to sample from the distribution prepared by measuring in the computational basis. More generally, tensor network methods often prove to be efficient when any or all of the following are satisfied: there is low entanglement; the quantum circuit maps directly onto a known tensor network structure (whose contraction is efficiently executable); the circuit is shallow, when expressed with swap gates included such that it can be exactly embedded in a two or three dimensional surface; when the *bond dimension* is low. Tensor network methods constitute an entire mathematical area in their own right, and hence are beyond the scope of this book – however, *further reading* highlights some useful resources for the interested reader. Table 6.1 summarises the main ways of classically simulating quantum circuits.

6.4 The road to quantum advantage

Having identified classes of quantum circuits that can be efficiently classically simulated as well as those for which it is expected that no efficient classical simulation algorithm exists, a natural question to ask is whether this is sufficient to establish when there is and is not a quantum advantage. It is possible to define sampling problems directly from classes of circuits that sample distributions that are expected to be hard to prepare classically, so in this sense the answer is a clear 'yes'; however beyond sampling problems, the picture is a little more nuanced. For instance, the *quantum approximate optimisation algorithm* (introduced in Chapter 12) employs quantum circuits that sample distributions that are hard to prepare classically, but once the 'end-to-end' complexity of the entire algorithm is taken into account, it is unclear that there is a quantum advantage available in most problem instances.

Nevertheless, as stated in the opening sentence of this chapter, *the* central question asked in the study of quantum computing is *when is there a quantum advantage?*, and to answer this more satisfactorily, it is necessary to first appreciate that demonstrating a quantum advantage may amount to any of the following (and possibly others) depending on the context:

1. Showing that BPP $\subsetneq$ BQP – or some other quantum complexity class is strictly more powerful than its classical counterpart.
2. Showing that there is a separation in classical and quantum computational complexity for some *query problem* (see Chapter 7).
3. Showing that a quantum algorithm asymptotically provides a super-polynomial advantage relative to the classical state-of-the-art (but that it has not been proved that this state-of-the-art algorithm is the best *possible* classical algorithm).
4. Showing that a quantum algorithm provides *any* asymptotic advantage relative to the classical state-of-the-art.
5. Showing that for some specific computational problem instance, a quantum algorithm running on real quantum hardware requires fewer computational resources than the best classical counterpart.

These notions of quantum advantage have been broadly ordered from 'of most theoretical interest' to 'of most practical interest'. Item 1 essentially amounts to a challenge to the *extended* Church–Turing thesis:

The extended Church–Turing thesis. *A probabilistic Turing machine can efficiently simulate any realistic model of computation.*

Whether quantum computing does indeed violate the extended Church–Turing thesis is the most important question in quantum computational complexity theory, whereas item 5 is essentially a general statement of the celebrated 'quantum supremacy' milestone achieved by the team at Google in 2019. In between, it is notable that Shor's algorithm for polynomial time factoring is an example of item 3 being achieved, whilst Grover's search – and its generalisations – exemplifies item 4. However, more than anything else, the hunt for quantum computational advantage began with the development of quantum algorithms for query problems, and this in turn provides a suitable starting point for our study of quantum algorithms, in the next chapter.

Chapter problems

1. The matrices defining probabilistic automata have the property that the entries in each column add up to 1. Prove that this property is preserved under matrix multiplication.

2. (a) What is the language accepted by the quantum automaton defined in Section 6.2.1 (i.e., with transition matrix in Eqn. (6.8))? Define the (range of) acceptance condition(s).
   (b) Prove that there is no two-state probabilistic automaton with this behaviour (i.e., with a single-letter alphabet).
   (c) Describe a probabilistic automaton (with more than two states) that exhibits this behaviour.
3. Consider a quantum automaton with initial state $|0\rangle$, a single accepting state, $|2\rangle$, and input letters, c and d, with transition matrices respectively:

$$M_c = \frac{1}{2}\begin{bmatrix} 1+i & 1-i & 0 \\ 1-i & 1+i & 0 \\ 0 & 0 & 2 \end{bmatrix}; \quad M_d = \frac{1}{2}\begin{bmatrix} 2 & 0 & 0 \\ 0 & 1+i & 1-i \\ 0 & 1-i & 1+i \end{bmatrix} \quad (6.28)$$

   (a) Verify that M_c and M_d are unitary as is required.
   (b) Give a four-letter input string containing two occurrences of c and two occurrences of d that is accepted with 100% probability.
   (c) Give an eight-letter input string containing both c and d that returns to the initial state with 100% probability.
4. Suppose that M is a quantum Turing machine that accepts a language L in the bounded probability sense: for each string $w \in L$, there is a probability $\geq \frac{2}{3}$ that M is observed in an accepting state after reading w and for each string $w \notin L$, there is a probability $\leq \frac{1}{3}$ that M is observed in an accepting state after reading w. We define a new machine M_0 that, on input w makes three independent runs of M on input w and decides acceptance by majority. What is the probability that M_0 accepts $w \in L$? What about $w \notin L$?
5. Consider the state $|\psi\rangle = \frac{1}{2}(|00\rangle + |01\rangle + |10\rangle - |11\rangle)$:
   (a) Express $|\psi\rangle$ as a quadratic form expansion.
   (b) Update the quadratic form expansion if a CNOT with first qubit as control and second as target is applied.
   (c) Simulate the CNOT applied to $|\psi\rangle$ directly, and confirm that the same answer is obtained as in (b).

Further reading

The counting argument that some states must require an exponentially large number of gates to prepare is due to Herbert et al [2024], which is similar to an earlier argument by Claude Shannon [1949]. The *generalised principle of deferred measurement* is well known in the community (as an easy result to obtain) and is stated explicitly by Herbert [2024]. Shende et al [2003] prove that {X, CNOT, Toffoli} can generate any permutation.

On computability and complexity theory in general, Hopcroft et al's [2006] book is a standard text and gives a suitable definition of a Turing machine, as used here. Scott Aaronson [2013b] provides a nice discussion of why complexity theory matters at a philosophical level, and Aaronson's [2013a] textbook is also an excellent resource for broaching quantum computing from a more complexity-theoretic direction. The assertion that quantum computing is primarily of interest for complexity rather than computability-theoretic reasons has one major exception: the result MIP*=RE by Ji *et al* [2021], which has far-reaching implications including for computability as discussed in his blog by Scott Aaronson [2020].

On classes of circuits that are expected to be hard to simulate classically, the result that IQP circuits are (assuming the non-collapse of the polynomial hierarchy) hard to simulate classically is due to Bremner *et al* [2010]; and Google's quantum supremacy experiment is documented by Arute *et al* [2019]. Another widely-celebrated quantum process whose efficient classical simulation would the collapse of the polynomial hierarchy is *Boson sampling*, originally proposed by Aaronson and Arkhipov [2011].

On classes of circuits that can be efficiently classically simulated, Guifré Vidal [2003] proves that slightly entangling circuits can still be efficiently simulated, and one other important class of circuits that can be efficiently classically simulated that we have not covered is *matchgate circuits* (Jozsa and Miyake [2008]; Brod [2016]). There is an abundance of literature on tensor network methods, Bridgeman and Chubb [2017] provide a nice pedagogical introduction, which shows how the bubbling method can be used to simulate shallow quantum circuits (when swap gates are included to make the interactions local), and Napp *et al* [2022] also address shallow (random) circuit simulation with tensor methods. The efficient classical simulation of stabiliser circuits was given by Daniel Gottesman [1997] in his PhD thesis, where he also credits Emanuel Knill for the discovery, and the state-of-the-art method of simulating such circuits with the stabiliser formalism is that of Aaronson and Gottesman [2004], whilst the method given herein is due to de Beaudrap and Herbert [2022] which Alec Edgington [2022] has implemented. Farhi and Harrow have shown that certain quantum approximate optimisation algorithm circuits prepare states that are hard to sample classically [2019].

7
Quantum algorithms for query problems

In 1985, David Deutsch proposed the original quantum algorithm, a simple yet remarkable demonstration of the enhanced computational power offered by quantum mechanics. Deutsch's algorithm provides a reduction in the number of *queries* to an unknown function required to discern some property of that function, relative to the best possible classical counterpart. This is known as an advantage in *query complexity*, and whilst the attention of the quantum computing community has, by and large, moved on to quantum algorithms that provide an advantage in *computational* rather than query complexity, it is still instructive to study some of the most important quantum algorithms for query problems. Such algorithms are historically important and, moreover, reveal the source of the computational power of quantum mechanics in a clear and easy to understand manner.

This chapter studies four algorithms for query problems: Deutsch's algorithm; the Deutsch–Jozsa algorithm, a generalisation of Deutsch's algorithm providing an exponential quantum advantage in the worst (classical) case; the Bernstein–Vazirani algorithm, which uses the same quantum circuit as the Deutsch–Jozsa algorithm to solve a problem in a way that provides a quantum advantage for the average case; and finally, Simon's algorithm, which provides an exponential separation between BPP and BQP in query complexity and served as an inspiration for Shor's algorithm.

A recap on binary numbers as quantum states

The algorithms in this chapter take as an input a superposition of computational basis states (as indeed do many other algorithms in the rest of the book), and it is convenient to analyse these by first expressing the action of the quantum circuit for any (single) computational basis state and then using linearity to express a sum of superposed terms. The means that each circuit is analysed for a general n-qubit computational basis state, $|x\rangle$:

$$|x\rangle = |x_1 x_2 \cdots x_n\rangle = |x_1\rangle \otimes |x_2\rangle \otimes \cdots \otimes |x_n\rangle \tag{7.1}$$

where $x \in \{0, 1\}^n$; thus, the natural correspondence between the computational basis state $|x\rangle$ and the n-bit binary number x (as spelled out in detail in Section 2.3.3) is leveraged to treat the input as a binary value.

7.1 Deutsch's algorithm

The unknown function in Deutsch's algorithm takes a single-bit input and returns a single-bit output and is encoded as a reversible circuit (see Section 4.4.1):

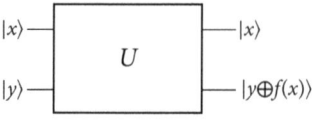

Figure 7.1

In the case of the quantum algorithm, this means that the reversible circuit can be used as a 'gate' in a quantum circuit, and the goal in each case (classical and quantum) is to use the unknown function as few times as possible to solve the problem posed. In much of the literature, terms including *black box* and *oracle* are often used synonymously in place of 'unknown function', although the latter has a slightly broader meaning within theoretical computer science, and so some care is sometimes needed to avoid ambiguity.

Deutsch's algorithm decides whether a function with a single input bit and single output bit is constant or balanced, that is, whether the output is always the same regardless of the input (constant) or the output is each of 0 and 1 for half of the inputs (balanced). There are four distinct functions with a single input bit and a single output bit:

Function	x	$f(x)$	Type	Example unitary
$f(x) = 0$	0	0	Constant	$\|x\rangle \longrightarrow \|x\rangle$ $\|y\rangle \longrightarrow \|y \oplus f(x)\rangle$
	1	0		
$f(x) = 1$	0	1	Constant	$\|x\rangle \longrightarrow \|x\rangle$ $\|y\rangle - \boxed{X} - \|y \oplus f(x)\rangle$
	1	1		
$f(x) = x$	0	0	Balanced	$\|x\rangle \longrightarrow \|x\rangle$ $\|y\rangle \longrightarrow \|y \oplus f(x)\rangle$
	1	1		
$f(x) = x \oplus 1$	0	1	Balanced	$\|x\rangle \longrightarrow \|x\rangle$ $\|y\rangle - \oplus \boxed{X} - \|y \oplus f(x)\rangle$
	1	0		

Classically, to decide whether an unknown function is constant or balanced, the function must be evaluated (queried) twice. Whichever input value, x, is selected, and whichever output, $f(x)$, is returned, there are always two possible functions that are consistent with this input–output pair: one balanced and one constant. Therefore, a second function evaluation is required to decide between the two possibilities.

Using Deutsch's algorithm, it is possible to solve the problem by querying the function just once quantumly.

Deutsch's algorithm incorporates the unknown (unitary) function into the following quantum circuit, applied to the initial state $|\psi_0\rangle = |01\rangle$ as shown:

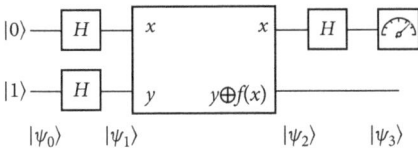

Figure 7.2

Deutsch's algorithm – even though relatively simple – reveals many of the important features of quantum algorithms in general, as can be seen by stepping through the algorithm.

Superposition

The first qubit corresponds to the input, x, of the function, and by setting it as $\frac{|0\rangle + |1\rangle}{\sqrt{2}}$, the function is thus queried for each of the possible inputs (in this case 0 and 1) *in superposition*. In many oversimplified popular 'explanations' of quantum computing, this is taken as the whole story – and it is implied that nothing more than superposition is required to obtain extraordinary quantum speed-ups; however, Deutsch's algorithm shows that this is emphatically not the case. Such fallacious explanations imply a circuit of the form,

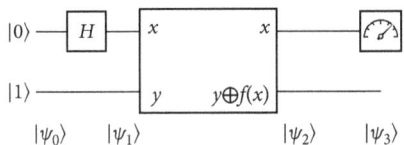

Figure 7.3

which is equivalent to running a randomised (classical) algorithm with the input being 0 or 1, each with 50% probability; and at least two circuit executions are still needed to decide if the function is constant or balanced.

Instead, there are two additional features of the quantum circuit that are crucial: first that the second qubit is initialised in the state $\frac{|0\rangle - |1\rangle}{\sqrt{2}}$, which facilitates *phase kickback*; and second that the Hadamard gates after the unknown unitary *interferes* the state, such that the correct decision is obtained with certainty.

Phase kickback

To see phase kickback in action, we initially treat the input on the first qubit as a computational basis state, denoted $|x\rangle$. In this case, the possibilities for the state immediately after the unknown function are: (i) when x is such that $f(x) = 0$,

$$|\psi_2'\rangle = |x\rangle\frac{|0 \oplus 0\rangle - |1 \oplus 0\rangle}{2} = |x\rangle\frac{|0\rangle - |1\rangle}{2} = (-1)^{f(x)}|x\rangle\frac{|0\rangle - |1\rangle}{2} \quad (7.2)$$

(using $|\psi_2'\rangle$ rather than $|\psi_2\rangle$ because the input is taken as a computational basis state, $|x\rangle$); and (ii) when x is such that $f(x) = 1$,

$$|\psi_2'\rangle = |x\rangle\frac{|0 \oplus 1\rangle - |1 \oplus 1\rangle}{2} = |x\rangle\frac{|1\rangle - |0\rangle}{2} = (-1)|x\rangle\frac{|0\rangle - |1\rangle}{2} = (-1)^{f(x)}|x\rangle\frac{|0\rangle - |1\rangle}{2} \quad (7.3)$$

Thus, the same expression, in terms of $f(x)$, is obtained in each case. In particular, $f(x)$ appears as the exponent of -1 in the coefficient multiplying the state, and whilst this is an unobservable *global* phase factor if the input is a computational basis state input, when the input is instead a superposition of computational basis states, by linearity, the corresponding terms are the *relative* phase factors. This is called *phase kickback*, and it can be used to write down the four possibilities for $|\psi_2\rangle$ when the first qubit is (prior to the unknown function) $\frac{|0\rangle + |1\rangle}{\sqrt{2}}$:

$$f(x) = 0 \quad |\psi_2\rangle = \left(\frac{(-1)^0|0\rangle + (-1)^0|1\rangle}{\sqrt{2}}\right)\left(\frac{|0\rangle - |1\rangle}{\sqrt{2}}\right) = \left(\frac{|0\rangle + |1\rangle}{\sqrt{2}}\right)\left(\frac{|0\rangle - |1\rangle}{\sqrt{2}}\right) \quad (7.4)$$

$$f(x) = 1 \quad |\psi_2\rangle = \left(\frac{(-1)^1|0\rangle + (-1)^1|1\rangle}{\sqrt{2}}\right)\left(\frac{|0\rangle - |1\rangle}{\sqrt{2}}\right) = \left(-\frac{|0\rangle + |1\rangle}{\sqrt{2}}\right)\left(\frac{|0\rangle - |1\rangle}{\sqrt{2}}\right) \quad (7.5)$$

$$f(x) = x \quad |\psi_2\rangle = \left(\frac{(-1)^0|0\rangle + (-1)^1|1\rangle}{\sqrt{2}}\right)\left(\frac{|0\rangle - |1\rangle}{\sqrt{2}}\right) = \left(\frac{|0\rangle - |1\rangle}{\sqrt{2}}\right)\left(\frac{|0\rangle - |1\rangle}{\sqrt{2}}\right) \quad (7.6)$$

$$f(x) = x \oplus 1 \quad |\psi_2\rangle = \left(\frac{(-1)^1|0\rangle + (-1)^0|1\rangle}{\sqrt{2}}\right)\left(\frac{|0\rangle - |1\rangle}{\sqrt{2}}\right) = \left(-\frac{|0\rangle - |1\rangle}{\sqrt{2}}\right)\left(\frac{|0\rangle - |1\rangle}{\sqrt{2}}\right) \quad (7.7)$$

from which it can be seen that (i) in the case where the function is constant, phase kickback means that both superposed terms accumulate the *same* phase factor; (ii) in the case where the function is balanced, phase kickback means that exactly one of the superposed terms is transformed to have coefficient equal to -1, and the other is unchanged. This allows the states of the two classes of function to be grouped:

$$|\psi_2\rangle = \begin{cases} \pm\left(\frac{|0\rangle + |1\rangle}{\sqrt{2}}\right)\left(\frac{|0\rangle - |1\rangle}{\sqrt{2}}\right) & \text{if } f(0) = f(1) \\ \pm\left(\frac{|0\rangle - |1\rangle}{\sqrt{2}}\right)\left(\frac{|0\rangle - |1\rangle}{\sqrt{2}}\right) & \text{if } f(0) \neq f(1) \end{cases} \quad (7.8)$$

That is, the two balanced cases differ only by an unobservable global phase (and likewise for the two constant cases).

Interference

The two types of functions (constant and balanced) are now grouped in a way that is observable; however, a computational basis measurement at this point is still insufficient to obtain the correct answer: simply measuring both qubits will yield one of the four possible bit strings, $\{00, 01, 10, 11\}$, such that each is obtained with equal probability. The final Hadamard gate is necessary as it interferes the superposition on the first qubit, giving:

$$|\psi_3\rangle = \begin{cases} \pm|0\rangle \left(\dfrac{|0\rangle - |1\rangle}{\sqrt{2}} \right) & \text{if } f(0) = f(1) \\ \pm|1\rangle \left(\dfrac{|0\rangle - |1\rangle}{\sqrt{2}} \right) & \text{if } f(0) \neq f(1) \end{cases} \quad (7.9)$$

Measuring the first qubit *after* this Hadamard gate returns the correct answer as the outcome will always be 0 if the function is constant, and 1 if balanced. Thus, Deutsch's algorithm has indeed determined whether the function is constant or balanced with just a single query, rather than the two queries needed in the classical case.

7.2 The Deutsch–Jozsa algorithm

In 1992, David Deutsch, together with Richard Jozsa, generalised Deutsch's algorithm to decide if an unknown function of *any* input size is constant or balanced. That is, the unknown function is now $f : \{0, 1\}^n \to \{0, 1\}$, with the promise that it is either constant or balanced.

Classically, the function must be queried $\frac{2^n}{2} + 1$ times to be sure whether the function is constant or balanced, because there are 2^n possible input bit strings, so in the worst case even if the same outcome is obtained for the first $\frac{2^n}{2}$ queries (with a different input bit string each time), it is still possible that the function could be either balanced or constant. By contrast, the Deutsch–Jozsa algorithm determines whether the function is constant or balanced with just a single query.

The circuit of the Deutsch–Jozsa algorithm closely resembles that of Deutsch's algorithm:

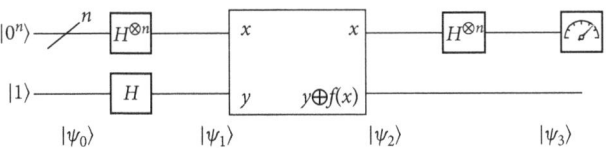

Figure 7.4

The initial state is:

$$|\psi_0\rangle = |0^n\rangle|1\rangle \tag{7.10}$$

which is then put into superposition:

$$|\psi_1\rangle = \sum_{x \in \{0,1\}^n} \frac{1}{\sqrt{2^{n+1}}} |x\rangle (|0\rangle - |1\rangle) \tag{7.11}$$

Phase kickback means that $|\psi_1\rangle$ is transformed to:

$$|\psi_2\rangle = \sum_{x \in \{0,1\}^n} \frac{1}{\sqrt{2^{n+1}}} (-1)^{f(x)} |x\rangle (|0\rangle - |1\rangle) \tag{7.12}$$

Considering the cases where the function is (i) constant and (ii) balanced in turn, when $f(x)$ is always the same value (constant), it can be written that:

$$|\psi_2\rangle = \pm \sum_{x \in \{0,1\}^n} \frac{1}{\sqrt{2^{n+1}}} |x\rangle (|0\rangle - |1\rangle) = \pm H^{\otimes n} |0^n\rangle |-\rangle \tag{7.13}$$

As the Hadamard gate is self-inverse, the final bank of Hadamards thus transforms the state to:

$$|\psi_3\rangle = \pm |0^n\rangle |-\rangle \tag{7.14}$$

and so the measurement of the first n qubits produces 0^n with certainty.

In the case where the function is instead balanced, it is necessary to explicitly express the interfering effect of the final bank of Hadamard gates. For this, it is useful to observe that for any computational basis state, $|x\rangle$, the following holds (see chapter problems):

$$H^{\otimes n}|x\rangle = \frac{1}{\sqrt{2^n}} \sum_{z \in \{0,1\}^n} (-1)^{x \cdot z} |z\rangle \tag{7.15}$$

where $x \cdot z$ is the bitwise modulo-2 dot-product, i.e., $x \cdot z = x_1 z_1 \oplus x_2 z_2 \oplus \cdots \oplus x_n z_n$, which can be used along with Eqn. (7.12) to obtain:

$$|\psi_3\rangle = \sum_{x \in \{0,1\}^n} \sum_{z \in \{0,1\}^n} \frac{1}{2^n} (-1)^{x \cdot z + f(x)} |z\rangle \frac{|0\rangle - |1\rangle}{\sqrt{2}} \tag{7.16}$$

To see how the Deutsch–Jozsa algorithm allows the cases where the function is constant and balanced to be distinguished, it is necessary to rewrite the expression for $|\psi_3\rangle$ with the term $|0^n\rangle$ written explicitly outside of the sum:

$$|\psi_3\rangle = \sum_{x \in \{0,1\}^n} \frac{1}{2^n}(-1)^{0+f(x)}|0^n\rangle\frac{|0\rangle - |1\rangle}{\sqrt{2}} + \sum_{x \in \{0,1\}^n} \sum_{z \in \{0,1\}^n \setminus 0^n} \frac{1}{2^n}(-1)^{x \cdot z + f(x)}|z\rangle\frac{|0\rangle - |1\rangle}{\sqrt{2}}$$

$$= \underbrace{\left(\frac{1}{2^n}\sum_{x \in \{0,1\}^n}(-1)^{0+f(x)}\right)}_{=0}|0^n\rangle\frac{|0\rangle - |1\rangle}{\sqrt{2}} + \sum_{x \in \{0,1\}^n} \sum_{z \in \{0,1\}^n \setminus 0^n} \frac{1}{2^n}(-1)^{x \cdot z + f(x)}|z\rangle\frac{|0\rangle - |1\rangle}{\sqrt{2}}$$

$$= \sum_{x \in \{0,1\}^n} \sum_{z \in \{0,1\}^n \setminus 0^n} \frac{1}{2^n}(-1)^{x \cdot z + f(x)}|z\rangle\frac{|0\rangle - |1\rangle}{\sqrt{2}} \qquad (7.17)$$

the final line following because $f(x)$ is equal to each of 0 and 1 for exactly half of the bit strings in the balanced case, meaning that exactly half of the terms in the sum in parentheses in the penultimate line are equal to −1, and the other half equal to 1. Thus, whereas the measurement outcome is *always* 0^n when the function is constant, the measurement outcome is *never* 0^n when the function is balanced. Therefore, measurement of the first n qubits distinguishes the two possibilities.

Whilst the Deutsch–Jozsa algorithm demonstrates a reduction from $2^{n-1} + 1$ classical evaluations to just one quantum query *in the worst case*, the classical algorithm does rather better *on average*. Let the function be evaluated just some m times classically, each time choosing a input bit string uniformly at random; then, if the function is balanced, the probability of obtaining the same outcome each time is $2^{-(m-1)}$. As soon as a different outcome is obtained, one can be certain that the function is balanced not constant; and conversely after a small number of queries, if the same output is returned each time, one can conclude that the function is highly likely to be constant. It follows that there exists a classical algorithm that decides whether the function is constant or balanced, which requires an average number of queries that does not grow with n and succeeds with high probability. Subsequent algorithms addressed this deficiency in the Deutsch–Jozsa algorithm as a demonstration of quantum advantage by proposing more elaborate unknown functions.

7.3 The Bernstein–Vazirani algorithm

The Bernstein–Vazirani algorithm tackles the following problem: an unknown function, $f_s : \{0, 1\}^n \to \{0, 1\}$, is promised to output the bitwise modulo-2 dot-product of the

---

Algorithm 7.1 Classical algorithm to discover s

1: Initialise $s = 0^n$
2: **for** $i = 1{:}n$ **do**
3: $\quad s_i \leftarrow f(e(i))$
4: **end for**
5: Return s

input bit string, x, and some hidden string s. That is, $f_s(x) = x_1 s_1 \oplus x_2 s_2 \oplus \cdots \oplus x_n s_n$ for some fixed s, and the goal is to obtain the string s by querying the unknown function.

The possible functions, $\{f_*(x)\}$, can be thought of as a statistical ensemble constructed by uniformly sampling s from all possible n-bit strings, and so, in an information-theoretic sense, n bits of information must be obtained to determine s. Any classical algorithm can obtain at most one bit of information from a single query to the unknown function, and thus, n queries are a lower-bound on the classical query complexity. Running Algorithm 7.1 (where $e(i)$ is the n-bit string of all zeros except for a single one in the ith element) produces s with n queries, thus achieving the lower-bound.

In the quantum case, the same circuit as that used in the Deutsch–Jozsa algorithm can be deployed to solve the problem with a single query. For this reason, the Bernstein–Vazirani algorithm is sometimes referred to as a variant of the Deutsch–Jozsa algorithm, although this is somewhat misleading, as the problem being computed is quite distinct. Nevertheless, the analysis of the evolution of the quantum state is identical, and so the state $|\psi_3\rangle$ can be expressed in a form that explicitly makes use of the form of the unknown function:

$$\begin{aligned}
|\psi_3\rangle &= \sum_{x\in\{0,1\}^n} \sum_{z\in\{0,1\}^n} \frac{1}{2^n} (-1)^{x\cdot z + f(x)} |z\rangle \frac{|0\rangle - |1\rangle}{\sqrt{2}} \\
&= \sum_{x\in\{0,1\}^n} \sum_{z\in\{0,1\}^n} \frac{1}{2^n} (-1)^{x\cdot z + x\cdot s} |z\rangle \frac{|0\rangle - |1\rangle}{\sqrt{2}} \\
&= \sum_{x\in\{0,1\}^n} \sum_{z\in\{0,1\}^n} \frac{1}{2^n} (-1)^{x\cdot (z\oplus s)} |z\rangle \frac{|0\rangle - |1\rangle}{\sqrt{2}} \\
&= \sum_{x\in\{0,1\}^n} \frac{1}{2^n} (-1)^{x\cdot (s\oplus s)} |s\rangle \frac{|0\rangle - |1\rangle}{\sqrt{2}} + \sum_{x\in\{0,1\}^n} \sum_{z\in\{0,1\}^n\setminus s} \frac{1}{2^n} (-1)^{x\cdot (z\oplus s)} |z\rangle \frac{|0\rangle - |1\rangle}{\sqrt{2}} \\
&= \sum_{x\in\{0,1\}^n} \frac{1}{2^n} (-1)^{0} |s\rangle \frac{|0\rangle - |1\rangle}{\sqrt{2}} + \sum_{x\in\{0,1\}^n} \sum_{z\in\{0,1\}^n\setminus s} \frac{1}{2^n} (-1)^{x\cdot (z\oplus s)} |z\rangle \frac{|0\rangle - |1\rangle}{\sqrt{2}} \\
&= |s\rangle \frac{|0\rangle - |1\rangle}{\sqrt{2}}
\end{aligned} \qquad (7.18)$$

The double sum in the penultimate line can be seen to equal 0 as it is possible to 'pair off' all of the bit strings x such that for one of the pair $x \cdot (z \oplus s) = 1$ and for the other $x \cdot (z \oplus s) = 0$. Thus, by measuring the first register, the Bernstein–Vazirani algorithm indeed returns the hidden bit string s with a single query. However, as the classical algorithm only requires n queries, the quantum advantage is not as large as in the worst (classical) case as for the Deutsch–Jozsa algorithm, which offers an exponential advantage. The final algorithm that we study in this chapter, Simon's algorithm, gives a quantum advantage that is exponential even in the average case.

7.4 Simon's algorithm

In the Deutsch–Jozsa and Bernstein–Vazirani algorithms, as well as using phase kickback, it is significant that the bitwise dot product, $x \cdot z$, also appears in the exponent of -1 in the superposition prepared by querying the unknown function. This is put to effect (without phase kickback) in Simon's algorithm.

Simon's algorithm tackles Simon's problem, which is: consider a function $f: \{0,1\}^n \to \{0,1\}^n$ with the promise that:

$$\forall x, y \in \{0,1\}^n, \ f(x) = f(y) \iff x \oplus y \in \{0^n, s\} \tag{7.19}$$

where s is some n-bit string that may or may not be 0^n. Note that $x \oplus y = 0^n \implies x = y$. Simon's problem may be posed in either of two slightly different forms: (i) the problem of *deciding* whether $s = 0^n$ or some other bit string; or (ii) the problem of *obtaining* the bit string s. The two are actually equivalent in terms of computational hardness, and Simon's algorithm resolves both – but we shall focus on the former here, for clarity of presentation. The key to solving the decision version of Simon's problem (both classically and quantumly) is to notice that the two possibilities are distinguished by the fact that:

1. when $s = 0^n$, the function is one–one: each input maps to a unique output;
2. when $s \neq 0^n$, there are exactly two inputs that map to each possible output, and hence, the function is two–one.

Classically, to solve Simon's problem, one must query $f(x)$ for various inputs and observe whether the output is unique in each case. This leads to a complexity of $\Theta(\sqrt{2^n}) = \Theta(2^{n/2})$ queries, and so any classical algorithm requires an exponential number of queries to solve the problem with high probability.

Simon's (quantum) algorithm tackles Simon's problem using the following circuit:

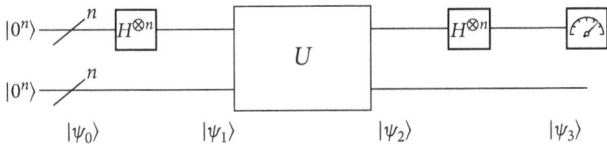

Figure 7.5

from which the state $|\psi_2\rangle$ can be immediately written down:

$$|\psi_2\rangle = \frac{1}{\sqrt{2^n}} \sum_{x \in \{0,1\}^n} |x\rangle |f(x)\rangle \tag{7.20}$$

In the case where the function is one–one, each $|x\rangle$ will be entangled with a different state in the second register, and so Eqn. (7.15) can be used to obtain the state after the final bank of Hadamard gates:

$$|\psi_3\rangle = \frac{1}{2^n} \sum_{x \in \{0,1\}^n} \sum_{z \in \{0,1\}^n} (-1)^{x \cdot z} |z\rangle |f(x)\rangle \tag{7.21}$$

Measurement of the first register will thus return a uniformly random bit string.

When $f(x)$ is two-one, this is no longer the case, as the states in the second register are such that each is entangled with two states in the first register. In this case, even though Eqn. (7.20) is still valid, it is more helpful to re-express $|\psi_2\rangle$ such that states that are entangled with the same value in the second register are now grouped.

$$|\psi_2\rangle = \frac{1}{\sqrt{2^n}} \sum_x (|x\rangle + |x \oplus s\rangle) |f(x)\rangle \tag{7.22}$$

The range of the summation is now ill-defined, as for each x that is included, $x \oplus s$ is not. However, this can be tolerated in the following analysis and allows the state after the final bank of Hadamard gates to be expressed

$$|\psi_3\rangle = \frac{1}{2^n} \sum_{x,z} ((-1)^{x \cdot z} + (-1)^{(x \oplus s) \cdot z}) |z\rangle |f(x)\rangle \tag{7.23}$$

For every x, z such that $(-1)^{x \cdot z} + (-1)^{(x \oplus s) \cdot z} = 0$, the corresponding $|z\rangle$ will not appear in the superposition, and hence, the bit string z will not be obtained as a measurement outcome. Thus, only states such that $x \cdot z = (x \oplus s) \cdot z$ will be present, which can be rearranged to give:

$$x \cdot z = (x \oplus s) \cdot z \tag{7.24}$$
$$x \cdot z = x \cdot z \oplus s \cdot z \tag{7.25}$$
$$\implies s \cdot z = 0 \tag{7.26}$$

and notably all such states will occur with equal probability. Simon's algorithm now relies on (efficient) classical post-processing to obtain s. It is necessary to run the quantum circuit a number of times to obtain n linearly independent outcome bit strings which we label $z^{(1)}, \ldots, z^{(n)}$ (as each possible outcome is equally probable, with high probability only a small number of runs is needed to achieve this). As for each, $s \cdot z^{(i)} = 0$ is sure to hold, a system of n independent linear equations,

$$s \cdot z^{(1)} = 0 \mod 2 \tag{7.27}$$
$$s \cdot z^{(2)} = 0 \mod 2 \tag{7.28}$$
$$\vdots$$
$$s \cdot z^{(n)} = 0 \mod 2 \tag{7.29}$$

can be solved to obtain s. Therefore, classically, $\Theta(2^{n/2})$ queries are needed, whereas Simon's algorithm solves the problem with just $\Theta(n)$ queries.

7.5 Conditions for exponential quantum speed-up in query problems

Of the four algorithms studied in this chapter, the first, Deutsch's algorithm, applies only to a single-input bit string size, whereas the others are defined for any input size. Indeed, in these terms, it is appropriate to think of Deutsch's algorithm as the smallest instance of the Deutsch–Jozsa algorithm. In this way, we can treat this chapter as the study of three rather than four distinct algorithms, and one common theme is that the particular problems tackled have been carefully crafted to separate the capabilities of the classical and quantum solutions. Moreover, there are two essential features that a query problem must have in order for the quantum algorithm to exhibit an exponential advantage:

1. The feature of the unknown function needs to be such that the function must be evaluated for a large fraction of the possible input strings to be discovered classically. This guarantees exponential run-time of the classical algorithm and is satisfied for the Deutsch–Jozsa algorithm (in the worst case classically) and for Simon's algorithm (on average, not just in the worst case), but not for the Bernstein–Vazirani algorithm, where only n queries are required classically.
2. Perhaps less obviously, it is necessary that the unknown function is restricted in some way. That is, that there is some *promise* on what the function should be. In the case of the Deutsch–Jozsa algorithm, there is the promise that the output is constant or balanced, whereas in the Bernstein–Vazirani and Simon's algorithms, the function encodes a hidden string, s, in different ways.

It is the necessity of both (1) and (2), which means that query problems are good for demonstrating the potential for quantum computation to obtain an exponential speed-up relative to the best possible classical counterpart. It is often possible to use elementary combinatorial arguments about black-box functions to lower-bound the classical query complexity as exponential in the problem size (that is, that no classical strategy can be much better than iterating through all of the possible inputs), whilst simultaneously restricting the function – that is, endowing it with a certain *structure* – such that when the unknown function is queried by a quantum mechanical superposition of the possible inputs, the solution can be extracted efficiently.

It is the structure of the problem that enables quantum mechanical interference to be used to concentrate probability mass on 'correct' answers, in a way that has no classical analogue. This idea is also present in non-query problems, where an actual explicitly-specified function takes the place of the black-box – and in one of the most spectacular examples, Shor's algorithm, the quantum Fourier transform may be thought of as an interfering step that efficiently extracts the period of a function (being periodic constituting the required structure of the function in this case). However, as it is not known that P ≠ NP, when the problem is posed in terms of an explicit mathematical function (not an unknown function to be queried), it cannot absolutely be precluded that an efficient classical algorithm for the problem does not exist, and hence, given the current state of computational complexity theory, an unconditional proof of exponential quantum speed-up for non-query problems is out of reach.

On the other hand, when the problem is *unstructured*, an exponential quantum advantage is completely off the table; however, there is still the opportunity for a practically very valuable lower-order quantum advantage. This is the subject of the next chapter, where the notion of structure is also more precisely defined.

Chapter problems

1. Use the generalised principle of deferred measurement, in Chapter 6, to explain why the circuit in Section 7.1 for Deutsch's algorithm with the interfering H (i.e., after the unknown unitary) omitted is equivalent to a randomised classical algorithm.
2. Show that:
   (a) $H^{\otimes n}|0\rangle^{\otimes n} = \frac{1}{\sqrt{2^n}} \sum_{x \in \{0,1\}^n} |x\rangle$.
   (b) When $|x\rangle$ is any n-qubit computational basis state, $H^{\otimes n}|x\rangle = \frac{1}{\sqrt{2^n}} \sum_{z \in \{0,1\}^n} (-1)^{x \cdot z} |z\rangle$.
3. For each of the following cases, work through the Deutsch–Jozsa algorithm explicitly for $n = 2$, and show the probabilities for each of the $2^2 = 4$ possible measurement outcomes of the two qubits.
   (a) $f(x) = 0$ if x is even or zero, and 1 otherwise.
   (b) $f(x) = 0$ for all x.
4. A Toffoli gate is to be used as the oracle in the Deutsch–Jozsa algorithm.
   (a) Why is this not a valid oracle for the Deutsch–Jozsa algorithm?
   (b) If the Deutsch–Jozsa algorithm is run anyway with a Toffoli gate as the oracle, what will the outcome be?
   (c) How can two Toffoli gates be used to construct an oracle that *is* valid for the Deutsch–Jozsa algorithm?
5. For the Bernstein–Vazirani algorithm with hidden two-bit string 01, write down the final state (before measurement) explicitly in full and show that the unwanted terms cancel to leave just the state $|01\rangle|-\rangle$.
6. If Simon's algorithm is executed for the single-bit hidden bit string:
   (a) $s = 1$
   (b) $s = 0$
   then in each case what is the state prior to measurement? Verify that in each case Simon's algorithm works as promised.

Further reading

Deutsch's algorithm is due to David Deutsch [1985], who together with Richard Jozsa generalised this as the Deutsch–Jozsa algorithm [1992]. Ethan Bernstein and Umesh Vazirani proposed the Bernstein–Vazirani algorithm [1997], and Simon's algorithm is due to Daniel Simon [1994; 1997]. The discussion about the structure needed for exponential speed-up is developed further in the next chapter, and the further reading for this topic can be found at the end of that chapter.

8
Quantum search

Quantum superposition is a powerful computational tool to have at one's disposal – in fact, a phrase along the lines *quantum computers evaluate a function for every possible input in superposition and hence achieve a massive speed-up* appears in virtually every popular and media article on the subject. The reality is, however, a little more complicated. In this chapter, we look at what is probably the closest quantum algorithm to the naive idea of 'evaluating for every possibility at once', namely Lov Grover's quantum search algorithm.

Grover's algorithm tackles an unstructured search problem (in fact, the problem it tackles is usually referred to as simply unstructured search, although this is a little misleading as by some common definitions of *structure*, it is possible to define other unstructured search problems). There is a 'database' of some N-elements, of which a small number M are *marked*. The goal is to find a marked element (any marked element), and here 'unstructured' means that there is no pattern that would allow (one of) the marked element(s) to be found quickly. Thus, classically, a brute-force search through the entire database is the only viable approach. In the worst case, this would entail checking $N - M$ unmarked elements before obtaining a marked element, and even the average case involves searching through some $\frac{N}{2M}$ elements before one can expect to obtain a marked element. We may think of the process of searching as *querying* the database for various elements $0, 1, \ldots, N$, and remarkably using quantum rather than classical computation, Grover's algorithm (sometimes referred to simply as *Grover search*) finds a marked element with only $\Theta(\sqrt{N/M})$ queries. This improvement from $\Theta(N/M)$ to $\Theta(\sqrt{N/M})$ queries is commonly referred to as the 'quadratic speed-up' or 'quadratic advantage'.

8.1 Grover's algorithm

To show how Grover's algorithm works, it is first necessary to be a little more specific about precisely what is meant by 'querying a database'. In the quantum algorithm, this is defined in terms of a 'search oracle'. The search oracle is similar in nature to the black boxes of Chapter 7, in that they 'mark' some M input binary strings as 1 (i.e., $f(x) = 1$) whilst leaving the others unchanged (i.e., $f(x) = 0$), and so are of the form:

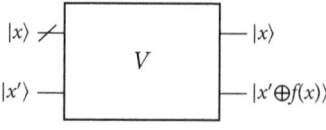

Figure 8.1

As Grover's algorithm addresses the problem of *unstructured search*, there is no promise that there is some pattern concerning which elements x are such that $f(x) = 1$. For example, consider a search through a database of $2^4 = 16$ elements, of which only one is marked (i.e., $M = 1$) – say that 0010 is the unique 'marked' binary string – then an appropriate oracle would be:

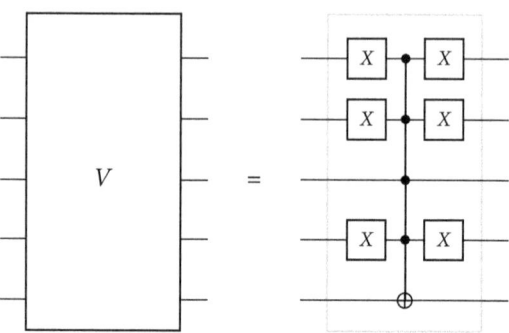

Figure 8.2

In Grover search, the search oracle is queried with the final qubit initialised in the state $|-\rangle$ to take advantage of phase kickback (see Section 7.1). If the oracle is queried for the binary string x (again using the strategy introduced in Chapter 7 of first analysing the circuit for a single computational basis state input), then the input to the oracle is the quantum state $|x\rangle|-\rangle$. The oracle then has the effect:

$$|x\rangle|-\rangle \xrightarrow{V} (-1)^{f(x)}|x\rangle|-\rangle \quad (8.1)$$

which can be illustrated as:

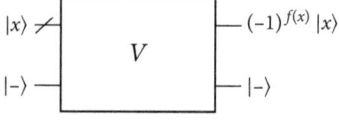

Figure 8.3

Decomposing the uniform superposition

Grover's algorithm does not apply the search oracle to a single computational basis state (input bit string), but rather to a superposition. For simplicity, it can be assumed that the number of elements in the search space, N, is a power of 2 (even if this is not already the case, it can always be achieved by *zero-padding*, that is, introducing additional unmarked 'dummy' elements so that the database is at most twice its original size), and the first n qubits are typically referred to as the 'search register'. The search register is then put in a uniform superposition of all 2^n computational basis states, a state that can be prepared by applying $H^{\otimes n}$ to the n qubits initially in the state $|0^n\rangle$. This state is denoted

$$|\psi\rangle = H^{\otimes n}|0^n\rangle = \frac{1}{\sqrt{N}} \sum_{x \in \{0,1\}^n} |x\rangle \qquad (8.2)$$

To analyse how Grover's algorithm works, it is helpful to decompose $|\psi\rangle$ into a sum of terms corresponding to (i) a uniform superposition of unmarked states (denoted $|y\rangle$) and (ii) a uniform superposition of marked states (denoted $|z\rangle$):

$$|\psi\rangle = \frac{1}{\sqrt{N}} \left(\sqrt{N-M} \underbrace{\frac{1}{\sqrt{N-M}} \sum_{x \text{ s.t. } f(x)=0} |x\rangle}_{|y\rangle} + \sqrt{M} \underbrace{\frac{1}{\sqrt{M}} \sum_{x \text{ s.t. } f(x)=1} |x\rangle}_{|z\rangle} \right) \qquad (8.3)$$

Visualising the action of the search oracle on $|\psi\rangle|-\rangle$

A key property of Grover search is that, once the uniform superposition has been prepared, the entire unitary evolution remains in the subspace spanned by $|z\rangle|-\rangle$ and $|y\rangle|-\rangle$, which are orthogonal by construction. As a result, even though the search register in Grover's algorithm includes some n qubits (for any n), the unitary evolution can be represented by vectors on a Bloch sphere with $|z\rangle|-\rangle$ and $|y\rangle|-\rangle$ as co-polar points. Furthermore, the evolution of the state occurs in just a single great circle of this Bloch sphere – namely that consisting of superpositions of $|z\rangle|-\rangle$ and $|y\rangle|-\rangle$ where the coefficients are real.

The most practically relevant cases for speeding up unstructured search occur when marked elements are rare (i.e., $N \gg M$), and so the following illustrations are suggestively drawn with the uniform superposition accordingly dominated by $|y\rangle$. To begin with, the uniform superposition itself (tensored with $|-\rangle$ on the final qubit) and the state prepared by a single application of the search oracle can be depicted:

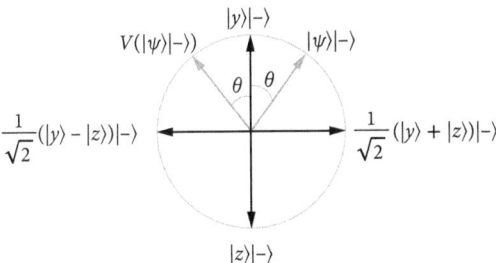

Figure 8.4

(Note that to keep the subsequent diagrams uncluttered, the horizontal axis is henceforth omitted.) As shown, the state after the search oracle has been applied to the uniform superposition amounts to reflection of $|\psi\rangle|-\rangle$ in the vertical axis, because:

$$V(|\psi\rangle|-\rangle)$$

$$= V \frac{1}{\sqrt{N}} \left(\sqrt{N-M} \underbrace{\frac{1}{\sqrt{N-M}} \sum_{x \text{ s.t. } f(x)=0} |x\rangle}_{|y\rangle} + \sqrt{M} \underbrace{\frac{1}{\sqrt{M}} \sum_{x \text{ s.t. } f(x)=1} |x\rangle}_{|z\rangle} \right) |-\rangle$$

$$= \frac{1}{\sqrt{N}} \left(\sqrt{N-M} \frac{1}{\sqrt{N-M}} \sum_{x \text{ s.t. } f(x)=0} (-1)^{f(x)} |x\rangle + \sqrt{M} \frac{1}{\sqrt{M}} \sum_{x \text{ s.t. } f(x)=1} (-1)^{f(x)} |x\rangle \right) |-\rangle$$

$$= \frac{1}{\sqrt{N}} \left(\sqrt{N-M} \frac{1}{\sqrt{N-M}} \sum_{x \text{ s.t. } f(x)=0} |x\rangle + \sqrt{M} \frac{1}{\sqrt{M}} \sum_{x \text{ s.t. } f(x)=1} -|x\rangle \right) |-\rangle$$

$$= \left(\sqrt{\frac{N-M}{N}} |y\rangle - \sqrt{\frac{M}{N}} |z\rangle \right) |-\rangle \tag{8.4}$$

The second step: reflecting about the original superposition

To find a marked element, it is necessary to evolve the state such that a superposition dominated by $|z\rangle|-\rangle$ is obtained. In such a case, measuring the search register will, with high probability, yield a binary string corresponding to a marked element. However, the above step of simply applying the search oracle has not helped in this regard: $V(|\psi\rangle|-\rangle)$ is no closer to $|z\rangle|-\rangle$ than $|\psi\rangle|-\rangle$ itself is. To make progress, it is necessary to perform a second step, namely a reflection in the line of the original superposition, $|\psi\rangle$. This reflection is achieved by the unitary operation $W = (2|\psi\rangle\langle\psi| - I)$ (as W acts only on the first n qubits, implicitly, the identity is applied to the final qubit and so the overall operation is $W \otimes I$) and can be illustrated:

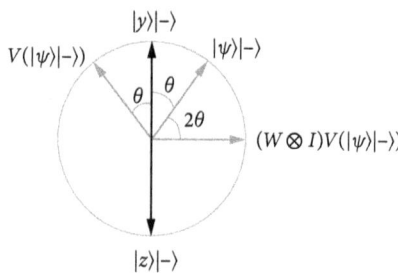

Figure 8.5

The operation W

The unitary W can be shown to perform the desired reflection by considering an arbitrary state $|\phi\rangle$, decomposed into a component in the direction of the uniform superposition, and a component perpendicular to the uniform superposition, denoted $|\psi^\perp\rangle$, i.e.:

$$|\phi\rangle = a|\psi\rangle + b|\psi^\perp\rangle \tag{8.5}$$

From this, it can be seen that:

$$\begin{aligned}
W|\phi\rangle &= (2|\psi\rangle\langle\psi| - I)|\phi\rangle \\
&= (2|\psi\rangle\langle\psi| - I)(a|\psi\rangle + b|\psi^\perp\rangle) \\
&= 2a|\psi\rangle \underbrace{\langle\psi|\psi\rangle}_{=1} - 2b|\psi\rangle \underbrace{\langle\psi|\psi^\perp\rangle}_{=0} - a|\psi\rangle - b|\psi^\perp\rangle \\
&= 2a|\psi\rangle - a|\psi\rangle - b|\psi^\perp\rangle \\
&= a|\psi\rangle - b|\psi^\perp\rangle
\end{aligned} \tag{8.6}$$

that is, the desired reflection about the uniform superposition.

It is further necessary to show that W can be implemented efficiently, i.e., with an acceptably small number of gates from a standard universal gateset. Doing so also suffices to verify that W is indeed unitary. The first step towards this goal is to pre- and post-multiply W by a bank of Hadamard gates:

$$\begin{aligned}
H^{\otimes n} W H^{\otimes n} &= 2|0^n\rangle\langle 0^n| - H^{\otimes n} I H^{\otimes n} \\
&= 2|0^n\rangle\langle 0^n| - I
\end{aligned} \tag{8.7}$$

which follows because H is self-inverse and also uses $|\psi\rangle = |+\rangle^{\otimes n}$ and $H|+\rangle = |0\rangle$. This state is next pre- and post-multiplied by a bank of Pauli-X gates. Using the fact that X is self-inverse, and that $X|0\rangle = |1\rangle$, this gives:

$$\begin{aligned}
X^{\otimes n}(H^{\otimes n} W H^{\otimes n})X^{\otimes n} &= X^{\otimes n}(2|0^n\rangle\langle 0^n| - I)X^{\otimes n} \\
&= 2|1^n\rangle\langle 1^n| - I
\end{aligned} \tag{8.8}$$

which is the matrix:

$$2\begin{bmatrix} 0 & 0 & \cdots & 0 \\ 0 & 0 & & \\ \vdots & & \ddots & \\ 0 & & & 1 \end{bmatrix} - \begin{bmatrix} 1 & 0 & \cdots & 0 \\ 0 & 1 & & \\ \vdots & & \ddots & \\ 0 & & & 1 \end{bmatrix} = -\begin{bmatrix} 1 & 0 & \cdots & 0 \\ 0 & 1 & & \\ \vdots & & \ddots & \\ 0 & & & -1 \end{bmatrix} \tag{8.9}$$

Therefore, $(-2|1^n\rangle\langle 1^n| + I)$ is just an n-qubit CZ gate, denoted $C^{n-1}Z$. A multi-controlled Z can be rewritten as a circuit consisting instead of a multi-controlled X gate and two Hadamard gates:

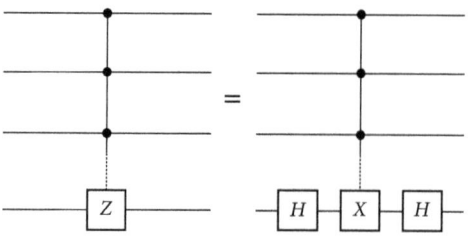

Figure 8.6

as $Z = HXH$, and so $C^{n-1}Z = (I \otimes H)(C^{n-1}X)(I \otimes H)$. $C^{n-1}X$ can be efficiently implemented using Toffoli gates with some extra ancilla qubits, as detailed in Chapter 4. Putting these steps together gives:

$$X^{\otimes n} H^{\otimes n} W H^{\otimes n} X^{\otimes n} = -(I \otimes H)(C^{n-1}X)(I \otimes H) \qquad (8.10)$$

which can be re-arranged to give:

$$W = -H^{\otimes n} X^{\otimes n} (I \otimes H)(C^{n-1}X)(I \otimes H) X^{\otimes n} H^{\otimes n} \qquad (8.11)$$

where the leading '-1' is a global phase factor that can safely be omitted from further consideration.

Thus, we have shown that W can be implemented using gates from a standard universal gateset, with a total gate count that grows linearly in n and, therefore, logarithmically in N.

The Grover iterate

The product of the two reflections, $(W \otimes I)V$, is known as the *Grover iterate*, and the circuit of Grover's algorithm simply consists of a suitable number of repeats of the Grover iterate. To see that this has the desired effect, consider a state, $|\psi'\rangle|-\rangle$, in the subspace spanned by $|y\rangle|-\rangle$ and $|z\rangle|-\rangle$ that is rotated some arbitrary $k\theta$ from the $|y\rangle|-\rangle$ axis:

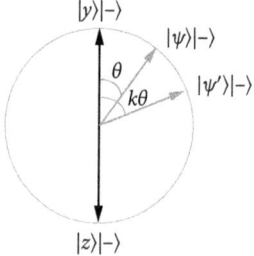

Figure 8.7

For any such $|\psi'\rangle|-\rangle$, first applying the search oracle, V, reflects the state in the vertical axis; and then applying the unitary $W \otimes I$ reflects the state in the line of the original superposition:

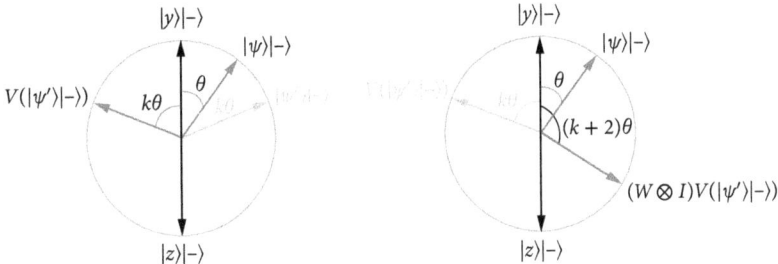

Figure 8.8

and so *each* occurrence of the Grover iterate, $(W \otimes I)V$, leads to a 2θ rotation.

It follows that the succession of Grover iterates rotates the uniform superposition $|\psi\rangle$ towards a superposition consisting of only (or dominated by) marked elements, $|z\rangle$:

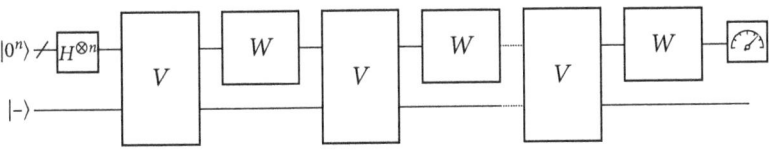

Figure 8.9

To find the right number of Grover iterates for any instance of unstructured search, θ can be expressed in terms of M and N. This requires the definition of how quantum states are represented on the Bloch sphere from Chapter 3, and also the decomposition of the uniform superposition in Eqn. (8.3), to give:

$$\sin(\theta/2) = \sqrt{\frac{M}{N}} \qquad (8.12)$$

Since Grover's algorithm will generally be employed when $M \ll N$ (that is, when marked elements are rare), $\sin(\theta/2)$ will be small, and hence, $\theta/2 \approx \sin(\theta/2) = \sqrt{\frac{M}{N}}$. As each Grover iterate rotates the superposition through an angle 2θ, and the total rotation needs to be $(\pi - \theta)$, this means that number, n_{it}, of required Grover iterates is given by:

$$\pi - \theta = n_{it} \times 2\theta$$

$$\implies n_{it} = \frac{\pi}{2\theta} - \frac{1}{2}$$

$$< \frac{\pi}{2\theta}$$

$$\approx \frac{\pi}{4}\sqrt{\frac{N}{M}} \qquad (8.13)$$

Therefore, only $\Theta(\sqrt{N})$ Grover iterates are required in the quantum algorithm, compared to checking $N - M$ elements in a classical exhaustive search.

One ostensible advantage of the classical algorithm is, however, is that it guarantees success. In Grover's algorithm, success is only guaranteed if the final superposition exactly aligns with the $|z\rangle|-\rangle$ axis, which, in general, will not be the case. However, the failure probability can be bounded, using the fact that it is always possible to choose a number of Grover iterates such that the state is rotated to within θ of $|z\rangle|-\rangle$:

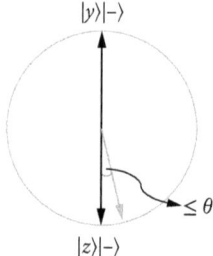

Figure 8.10

It follows that a marked state will be measured with probability at least:

$$(\cos(\theta/2))^2 = \frac{N-M}{N} = 1 - \frac{M}{N} \qquad (8.14)$$

and so for 'hard' searches, where the ratio of marked elements to total elements, $\frac{M}{N}$, is small, Grover search finds a marked element with high probability. As it is easy to check if the bit string returned by Grover's algorithm corresponds to a marked element (by querying the search oracle with that bit string as the input), a repeat until success strategy can be deployed, which still has $\Theta(\sqrt{N})$ expected number of search oracle uses.

Another feature of Grover's algorithm is that it is not convergent, in the sense that Eqn. (8.13) specifies the number of Grover iterates required and not a lower-bound thereon. Executing a number of Grover iterates greater than n_{it} will cause an 'over-rotation', and the probability of successfully measuring a marked element will decrease. Needing to specify the exact number of Grover iterates is not necessarily a problem, per se, except that to calculate n_{it} requires the number of marked elements, M, to be known in advance. Requiring M to be known is quite a restrictive condition to impose, but fortunately it is possible to combine Grover's algorithm with the quantum phase estimation subroutine (Chapter 9) to 'approximately count' the number of marked elements when not known. Moreover, the complexity of this approximate quantum counting algorithm is itself $\Theta(\sqrt{N})$ (see Section 9.5), and so it is possible to find a marked element even when the number of marked elements is unknown with overall complexity $\Theta(\sqrt{N})$.

8.1.1 The source of quantum advantage in Grover's algorithm

Grover's algorithm shows that there is a quantum speed-up available for unstructured search, and an important question to ask (in general with regard to quantum algorithms) is *what is the source of the quantum advantage?* In the case of Grover search, the advantage stems directly from the fact that quantum states have unit ℓ^2 norm, whereas probability distributions have unit ℓ_1 norm.

Focusing for simplicity on the case where $M = 1$, each element in the database has a $1/N$ probability of being the marked element. For a classical exhaustive search, each new element that is 'checked' increases the overall probability of the marked element having been found by $1/N$. Thus, it takes $1/(1/N) = N$ iterations to find the marked state with certainty (by a process of elimination, one can actually be certain about the identity of the marked element if $N - 1$ unmarked elements have been checked, but no earlier). Quantumly, it is possible to put all elements in superposition and then 'move the superposition around', as Grover search indeed does. Crucially, the modulus of the coefficient of a term in the superposition is the *square root* of its probability, and so the initial uniform superposition is initially at an angle that is approximately $1/\sqrt{N}$. As the superposition is rotated such that the angle increases in constant increments proportional to its initial value, only $\Theta(1/(1/\sqrt{N}) = \Theta(\sqrt{N})$ iterations are required to rotate to a state dominated by the marked state.

8.1.2 Grover's algorithm for NP-complete problems

Turning to the question of whether the quadratic speed-up offered by Grover's algorithm is useful in practice, an ostensibly obvious starting point is to consider NP-complete problems, for which no polynomial time classical algorithm is known to exist (of course, P = NP has not been formally precluded, and so it cannot be said with certainty that no such algorithm does exist). Focusing on the oracle as something that can *recognise* an answer by outputting a bit which is unflipped (recognised as 'unmarked') or flipped (recognised as 'marked'), Grover search can be applied directly to NP-complete problems. By definition, an NP (decision) problem is one for which a polynomial-sized circuit (in the problem size) can be constructed that decides whether any input is part of the accepted language or not. This circuit can thus be used directly as the search oracle, and Grover's algorithm then provides a quadratic speed-up for the NP-complete problem, relative to a classical strategy of exhaustive search.

However, an important caveat is that, even though no polynomial time algorithm is known for NP-complete problems, it is usually the case that the best classical strategy will still be much better than exhaustive search. For this reason, there is a degree of subtlety needed when exploring the real-world uses of Grover's algorithm, and quantum algorithms that use Grover iterate-like primitives to obtain a quadratic advantage over the classical state of the art are usually tackling estimation rather than search problems (see, e.g., Section 9.4.3).

8.2 Grover's algorithm decides the OR problem

Grover's algorithm, complemented with the ability to estimate the number of marked elements, achieves the following for a database with some N-elements in total: either (i) a marked element is returned if there is at least one such element; or (ii) if there are no marked elements, then this is explicitly indicated as the output. The total complexity is $\Theta(\sqrt{N})$. This clearly generalises the problem of *deciding* whether an unstructured database has at least one marked element or no marked elements.

Deciding if there is a marked element is a decision problem that applies to the *outputs* of the oracle queries and is usually stated as the OR problem: the OR problem is the problem of deciding if the logical OR of the N outputs obtained by querying the search oracle for every input is equal to 0 or 1.

Defining a decision problem such that the inputs of the decision problem are obtained from the outputs of the oracle queries allows *structure* to be precisely defined (at least in this context): if the decision problem is a *total function* – that is, the correct output is specified as either 0 or 1 for each of the 2^N possible inputs – then the problem is said to be unstructured. Conversely, if there are some inputs for which the output is undetermined (an algorithm can return either 0 or 1), then the function is not total, and the problem is said to be structured. It has been shown that unstructured decision problems have at best a polynomial quantum-classical separation in query complexity. The function OR is clearly total, as is every Boolean function, and it is possible to prove that $\Omega(\sqrt{N})$ queries is a lower-bound for deciding the OR problem. As any algorithm that searches an unstructured database will automatically solve the OR problem, this result immediately implies the asymptotic optimality of Grover's algorithm. (As an interesting aside, it is possible to define an analogous notion of structure for search rather than decision problems, and it came as something of a surprise when an unstructured search problem with an exponential quantum-classical query complexity was proposed; this result reminds us of the importance of not speaking too loosely of Grover's algorithm solving – and being optimal for – *unstructured search* without further qualification!)

8.2.1 $\Omega(\sqrt{N})$ is a lower-bound for deciding the OR problem

There are two main approaches to lower-bounding query complexity of deciding the OR problem: the *adversary method*, and the *polynomial method*, the latter of which we will now go through. The polynomial method uses the fact that *any* quantum algorithm must alternate between unitary operations and queries to the oracle and, thus, bounds how much the state can be altered by each oracle use. This setting maintains full generality, as even an algorithm that uses classical computation can be treated as a single quantum circuit, owing to the computational universality of the quantum circuit model (so the classical computation can be encoded in quantum circuitry) with the principle of deferred measurement employed to ensure that this circuit is unitary. The structure of the circuit can further be restricted by treating the qubits as grouped into three registers: a first n-qubit register corresponding to the input of the oracle; a second single-qubit register corresponding to the output of the oracle; and finally a third register of any number of qubits (in the following, this number is denoted m). As the oracle queries are interleaved with unitary blocks – which can thus include swap gates – there is no loss of generality incurred by enforcing this structure. Thus, a general form of a quantum circuit for deciding the OR problem with d uses of the search oracle, V, is:

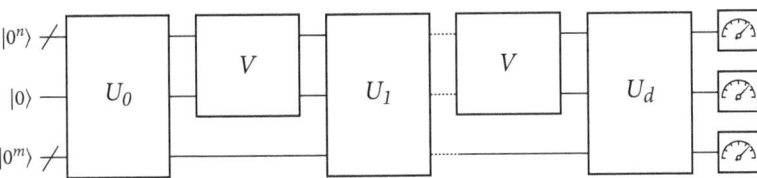

Figure 8.11

where $U_0 \ldots U_d$ can be any unitary operations. (Note that if some particular algorithm requires an initial state other than $|0^{n+m+1}\rangle$, this can be treated as a state preparation unitary applied to $|0^{n+m+1}\rangle$, which is, therefore, absorbed into U_0; and similarly if the measurement is not a computational basis measurement, the unitary U_d can be adjusted accordingly so that the measurement is then a computational basis measurement.) At any point during the quantum circuit, the state may be expressed:

$$|\psi\rangle = \sum_{i\in\{0,1\}^n} \sum_{j\in\{0,1\}} \sum_{k\in\{0,1\}^m} \alpha_{i,j,k} |i\rangle |j\rangle |k\rangle \tag{8.15}$$

To keep track of the relevant terms as the quantum circuit unfolds, it is necessary to introduce some notation: y_i is used to denote the output when the input to the oracle is $|i\rangle$ for $i \in \{0, \ldots, N-1\}$, and so $\{y_i\}_{i\in\{0,\ldots,N-1\}}$, abbreviated as $\{y_*\}$, is the set of all possible outputs.

Before the oracle has been used at all, each coefficient $\alpha_{i,j,k}$ is necessarily independent of $\{y_*\}$, which provides the base case to prove by induction that after some d uses of the oracle, each $\alpha_{i,j,k}$ is a polynomial in $\{y_*\}$ of degree at most d. For the inductive step, let the oracle have been used d times, and $P^{(d)}(\cdot)$ be some polynomial of degree d; assuming that before the $(d+1)$th use of the oracle, the state is of the form:

$$|\psi\rangle = \sum_{i\in\{0,1\}^n} \sum_{j\in\{0,1\}} \sum_{k\in\{0,1\}^m} P^{(d)}_{i,j,k}(\{y_*\}) |i\rangle |j\rangle |k\rangle \tag{8.16}$$

when the oracle, V, is applied, the state transforms to:

$$V|\psi\rangle = \sum_{i\in\{0,1\}^n} \sum_{j\in\{0,1\}} \sum_{k\in\{0,1\}^m} P^{(d)}_{i,j,k}(\{y_*\})|i\rangle|j\oplus y_i\rangle|k\rangle$$

$$= \sum_{i\in\{0,1\}^n} \sum_{j'\in\{0,1\}} \sum_{k\in\{0,1\}^m} \left(\sum_{a\in\{0,1\}} (1\oplus a \oplus y_i) P^{(d)}_{i,j'\oplus a,k}(\{y_*\})\right)|i\rangle|j'\rangle|k\rangle$$

$$= \sum_{i\in\{0,1\}^n} \sum_{j\in\{0,1\}} \sum_{k\in\{0,1\}^m} P^{(d+1)}_{i,j,k}(\{y_*\})|i\rangle|j\rangle|k\rangle \tag{8.17}$$

where the second line makes the substitution $j' = j \oplus y_i$ and follows because the parenthesised term is equal to 1 if $a = y_i$ and 0 otherwise, meaning that $P^{(d)}_{i,j'\oplus a,k}$ may be used rather than the equivalent $P^{(d)}_{i,j'\oplus y_i,k}$. The final line explicitly identifies the term $1 \oplus a \oplus y_u$ as a polynomial of degree $d+1$, and the inductive proof is completed by noting that the interleaved unitary blocks (which are independent of $\{y_*\}$) are linear operations whose effect is, therefore, to (linearly) combine computational basis state coefficients and as such do not change the maximum degree of the polynomial. Therefore, for d uses of the oracle, the coefficient of any computational basis state is a polynomial of degree at most d.

As $y_i \in \{0, 1\}$ and therefore $y_i^b = y_i$ for all i and any positive integer b, P is not just a polynomial, but a *multi-linear* function of degree at most d. The Born rule states that the probability of obtaining a certain computational basis measurement outcome is equal to the corresponding coefficient's modulus squared, and so it follows that the probability of any given (computational basis) measurement outcome is a polynomial (in particular a multi-linear function) of degree at most $2d$. Moreover, as the probability of obtaining any one of some set of measurement outcomes is simply the sum of the probabilities of the individual outcomes (as computational basis state measurement outcomes are mutually exclusive events), this is also a polynomial (multi-linear function) of degree at most $2d$.

To use this property to lower-bound the required number of quantum queries, it is necessary to be a little more precise about what it means for a quantum algorithm to successfully compute the OR problem. As the OR function itself may be stated

$$\text{OR}(y_0, y_1, \ldots, y_{N-1}) = \begin{cases} 0 & \text{if } \forall i, y_i = 0 \\ 1 & \text{otherwise} \end{cases} \tag{8.18}$$

it follows that the quantum algorithm computing the OR problem can be treated as a function with N inputs that are the outputs when the oracle is queried for all of N possible bit strings, defined $A_{\text{OR}}(y_0, y_1, \ldots, y_{N-1})$. A_{OR} returns a single bit, obtained by a computational basis measurement with certain outcomes interpreted as 1 and others 0. A_{OR} is deemed to have correctly computed (decided) the OR problem if it gives the correct answer with at least 2/3 probability (as detailed in Chapter 6):

$$A_{\text{OR}}(y_0, y_1, \ldots, y_{N-1}) \begin{cases} \leq 1/3 & \text{if } \forall i, y_i = 0 \\ \geq 2/3 & \text{otherwise} \end{cases} \tag{8.19}$$

As shown above, the sum of any subset of computational basis measurement outcome probabilities is a polynomial in $\{y_*\}$ of degree at most $2d$, and so A_{OR} is itself a polynomial of its inputs of degree at most $2d$. The following lower-bound is essentially a proof of how rapidly such a function can rise to distinguish the two cases in Eqn. (8.19).

To prove the polynomial lower-bound, a univariate polynomial, $\tilde{A}$, that inherits the required properties of A_{OR} is used. Specifically, $\tilde{A}(x)$ is obtained by averaging the value of A_{OR} when exactly x input bits are set to 1. To define this formally, we may think of $y_1 y_2 \ldots y_N$ as a binary string, y, with Hamming weight $|y|$, which gives

$$\tilde{A}(x) = \mathbb{E}_{|y|=x}[A_{OR}(y)] \tag{8.20}$$

$\tilde{A}(x)$ is a polynomial in x whose degree is itself at most $2d$ – i.e., the degree of A_{OR}. To see this, first we express the multi-linear function A_{OR} as:

$$A_{OR} = \sum_{j=1}^{N_A} c_j \prod_{i \in S_j} y_i \tag{8.21}$$

where c_j are constants, the multi-linear function has N_A terms, and each set S_j contains the indices for y_i that are included in the corresponding term of the sum. From this, $\tilde{A}(x)$ can be expressed as:

$$\tilde{A}(x) = \mathbb{E}_{|y|=x}\left[\sum_{j=1}^{N_A} c_j \prod_{i \in S_j} y_i\right] = \sum_{j=1}^{N_A} c_j \mathbb{E}_{|y|=x}\left[\prod_{i \in S_j} y_i\right] \tag{8.22}$$

where the second equality follows by the linearity of expectation. We continue by noticing that of the 2^N binary strings y, there are some $\binom{N}{x}$ that have exactly x ones, and furthermore, of these strings with exactly x ones, for a term $\prod_{i \in S_j} y_i$, there are some $\binom{N-|S_j|}{x-|S_j|}$ binary strings that evaluate to one (i.e., that are such that every included y_i is equal to 1). From this, the following can be derived:

$$\mathbb{E}_{|y|=x}\left[\prod_{i \in S_j} y_i\right] = \Pr\left(\forall i \in S_j, y_i = 1 \Big| |y| = x\right)$$

$$= \frac{\binom{N-|S_j|}{x-|S_j|}}{\binom{N}{x}}$$

$$= \frac{(N-|S_j|)! x! (N-x)!}{(x-|S_j|)!(N-x)!N!}$$

$$= \frac{(N-|S_j|)!}{N!} x(x-1)\ldots(x-|S_j|+1) \tag{8.23}$$

which is a polynomial in x of degree $|S_j|$. However, recall that $|S_j|$ is the number of values multiplied together in the corresponding term of the original multi-linear function, and hence is upper-bounded by the upper-bound on the degree of the original polynomial, $2d$. As $\tilde{A}(x)$ is a sum of such polynomials, it therefore also has degree at most $2d$.

The existence of an algorithm A_{OR} deciding the OR problem with some d uses of the search oracle, therefore, implies the existence of a degree $2d$ univariate polynomial, $\tilde{A}(x)$, with the properties:

$$0 \leq \tilde{A}(x) \leq \frac{1}{3} \quad \text{if } k = 0 \tag{8.24}$$

$$\frac{2}{3} \leq \tilde{A}(x) \leq 1 \quad \text{otherwise} \tag{8.25}$$

for integers k between 0 and $N-1$, inclusive. These follow from the symmetric nature of the OR problem: for every input binary string with $x \neq 0$ ones, A_{OR} returns a number between $\frac{2}{3}$ and 1, and so the average of these output values must itself be between $\frac{2}{3}$ and 1.

The final part of the proof follows from a standard bound on how rapidly a polynomial function can rise. For this, it is necessary to specify the 'box' in which the polynomial $\tilde{A}(x)$ exists. For our purposes, we are concerned only with x in the range $[0, N]$, and the maximum and minimum values that $\tilde{A}$ can take have not yet been specified (as the function value has only be defined for integer x) and so we let these be $\tilde{A}_{max}$ and $\tilde{A}_{min}$. A simple result due to Markov can be used to relate the maximum derivative of a polynomial to its degree:

$$\left|\frac{d\tilde{A}(x)}{dx}\right| \leq (\deg(\tilde{A}(x)))^2 \frac{\tilde{A}_{max} - \tilde{A}_{min}}{N} \tag{8.26}$$

In order for $\tilde{A}(x)$ to rise from a value at most $1/3$ at $x = 0$ to a value at least $2/3$ at $x = 1$, it is necessary that $\left|\frac{d\tilde{A}(x)}{dx}\right|$ must be at least $1/3$ at some point between $x = 0$ and $x = 1$, and this must hold regardless of the values of $\tilde{A}_{max}$ and $\tilde{A}_{min}$.

A simple analysis can be completed by noting that $\tilde{A}_{max} - \tilde{A}_{min} \geq 1/3$ and assuming that $\tilde{A}_{max} - \tilde{A}_{min}$ is also bounded above by some constant. Substituting $\left|\frac{d\tilde{A}(x)}{dx}\right| \geq 1/3$ and $\deg(\tilde{A}(x)) = 2d$ into Eqn. (8.26) gives:

$$\frac{1}{3} \leq (2d)^2 \frac{\tilde{A}_{max} - \tilde{A}_{min}}{N}$$

$$\Rightarrow d \geq \sqrt{\frac{N}{12(\tilde{A}_{max} - \tilde{A}_{min})}}$$

$$\in \Omega(\sqrt{N}) \tag{8.27}$$

However, this is to oversimplify the situation, as a family of polynomials in which the range $\tilde{A}_{max} - \tilde{A}_{min}$ grows with N has not been precluded. A complete, general analysis uses the fact that the construction of $\tilde{A}$ is such that at every integer $x \in [0, N]$, the polynomial is equal to a value between 0 and 1, as this corresponds to an average of measurement outcome probabilities. To use this property, we define

$$\tilde{A}_{\lim} = \max(\tilde{A}_{\max} - 1, -\tilde{A}_{\min}) \tag{8.28}$$

and for sufficiently large $\tilde{A}_{\lim}$, the maximum derivative will be given by the fact that $\tilde{A}$ necessarily takes a value between 0 and 1 at the integers either side of its extremal value. This gives the inequality:

$$\left|\frac{d\tilde{A}(x)}{dx}\right| \geq \frac{\tilde{A}_{\lim}}{1/2} \tag{8.29}$$

where the value of 1/2 in the denominator occurs because, within an interval of width one, the function must rise or fall from a value between 0 and 1 to obtain its extremal value and then return to a value between 0 and 1. Using Eqn. (8.29) along with $\tilde{A}_{\max} - \tilde{A}_{\min} \leq 2\tilde{A}_{\lim} + 1$, which follows directly from the definition in Eqn. (8.28), and substituting into Eqn. (8.26) gives:

$$2\tilde{A}_{\lim} \leq (2d)^2 \frac{2\tilde{A}_{\lim} + 1}{N}$$

$$\implies d \geq \sqrt{\frac{\tilde{A}_{\lim} N}{2(2\tilde{A}_{\lim} + 1)}}$$

$$\in \Omega(\sqrt{N}) \tag{8.30}$$

Therefore, the same lower-bound holds when there is a family of polynomials such that the height of the box containing this polynomial grows with N. As d is the number of uses of the search oracle, this completes the lower-bound that $\Omega(\sqrt{N})$ queries are needed for any quantum algorithm that decides the OR problem, and hence any quantum algorithm for unstructured search. As Grover's algorithm uses $\Theta(\sqrt{N})$ queries for unstructured search, it is, therefore, *asymptotically optimal* – no other algorithm can do better by more than a constant factor.

Chapter problems

1. Grover's algorithm is performed for an oracle that operates on three input bits and is such that only the state $|010\rangle$ is marked. Calculate the probability of measuring the marked state after applying the Grover iterate 0, 1, 2, and 3 times.
   (a) by directly simulating the matrix operations;
   (b) using the fact that the rotation angle is constant for each Grover iteration. Verify that the same answer is obtained in each case.
2. Let there be a database containing 32 elements, indexed by the binary numbers 00000 to 11111. A single element 00110 is marked.

(a) Give an oracle circuit that identifies the marked element.
(b) If Grover's search algorithm is applied to find the marked element, what should the initial state be set to, and what is the state after a single Grover iterate has been applied?
(c) Find the marked element with maximum probability requires N iterates in total. What is the value of N, and what is the probability of correctly finding the marked element?
(d) If the algorithm is instead run with $3N$ iterates in total, what is the probability of correctly finding the marked element? Comment on your answer.

3. Let V be an oracle circuit that marks one or more elements, acting as follows:

$$V(|x\rangle|a\rangle) = |x\rangle|a \oplus f(x)\rangle$$

Here, a takes the values 0 or 1, and we have $f(x) = 1$ when x is the index of a marked element, and $f(x) = 0$ otherwise. How could V be altered to allow Grover's search to find an *unmarked* element?

4. A Grover iterate consists of the oracle circuit, typically denoted V, followed by the circuit W. What would happen if V and W were swapped, such that Grover's algorithm is run with V following W as the Grover iterate?

Further reading

Lov Grover [1996] proposed his eponymous algorithm. The polynomial method for lower-bounding query complexity was due to Beals *et al* [2001], and the adversary method was proposed by Andris Ambainis [2002] and further developed by Hoyer *et al* [2007]. The 'surprising' *unstructured search* problem with an exponential quantum speed-up is due to Yamakawa and Zhandry [2022], and more generally, Scott Aaronson [2022] provides a nice and accessible discussion about structure and quantum advantage.

9
Quantum phase estimation and quantum amplitude estimation

This chapter introduces two important subroutines that are deployed in myriad quantum algorithms. The first is *quantum phase estimation*, which (in the version presented here) uses the *quantum Fourier transform*. Quantum phase estimation is then combined with the general idea underpinning Grover's algorithm, *quantum amplitude amplification*, to give the second subroutine, *quantum amplitude estimation*.

9.1 The discrete Fourier transform

In the early 1800s, French mathematician Joseph Fourier discovered the Fourier transform, which extracts the frequency components of a time-varying signal and is still at the heart of modern day digital signal processing. There are a few variations of the Fourier transform, of which the *discrete Fourier transform* is the one which has the quantum analogue. The discrete Fourier transform takes an N-element input vector x and transforms it to an N-element output vector y such that:

$$y_k = \frac{1}{\sqrt{N}} \sum_{j=0}^{N-1} x_j e^{2\pi i jk/N} \qquad (9.1)$$

In the discrete Fourier transform, each element of the output vector is a weighted sum of the elements of the input vector, and hence, it may be represented as a matrix, W_N such that $y = W_N x$ and

$$W_N = \frac{1}{\sqrt{N}} \begin{bmatrix} (e^{2\pi i/N})^0 & (e^{2\pi i/N})^0 & (e^{2\pi i/N})^0 & (e^{2\pi i/N})^0 & \cdots & (e^{2\pi i/N})^0 \\ (e^{2\pi i/N})^0 & (e^{2\pi i/N})^1 & (e^{2\pi i/N})^2 & (e^{2\pi i/N})^3 & \cdots & (e^{2\pi i/N})^{N-1} \\ (e^{2\pi i/N})^0 & (e^{2\pi i/N})^2 & (e^{2\pi i/N})^4 & (e^{2\pi i/N})^6 & \cdots & (e^{2\pi i/N})^{2(N-1)} \\ (e^{2\pi i/N})^0 & (e^{2\pi i/N})^3 & (e^{2\pi i/N})^6 & (e^{2\pi i/N})^9 & \cdots & (e^{2\pi i/N})^{3(N-1)} \\ \vdots & \vdots & \vdots & \vdots & \ddots & \\ (e^{2\pi i/N})^0 & (e^{2\pi i/N})^{N-1} & (e^{2\pi i/N})^{2(N-1)} & (e^{2\pi i/N})^{3(N-1)} & & (e^{2\pi i/N})^{(N-1)(N-1)} \end{bmatrix}$$

$$(9.2)$$

It is interesting to notice that the 2×2 discrete Fourier transform matrix is the familiar Hadamard matrix:

$$W_2 = \frac{1}{\sqrt{2}} \begin{bmatrix} (e^{2\pi i/2})^0 & (e^{2\pi i/2})^0 \\ (e^{2\pi i/2})^0 & (e^{2\pi i/2})^1 \end{bmatrix} = \frac{1}{\sqrt{2}} \begin{bmatrix} 1 & 1 \\ 1 & -1 \end{bmatrix} \qquad (9.3)$$

More generally, the discrete Fourier transform matrix is always unitary – a fact that is implicitly proven by its quantum circuit form: the quantum Fourier transform. (Note that in the above, boldface font has been used to denote vectors in order to emphasise their classicality; however, these will naturally be replaced by quantum states in the analysis of the quantum Fourier transform.)

9.2 The quantum Fourier transform

To give an intuitive introduction to the quantum Fourier transform, and to see how the quantum circuit works, we begin with the three-qubit quantum Fourier transform as a simple example, which has the following circuit:

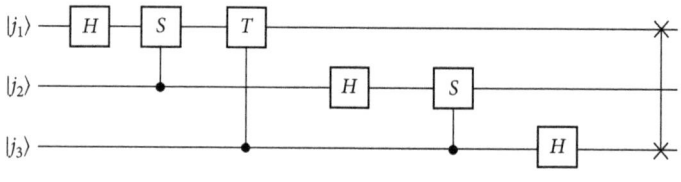

Figure 9.1

Stepping through the circuit for any computational basis state $|j_1 j_2 j_3\rangle$, and taking the evolution of each qubit in turn (but not yet including the swap gate):

$$|j_1\rangle \xrightarrow{H} \frac{|0\rangle + e^{\pi i j_1}|1\rangle}{\sqrt{2}} \xrightarrow{CS} \frac{|0\rangle + e^{\pi i j_1} e^{\pi i j_2/2}|1\rangle}{\sqrt{2}} \xrightarrow{CT} \frac{|0\rangle + e^{\pi i j_1} e^{\pi i j_2/2} e^{\pi i j_3/4}|1\rangle}{\sqrt{2}} \qquad (9.4)$$

$$|j_2\rangle \xrightarrow{H} \frac{|0\rangle + e^{\pi i j_2}|1\rangle}{\sqrt{2}} \xrightarrow{CS} \frac{|0\rangle + e^{\pi i j_2} e^{\pi i j_3/2}|1\rangle}{\sqrt{2}} \qquad (9.5)$$

$$|j_3\rangle \xrightarrow{H} \frac{|0\rangle + e^{\pi i j_3}|1\rangle}{\sqrt{2}} \qquad (9.6)$$

After the SWAP gate, the state is:

$$\frac{1}{2\sqrt{2}} \left(|0\rangle + e^{\pi i j_3}|1\rangle\right) \left(|0\rangle + e^{\pi i(j_2 + (j_3/2))}|1\rangle\right) \left(|0\rangle + e^{\pi i(j_1 + (j_2/2) + (j_3/4))}|1\rangle\right) \qquad (9.7)$$

The three-qubit quantum Fourier transform uses controlled S and T gates, which are (up to irrelevant global phase factors) members of the $R_z(\theta)$ family of parameterised gates. In particular, $S = R_z(\pi/2)$ and $T = R_z(\pi/4)$. This pattern of using controlled R_z rotations such that each successive gate has rotation angle half of that of the preceding gate continues in the general n-qubit quantum Fourier transform circuit:

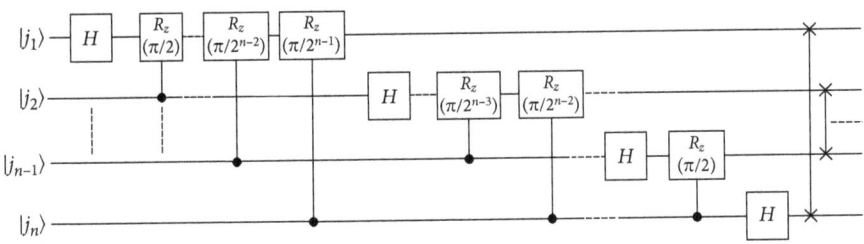

Figure 9.2

Each controlled rotation angle is a negative power of 2 multiplied by π, and the global phase can be factored out such that

$$R_z(2^{-k}\pi) = \begin{bmatrix} e^{-i2^{-k}\pi/2} & 0 \\ 0 & e^{i2^{-k}\pi/2} \end{bmatrix} = e^{-i2^{-k}\pi/2} \begin{bmatrix} 1 & 0 \\ 0 & e^{i2^{-k}\pi} \end{bmatrix} \qquad (9.8)$$

The coefficient $e^{-i2^{-k}\pi/2}$ (the global phase factor) is omitted in the following analysis. With this omission of global phase factors, the general quantum Fourier transform circuit has the following action on a computational basis state $|j\rangle = |j_1\rangle|j_2\rangle\cdots|j_n\rangle$:

$$|j\rangle \xrightarrow{\text{QFT}} \frac{1}{\sqrt{N}}\left(|0\rangle + e^{\pi i j_n}|1\rangle\right) \otimes \cdots \otimes \left(|0\rangle + e^{\pi i(j_2 + (j_3/2) + \cdots + (j_n/(2^{n-2})))}|1\rangle\right)$$

$$\otimes \left(|0\rangle + e^{\pi i(j_1 + (j_2/2) + \cdots + (j_n/(2^{n-1})))}|1\rangle\right) \qquad (9.9)$$

(where $N = 2^n$). This result can be re-expressed using *binary fractions*, that is, expressing a sum of terms that are integer multiples of successive negative powers of 2 using a *binary point*: $\frac{a_1}{2} + \frac{a_2}{4} + \cdots + \frac{a_n}{2^n} = 0 \cdot a_1 a_2 \cdots a_n$. The action of the quantum Fourier transform using this notation is:

$$|j\rangle \xrightarrow{\text{QFT}} \frac{1}{\sqrt{N}}\left(|0\rangle + e^{2\pi i(0 \cdot j_n)}|1\rangle\right) \otimes \cdots \otimes \left(|0\rangle + e^{2\pi i(0 \cdot j_2 j_3 \cdots j_n)}|1\rangle\right)\left(|0\rangle + e^{2\pi i(0 \cdot j_1 j_2 \cdots j_n)}|1\rangle\right)$$

$$(9.10)$$

Furthermore, as $e^{2m\pi i} = 1$ for any integer m, any number ahead of the binary point can be introduced without changing the value, and so the right-hand side of Eqn. (9.10) equals $\frac{1}{\sqrt{N}}\bigotimes_{l=1}^{n}\left(|0\rangle + e^{2\pi i j 2^{-l}}|1\rangle\right)$. ($\otimes$ represents the tensor product of a number of indexed terms in a manner analogous to $\sum$ for summation and $\prod$ for the regular product.) From this, the quantum Fourier transform can be arranged into standard form:

$$|j\rangle \xrightarrow{\text{QFT}} \frac{1}{\sqrt{N}} \bigotimes_{l=1}^{n} \left(|0\rangle + e^{2\pi i j 2^{-l}}|1\rangle\right) = \frac{1}{\sqrt{N}} \bigotimes_{l=1}^{n} \left(\sum_{k_l=0}^{1} e^{2\pi i j k_l 2^{-l}}|k_l\rangle\right)$$

$$= \frac{1}{\sqrt{N}} \sum_{k_1=0}^{1} \cdots \sum_{k_n=0}^{1} \bigotimes_{l=1}^{n} e^{2\pi i j k_l 2^{-l}}|k_l\rangle$$

$$= \frac{1}{\sqrt{N}} \sum_{k_1=0}^{1} \cdots \sum_{k_n=0}^{1} e^{2\pi i j (\sum_{l=1}^{n} k_l 2^{-l})}|k_1 \cdots k_n\rangle$$

$$= \frac{1}{\sqrt{N}} \sum_{k=0}^{N-1} e^{2\pi i j k/N}|k\rangle \tag{9.11}$$

For an arbitrary state, represented as a superposition of computational basis states, by linearity, the quantum Fourier transform has the following action:

$$\sum_{j=0}^{N-1} x_j |j\rangle \xrightarrow{\text{QFT}} \sum_{j=0}^{N-1} x_j \frac{1}{\sqrt{N}} \sum_{k=0}^{N-1} e^{2\pi i j k/N}|k\rangle$$

$$= \sum_{k=0}^{N-1} \left(\frac{1}{\sqrt{N}} \sum_{j=0}^{N-1} x_j e^{2\pi i j k/N}\right)|k\rangle$$

$$= \sum_{k=0}^{N-1} y_k |k\rangle \tag{9.12}$$

with y_k as defined in Eqn. (9.1). Treating the input and output quantum states as, respectively, the vectors $\boldsymbol{x}$ and $\boldsymbol{y}$, the quantum Fourier transform circuit *is* the matrix W_N, and the fact that it has a representation as a quantum circuit demonstrates its unitarity.

As noted in the introductory text of this chapter, the quantum Fourier transform is central to quantum phase estimation, which actually uses not the quantum Fourier transform itself, but the inverse quantum Fourier transform. The action of the inverse quantum Fourier transform can be expressed by simply reversing the direction of the arrow in Eqn. (9.11)

$$\frac{1}{\sqrt{N}} \sum_{k=0}^{N-1} e^{2\pi i j k/N}|k\rangle \xrightarrow{\text{QFT}^{\dagger}} |j\rangle \tag{9.13}$$

and for future reference, we also give the action of the inverse quantum Fourier transform on a single computational basis state:

$$|k\rangle \xrightarrow{\text{QFT}^{\dagger}} \frac{1}{\sqrt{N}} \sum_{j=0}^{N-1} e^{-2\pi i j k/N}|j\rangle \tag{9.14}$$

which comes from the fact that the quantum Fourier transform in unitary matrix form Eqn. (9.2) is symmetric, and so all that is required to find its inverse is to take its complex conjugate by negating the exponent of each element.

As with any quantum circuit, to invert the quantum Fourier transform, the circuit must be reversed, with each gate replaced by its inverse. The Hadamard gate is self-inverse, as is the SWAP gate, and the inverse of the rotations gate R_z is given by:

$$R_z^\dagger(\theta) = R_z(-\theta) \tag{9.15}$$

Therefore, the inverse quantum Fourier transform circuit is:

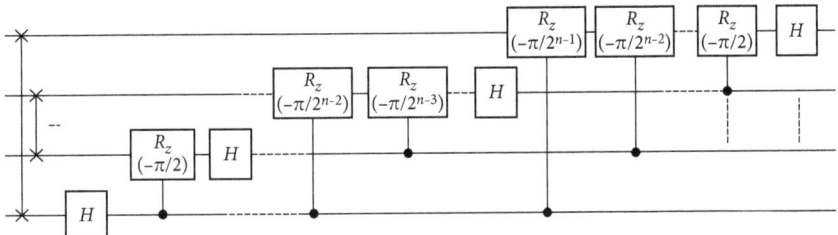

Figure 9.3

The gate complexity of the quantum Fourier transform

The quantum Fourier transform on n qubits requires n Hadamard gates, $\frac{n(n-1)}{2}$ 2-qubit conditional rotation gates, and $\lfloor \frac{n}{2} \rfloor$ SWAP gates, i.e., $\Theta(n^2)$ gates in total (this is also the case for the inverse quantum Fourier transform circuit). As n^2 is a polynomial in n, this is taken as an efficient subroutine, whereas, classically, the most efficient version of the discrete Fourier transform, the *fast Fourier transform (FFT)*, requires $\Theta(n2^n)$ operations and so is inefficient (exponential in n). As the FFT is ubiquitous in signal processing, the existence of the exponentially-faster quantum Fourier transform is extremely exciting, as it hints at the possibility of speeding up many digital signal processing tasks. However, digging a little deeper, this turns out not to be the case.

The quantum Fourier transform operates on a quantum state, which may be thought of as an 'encoding' of an $N = 2^n$ element vector, producing a similarly encoded version of the output vector. Even if one neglects any complication with preparing a suitable encoding of the signal (input) itself, typically in digital signal processing, one wishes to have direct access to the elements of the output vector, as these are the values of the various frequency components. In the case of the quantum Fourier transform, however, one must measure the prepared output state to extract classical information, and by the Born rule, this returns a single computational basis state with probability equal to the square of the magnitude of the corresponding element in the N-element vector. For this reason, the quantum Fourier transform cannot be used to (exponentially) speed-up signal processing tasks, but it is at the heart of many applications of quantum computing and simulation that demonstrate exponential speed-ups (compared to the best-known classical counterparts), as a central component of the *quantum phase estimation* subroutine.

9.3 Quantum phase estimation

Let U be a unitary matrix, with an eigenvector, $|u\rangle$ and associated eigenvalue, $e^{2\pi i\phi}$. Quantum phase estimation solves the problem of estimating to some t bits of precision the phase, ϕ, of the eigenvalue, when U is available for use in a quantum circuit in the form of a controlled-U unitary. To get started with quantum phase estimation, consider the action of controlled-U on $|u\rangle$ when controlled by $|+\rangle$:

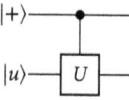

Figure 9.4

which gives the transformation:

$$\frac{1}{\sqrt{2}}(|0\rangle + |1\rangle)|u\rangle \longrightarrow \frac{1}{\sqrt{2}}(|0\rangle|u\rangle + |1\rangle e^{2\pi i\phi}|u\rangle) = \frac{1}{\sqrt{2}}(|0\rangle + e^{2\pi i\phi}|1\rangle)|u\rangle \quad (9.16)$$

The quantum phase estimation circuit repeatedly applies controlled-U, and so it is important to note that:

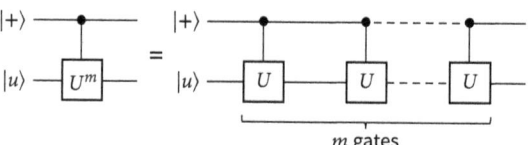

Figure 9.5

which has the effect:

$$\frac{1}{\sqrt{2}}(|0\rangle + |1\rangle)|u\rangle \longrightarrow \frac{1}{\sqrt{2}}(|0\rangle + e^{2m\pi i\phi}|1\rangle)|u\rangle \quad (9.17)$$

Having addressed these simple preliminaries, we can give the first part of the quantum phase estimation circuit:

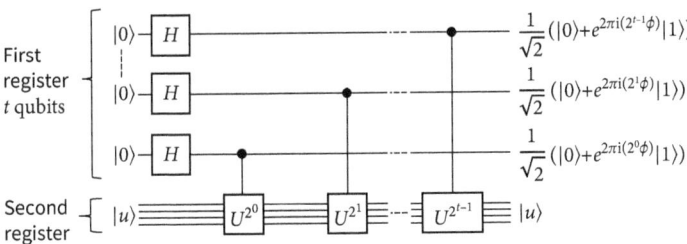

Figure 9.6

The state prepared by this circuit, applied to the initial state as shown in Fig. 9.6, can be expressed using analysis analogous to that leading to Eqn. (9.11), with 2^t in place of N and $2^t \phi$ in place of j:

$$\frac{1}{\sqrt{2^t}} \left(\bigotimes_{l=1}^{t} \left(|0\rangle + e^{2\pi i (2^t \phi) 2^{-l}} |1\rangle \right) \right) |u\rangle = \frac{1}{\sqrt{2^t}} \sum_{k=0}^{2^t-1} e^{2\pi i (2^t \phi) k / 2^t} |k\rangle |u\rangle \qquad (9.18)$$

The final component of the quantum phase estimation circuit is the inverse quantum Fourier transform applied to the first register.

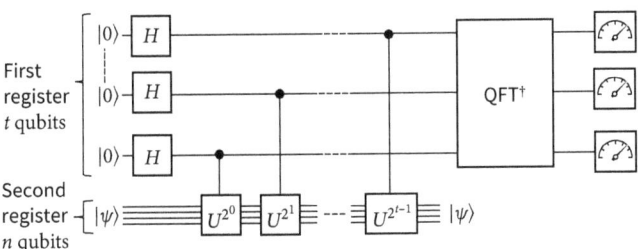

Figure 9.7

To express the state after the inverse quantum Fourier transform, we first assume that ϕ is exactly representable in t bits, such that the effect of the inverse quantum Fourier transform follows immediately from Eqn. (9.13):

$$\frac{1}{\sqrt{2^t}} \sum_{k=0}^{2^t-1} e^{2\pi i (2^t \phi) k / 2^t} |k\rangle |u\rangle \xrightarrow{\text{QFT}^\dagger} |2^t \phi\rangle |u\rangle \qquad (9.19)$$

Measuring the first register gives the eigenvalue phase, ϕ multiplied by 2^t. As the phase is itself less than one, the bit string obtained by the measurement can simply be thought as the value coming after the binary point in a binary fraction representation of ϕ.

In the case where ϕ is not exactly representable in t bits, this procedure still produces a *good* approximation of ϕ with high probability. (In the general case, it is necessary to use Eqn. (9.14) rather than Eqn. (9.13)):

$$\frac{1}{\sqrt{2^t}} \sum_{k=0}^{2^t-1} e^{2\pi i (2^t \phi) k / 2^t} |k\rangle |u\rangle \xrightarrow{\text{QFT}^\dagger} \frac{1}{2^t} \sum_{k=0}^{2^t-1} \sum_{j=0}^{2^t-1} e^{2\pi i (2^t \phi) k / 2^t} e^{-2\pi i j k / 2^t} |j\rangle |u\rangle$$

$$= \frac{1}{2^t} \sum_{k=0}^{2^t-1} \sum_{j=0}^{2^t-1} e^{2\pi i k (2^t \phi - j)/2^t} |j\rangle |u\rangle \quad (9.20)$$

which is a superposition dominated by the term(s) where $2^t \phi - j \approx 0 \implies j \approx 2^t \phi$. Therefore, measurement of the first register is likely to return a bit string that is a close approximation of $2^t \phi$ as desired. A full analysis can be undertaken to show that a first register size of $t' = t + \lceil \log(2 + 1/2\epsilon) \rceil$ qubits is sufficient to obtain a t-bit approximation of ϕ with at least probability $1 - \epsilon$. For practical and theoretical purposes, this constitutes a sufficiently modest uplift from the 'ideal' case of just t qubits.

A simple example of quantum phase estimation

In practical situations, quantum phase estimation offers a useful quantum advantage when applied to large unitary operators. However, for pedagogical purposes, it is instructive to explicitly express and work through a very simple quantum phase estimation circuit, for a single-qubit unitary:

$$U = \begin{bmatrix} 0 & -i \\ i & 0 \end{bmatrix} \quad (9.21)$$

which has an eigenvector $|u\rangle = \frac{1}{\sqrt{2}} \begin{bmatrix} 1 \\ -i \end{bmatrix}$ associated with eigenvalue -1, (i.e., $\phi = \frac{1}{2}$). To estimate the phase to two bits of precision, the circuit is (noting that, in this case, ϕ is exactly representable as a binary fraction):

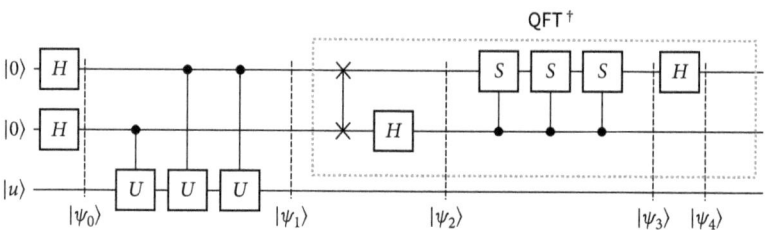

Figure 9.8

which can be simulated 'by hand' to give:

$$|\psi_0\rangle = \frac{1}{2}(|0\rangle + |1\rangle)(|0\rangle + |1\rangle)|u\rangle \quad (9.22)$$

$$|\psi_1\rangle = \frac{1}{2}(|0\rangle + e^{2\pi i}|1\rangle)(|0\rangle + e^{\pi i}|1\rangle)|u\rangle \quad (9.23)$$

$$|\psi_2\rangle = \frac{1}{\sqrt{2}}(|0\rangle - |1\rangle)|0\rangle|u\rangle \qquad (9.24)$$

$$|\psi_3\rangle = \frac{1}{\sqrt{2}}(|0\rangle - |1\rangle)|0\rangle|u\rangle \qquad (9.25)$$

$$|\psi_4\rangle = |1\rangle|0\rangle|u\rangle \qquad (9.26)$$

Therefore, the measurement outcome is 10, which corresponds to the value $1 \times \frac{1}{2} + 0 \times \frac{1}{4} = \frac{1}{2}$, as expected.

9.4 Quantum amplitude estimation

Quantum phase estimation is deployed as a subroutine in a number of quantum algorithms and is also used in quantum amplitude estimation, another important quantum subroutine. To define what it means to estimate the amplitude of a quantum state, and to show that quantum amplitude estimation indeed achieves this, first note that *every* n-qubit quantum state can be expressed in the form:

$$|\psi\rangle = \cos\frac{\theta}{2}|\Phi_0\rangle|0\rangle + \sin\frac{\theta}{2}|\Phi_1\rangle|1\rangle \qquad (9.27)$$

where $|\Phi_0\rangle$ and $|\Phi_1\rangle$ are $(n-1)$-qubit states and θ is some real number between 0 and π which may be interpreted as an angle. We further let A be the circuit that prepares, $|\psi\rangle$, that is:

$$|\psi\rangle = A|0^n\rangle \qquad (9.28)$$

Quantum amplitude estimation returns an estimate of the amplitude, $a = \sin^2\frac{\theta}{2}$, which is obtained using *quantum amplitude amplification* – which can be thought of as a generalisation of the Grover iterate.

9.4.1 Quantum amplitude amplification

The goal of quantum amplitude amplification is to build a unitary operator that has the action $\cos\frac{\theta}{2}|\Phi_0\rangle|0\rangle + \sin\frac{\theta}{2}|\Phi_1\rangle|1\rangle \to \cos\frac{3\theta}{2}|\Phi_0\rangle|0\rangle + \sin\frac{3\theta}{2}|\Phi_1\rangle|1\rangle$ or, more generally, to perform a rotation through an angle 2θ on any state in the two-dimensional subspace spanned by $|\Phi_0\rangle|0\rangle$ and $|\Phi_1\rangle|1\rangle$ when depicted on the corresponding Bloch sphere.

The premise of quantum amplitude amplification and estimation is that one has access to the circuit, A (as well as its inverse), and hence, it is possible to construct the following operator:

$$Q = -AS_0 A^\dagger S_\chi \qquad (9.29)$$

where

$$S_0 = X^{\otimes n}(C^{n-1}Z)X^{\otimes n} \tag{9.30}$$

$$S_\chi = I_{2^{n-1}} \otimes Z \tag{9.31}$$

i.e., S_χ is simply a Z gate applied to the final qubit. Therefore, all of the components of the operator Q can easily be constructed: S_0 and S_χ are simple circuits that do not vary with the particular state whose amplitude is being amplified, and by the premise, A and $A^\dagger$ are available for use.

The operator Q can be decomposed as the product of two reflections. The first reflection, S_χ, reflects the state in the $|\Phi_0\rangle|0\rangle$ axis (as it is just a Pauli-Z gate), which can be shown by considering a general state $|\omega\rangle$ in the subspace spanned by $|\Phi_0\rangle|0\rangle$ and $|\Phi_1\rangle|1\rangle$,

$$|\omega\rangle = \cos\frac{\omega}{2}|\Phi_0\rangle|0\rangle + e^{i\phi}\sin\frac{\omega}{2}|\Phi_1\rangle|1\rangle \tag{9.32}$$

for some ϕ and ω. The action of S_χ is thus:

$$|\omega\rangle \xrightarrow{S_\chi} \cos\frac{\omega}{2}|\Phi_0\rangle|0\rangle - e^{i\phi}\sin\frac{\omega}{2}|\Phi_1\rangle|1\rangle \tag{9.33}$$

If we now consider another state, $|\omega'\rangle$, again in the subspace spanned by $|\Phi_0\rangle|0\rangle$ and $|\Phi_1\rangle|1\rangle$, which we express (without loss of generality) as a term in the direction, $|\psi\rangle$, and a term orthogonal to $|\psi\rangle$:

$$|\omega'\rangle = \cos\frac{\omega'}{2}|\psi\rangle + e^{i\phi'}\sin\frac{\omega'}{2}|\psi^\perp\rangle \tag{9.34}$$

then it can be shown that the action of the operator $-AS_0A^\dagger = (-I)AS_0A^\dagger$ is a reflection about the direction of the state $|\psi\rangle$. Stepping through each operation in the product, $(-I)AS_0A^\dagger$, in turn:

$$|\omega'\rangle \xrightarrow{A^\dagger} \cos\frac{\omega'}{2}|0^n\rangle + e^{i\phi'}\sin\frac{\omega'}{2}|0^\perp\rangle \tag{9.35}$$

where the first term on the right-hand side follows from the definition $|\psi\rangle = A|0^n\rangle$, and the second term because $A^\dagger$ is, therefore, guaranteed to send a state orthogonal to $|\psi\rangle$ to a state orthogonal to $|0^n\rangle$; next

$$\cos\frac{\omega'}{2}|0^n\rangle + e^{i\phi'}\sin\frac{\omega'}{2}|0^\perp\rangle \xrightarrow{S_0} -\cos\frac{\omega'}{2}|0^n\rangle + e^{i\phi'}\sin\frac{\omega'}{2}|0^\perp\rangle \tag{9.36}$$

because S_0 'picks out' the all zero state and applies a phase of -1, leaving all states orthogonal to $|0^n\rangle$ unaffected; A undoes the action of the initial $A^\dagger$:

$$-\cos\frac{\omega'}{2}|0^n\rangle + e^{i\phi'}\sin\frac{\omega'}{2}|0^\perp\rangle \xrightarrow{A} -\cos\frac{\omega'}{2}|\psi\rangle + e^{i\phi'}\sin\frac{\omega'}{2}|\psi^\perp\rangle \tag{9.37}$$

and the final operation $-I$ adjusts the global phase:

$$-\cos\frac{\omega'}{2}|\psi\rangle + e^{i\phi'}\sin\frac{\omega'}{2}|\psi^\perp\rangle \xrightarrow{-I} \cos\frac{\omega'}{2}|\psi\rangle - e^{i\phi'}\sin\frac{\omega'}{2}|\psi^\perp\rangle \quad (9.38)$$

the right-hand side of which is indeed the intended reflection about the line of the state $|\psi\rangle$. Note that this final 'operation' $-I$ is only included for illustrative purposes and is not actually required in practice as all is does is to change the global phase factor.

The two reflections that compose Q, S_χ, and $-AS_0A^\dagger$ together give a rotation, which is such that a state in the subspace spanned by $|\Phi_0\rangle|0\rangle$ and $|\Phi_1\rangle|1\rangle$ is transformed to another state in the same subspace. Moreover, under the action of Q, a state in the real great circle of the Bloch sphere corresponding to the subspace spanned by $|\Phi_0\rangle|0\rangle$ and $|\Phi_1\rangle|1\rangle$ remains in the real great circle, which confirms the analogy with the Grover iterate and enables the action of Q to be depicted:

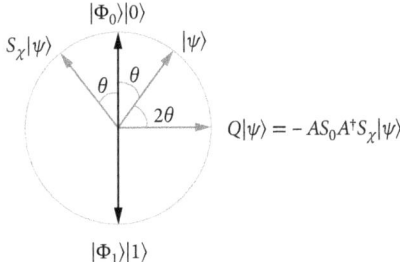

Figure 9.9

Having shown how the amplitude of a quantum state can be *amplified*, quantum amplitude *estimation* can be derived.

9.4.2 Estimating the amplitude with phase estimation

The fact that the action of Q is contained within the subspace spanned by $|\Phi_0\rangle|0\rangle$ and $|\Phi_1\rangle|1\rangle$ is not just convenient for illustrative purposes, but is also central to how it enables the amplitude to be estimated. Explicitly expressing Q as a rotation in this subspace gives:

$$Q = \begin{bmatrix} \cos\theta & -\sin\theta \\ \sin\theta & \cos\theta \end{bmatrix} \quad (9.39)$$

which is a matrix acting on a two-element complex unit vector, the first element being the coefficient of $|\Phi_0\rangle|0\rangle$ and the second element being that of $|\Phi_1\rangle|1\rangle$. As a simple verification, consider the action of Q, as expressed in Eqn. (9.39), on the initial state

166 Quantum phase estimation and quantum amplitude estimation

$|\psi\rangle$ expressed as a two-element vector in this subspace:

$$|\psi\rangle = \cos\frac{\theta}{2}|\Phi_0\rangle|0\rangle + \sin\frac{\theta}{2}|\Phi_1\rangle|1\rangle = \begin{bmatrix} \cos\frac{\theta}{2} \\ \sin\frac{\theta}{2} \end{bmatrix} \tag{9.40}$$

$$\Rightarrow Q|\psi\rangle = \begin{bmatrix} \cos\theta & -\sin\theta \\ \sin\theta & \cos\theta \end{bmatrix} \begin{bmatrix} \cos\frac{\theta}{2} \\ \sin\frac{\theta}{2} \end{bmatrix} = \begin{bmatrix} \cos\theta\cos\frac{\theta}{2} - \sin\theta\sin\frac{\theta}{2} \\ \sin\theta\cos\frac{\theta}{2} + \cos\theta\sin\frac{\theta}{2} \end{bmatrix} = \begin{bmatrix} \cos\frac{3\theta}{2} \\ \sin\frac{3\theta}{2} \end{bmatrix} \tag{9.41}$$

The idea at the heart of quantum amplitude estimation is to show that θ appears in the exponent of an eigenvalue, and hence, quantum *phase* estimation can be used to estimate the amplitude. The two relevant eigenvalues (and corresponding eigenvectors) of Q are, first:

$$Q\begin{bmatrix} 1 \\ -i \end{bmatrix} = \begin{bmatrix} \cos\theta & -\sin\theta \\ \sin\theta & \cos\theta \end{bmatrix}\begin{bmatrix} 1 \\ -i \end{bmatrix} = \begin{bmatrix} \cos\theta + i\sin\theta \\ \sin\theta - i\cos\theta \end{bmatrix} = \begin{bmatrix} \cos\theta + i\sin\theta \\ -i(\cos\theta + i\sin\theta) \end{bmatrix} = e^{i\theta}\begin{bmatrix} 1 \\ -i \end{bmatrix} \tag{9.42}$$

and so one eigenvalue is $e^{i\theta} = e^{2\pi i(\theta/2\pi)}$, with normalised eigenvector $(1/\sqrt{2})[1, -i]^T = (1/\sqrt{2})(|\Phi_0\rangle|0\rangle - i|\Phi_1\rangle|1\rangle)$. Second,

$$Q\begin{bmatrix} 1 \\ i \end{bmatrix} = \begin{bmatrix} \cos\theta & -\sin\theta \\ \sin\theta & \cos\theta \end{bmatrix}\begin{bmatrix} 1 \\ i \end{bmatrix} = \begin{bmatrix} \cos\theta - i\sin\theta \\ \sin\theta + i\cos\theta \end{bmatrix} = \begin{bmatrix} \cos(-\theta) + i\sin(-\theta) \\ i(\cos(-\theta) + i\sin(-\theta)) \end{bmatrix} = e^{-i\theta}\begin{bmatrix} 1 \\ i \end{bmatrix} \tag{9.43}$$

and so another eigenvalue is $e^{-i\theta} = e^{2\pi i(1-\theta/2\pi)}$, with normalised eigenvector $(1/\sqrt{2})[1, i]^T = (1/\sqrt{2})(|\Phi_0\rangle|0\rangle + i|\Phi_1\rangle|1\rangle)$.

Any state in the subspace spanned by $|\Phi_0\rangle|0\rangle$ and $|\Phi_1\rangle|1\rangle$ can, therefore, be written as a linear combination of these two eigenvectors, and hence, applying quantum phase estimation to such a state (loaded into the second register) provides an estimate of $\theta/2\pi$ or $1 - \theta/2\pi$. For example, the state prepared by applying A to $|0^n\rangle$, $|\psi\rangle = \cos\frac{\theta}{2}|\Phi_0\rangle|0\rangle + \sin\frac{\theta}{2}|\Phi_1\rangle|1\rangle$, is a suitable initial state. Furthermore, as $|\psi\rangle$ (as defined in Eqn. (9.27)) is such that $0 \le \theta \le \pi$, there is no ambiguity in the returned estimate: a value between 0 and 0.5 means that the estimated phase is $\theta/2\pi$, whereas a value between 0.5 and 1 means that the estimated phase is $1 - \theta/2\pi$.

Showing that quantum phase estimation can be deployed to estimate the amplitude of a quantum state says nothing about the computational efficiency of this procedure. For this, it is necessary to relate the error of the phase estimate to that of the corresponding amplitude estimate. From Section 9.3, quantum phase estimation with some $t' = t + \lceil \log(2 + 1/2\epsilon) \rceil$ qubits in the first register suffices to estimate the phase to t bits with probability at least $1 - \epsilon$. In the case where the measurement collapses to the eigenstate for which the eigenvalue is $e^{i\theta}$, this means that quantum phase estimation estimates the quantity $\theta/2\pi$ to t bits of precision. Letting the error in the estimate of θ be ϵ_θ, the error in the estimate of the amplitude, ϵ_a, can be bounded:

$$\epsilon_a = \sin^2 \frac{\theta + \epsilon_\theta}{2} - \sin^2 \frac{\theta}{2}$$

$$= \left(\sin \frac{\theta + \epsilon_\theta}{2} + \sin \frac{\theta}{2}\right)\left(\sin \frac{\theta + \epsilon_\theta}{2} - \sin \frac{\theta}{2}\right)$$

$$\implies |\epsilon_a| = \left|\sin \frac{\theta + \epsilon_\theta}{2} + \sin \frac{\theta}{2}\right| \times \left|\sin \frac{\theta + \epsilon_\theta}{2} - \sin \frac{\theta}{2}\right| \quad (9.44)$$

which can be further bounded by using the following properties/facts:

1. the magnitude of the gradient of $\sin \frac{\theta}{2}$ is at most one, so $\left|\sin \frac{\theta + \epsilon_\theta}{2} - \sin \frac{\theta}{2}\right|$ can be upper-bounded by $\frac{|\epsilon_\theta|}{2}$;
2. the modulus of the sum of any two real numbers is at most the sum of their moduli;
3. $\left|\sin \frac{\theta + \epsilon_\theta}{2}\right| \leq \sin \frac{\theta}{2} + \frac{|\epsilon_\theta|}{2}$ (a result from trigonometry);
4. $|\sin(\cdot)|$ is at most one;
5. the range of θ is known to be $0 \leq \theta \leq \pi$, so the error can also be bounded $|\epsilon_\theta| \leq \pi$.

Substituting these into Eqn. (9.44) gives (note that the below uses the fact that θ is between 0 and π by definition, and so $\sin \frac{\theta}{2}$ is positive):

$$|\epsilon_a| \leq \left(\left|\sin \frac{\theta + \epsilon_\theta}{2}\right| + \left|\sin \frac{\theta}{2}\right|\right) \frac{|\epsilon_\theta|}{2}$$

$$\leq \left(2 \sin \frac{\theta}{2} + \frac{|\epsilon_\theta|}{2}\right) \frac{|\epsilon_\theta|}{2} \quad (9.45)$$

$$\leq \left(2 + \frac{|\epsilon_\theta|}{2}\right) \frac{|\epsilon_\theta|}{2}$$

$$\leq \left(1 + \frac{\pi}{4}\right) |\epsilon_\theta| \quad (9.46)$$

Analogous analysis can be performed when the measurement outcome is such that, instead, the collapse is to the eigenstate associated with eigenvalue $e^{-i\theta}$. Therefore, the error in the estimate of the amplitude is proportional to the error in the estimate of the phase, and it turns out that this amounts to a quadratic advantage in estimation accuracy compared to (comparable means of) classical estimation. To see why, we turn our attention to *quantum Monte Carlo integration*.

9.4.3 Quantum Monte Carlo integration

The set-up for quantum Monte Carlo integration is as follows: as before, A is a circuit preparing the state, $|\psi\rangle = \cos\frac{\theta}{2}|\Phi_0\rangle|0\rangle + \sin\frac{\theta}{2}|\Phi_1\rangle|1\rangle$, and the amplitude, $a = \sin^2\frac{\theta}{2}$, is estimated in each of two ways:

1. using a classical algorithm, that is fed only samples obtained by measuring the final qubit of the state $|\psi\rangle = A|0^n\rangle$;
2. using a quantum algorithm that can make use of the circuits A and $A^\dagger$ to perform quantum computations.

The objective is to estimate a to a certain accuracy with as few uses of A (and $A^\dagger$) as possible, and a suitable measure of accuracy is the root-mean-squared error (RMSE). If $\hat{a}$ is the estimate of a, then the RMSE is defined as:

$$\text{RMSE} = \sqrt{\mathbb{E}((a - \hat{a})^2)} \tag{9.47}$$

$\hat{a}$ is itself a random variable (as each individual run of the estimation algorithm will return a different estimate, depending on the random samples), and it is over this that the expectation is taken.

The first (classical) case amounts to estimating the mean of a Bernoulli random variable, and the best strategy is to simply average the samples – a simple instance of a process known as *Monte Carlo integration* – in which case the error is:

$$\text{RMSE}_{\text{classical}} \propto \sqrt{\frac{1}{N_A}} \tag{9.48}$$

where N_A is the number of uses of A – in the classical case, the number of samples.

In the second (quantum) case, quantum amplitude estimation can be deployed, which has two types of error: (i) quantum amplitude estimation fails with probability ϵ; and (ii) if quantum amplitude estimation does not fail, the output is guaranteed to be within ϵ_a of the true value. How the possibility of failure (i.e., (i)) is handled varies on a case-by-case basis, for example in some applications (such as quantum counting, in Section 9.5), an incorrect amplitude estimate may become apparent in the 'downstream' processing, and so a simple retry-until-success strategy (with the failure probability set to a small constant) suffices. In other cases, quantum amplitude estimation may itself be run a few times with statistical techniques used to obtain a successful ('non-failed') estimate with high probability. To give a simple sketch of how quantum Monte Carlo integration offers a quadratic advantage, we assume that ϵ is a constant and that the associated failed runs can be detected or dealt with by some other means (i.e., for the following analysis, neglected).

In this case, the RMSE of the amplitude estimate is upper-bounded by $|\epsilon_a|$ because the maximum discrepancy upper bounds the RMSE. From Eqn. (9.46), $|\epsilon_a| \propto |\epsilon_\theta|$, and so if the first register has $t + \lceil \log(2 + 1/2\varepsilon) \rceil$ qubits such that the estimate of the phase, $|\varepsilon_\theta|/2\pi$, is accurate to t bits, then at worst $|\varepsilon_\theta| \propto 2^{-t}$. Putting this together gives

$$\text{RMSE}_{\text{quantum}} \leq |\varepsilon_a| \propto |\varepsilon_\theta| \propto 2^{-t} \tag{9.49}$$

For quantum phase estimation with $t + \lceil \log(2 + 1/2\varepsilon) \rceil$ qubits in the first register, Q is used $2^{t + \lceil \log(2 + 1/2\varepsilon) \rceil} - 1 = \Theta(2^t) = \Theta(\frac{1}{\text{RMSE}_{\text{quantum}}})$ times in total, and each time the operator Q is constructed, the circuits A and $A^\dagger$ are each used once. Therefore, the complexity can be rearranged to give the RMSE as a function of the total number of uses of A and $A^\dagger$:

$$\text{RMSE}_{\text{quantum}} \in \Theta\left(\frac{1}{N_A}\right) \tag{9.50}$$

Comparing Eqn. (9.50) with Eqn. (9.48) shows that, as claimed, the RMSE is suppressed quadratically faster (in the number of uses of A and $A^\dagger$) in the quantum case relative to its classical counterpart. From a practical perspective, quantum Monte Carlo integration would not be of much interest if it only offered a speed-up for the estimation of the mean of Bernoulli random variables; however, it is possible to encode, in the amplitude of a qubit, statistical quantities pertaining to complex, high-dimensional random processes – and so in fact, quantum Monte Carlo integration *is* one of the most practically interesting quantum algorithms.

9.4.4 Amplitude estimation without phase estimation

Why Monte Carlo integration enjoys a quadratic quantum advantage is not immediately obvious – certainly not when quantum amplitude estimation is achieved by quantum phase estimation. However, since the original quantum amplitude estimation algorithm was discovered, other means of performing quantum amplitude estimation have been proposed, which do not use quantum phase estimation. Whilst these have been mainly championed for their relatively shallow circuit depth (an essential feature for near-term quantum computation), they also have the attractive feature that the source of the quantum advantage is much more apparent and can be conveyed in a simple example. (The following explanation places a premium on providing a good intuitive explanation of the quadratic advantage, at the cost of some lack of rigour.)

Consider again a circuit A, which prepares the state $|\psi\rangle$ as defined in Eqn. (9.28), but that it is now promised that $\theta < \pi/3$. Further let some N_A samples, obtained by measuring the final qubit of $\cos\frac{\theta}{2}|\Phi_0\rangle|0\rangle + \sin\frac{\theta}{2}|\Phi_1\rangle|1\rangle$, be averaged to obtain an estimate of $\sin^2\theta/2$ with RMSE proportional to $\frac{1}{\sqrt{N_A}}$. The estimate of $\sin^2\theta/2$ can itself be thought of as a random variable, and if N_A is sufficiently large, this will have a distribution that is approximately normal, and be such that virtually all of the probability is concentrated into an approximately linear region of the function

$\sin^2(\cdot)$. It follows that if instead of estimating the amplitude ($\sin^2 \frac{\theta}{2}$), an estimate, $\frac{\hat{\theta}}{2}$, of $\frac{\theta}{2}$ is required, then $\frac{\hat{\theta}}{2}$ will itself be approximately normally distribution with RMSE proportional to $\frac{1}{\sqrt{N_A}}$. In the following, we let the constant of proportionality be k.

If the number of samples (that is, the number of times A and $A^\dagger$ are used in total) is tripled, then following Section 9.4.3, there are two alternatives for how the accuracy of the estimate, $\frac{\hat{\theta}}{2}$, can be improved:

1. $3N_A$ samples can now be obtained by measuring the final qubit of $\cos\frac{\theta}{2}|\Phi_0\rangle|0\rangle + \sin\frac{\theta}{2}|\Phi_1\rangle|1\rangle$, and averaging these leads to $\frac{\hat{\theta}}{2}$ with RMSE = $\frac{k}{\sqrt{3N_A}}$;
2. alternatively, the state $QA|0^n\rangle = \cos\frac{3\theta}{2}|\Phi_0\rangle|0\rangle + \sin\frac{3\theta}{2}|\Phi_1\rangle|1\rangle$ can be prepared N_A times (as Q requires one use of each of A and $A^\dagger$), and averaging samples obtained by measuring the final qubit of this state would give an estimate of $\frac{3\theta}{2}$ with RMSE = $\frac{k}{\sqrt{N_A}}$. (The promise that $\theta < \pi/3$ is important as it means that the conversion from $\sin^2\frac{3\theta}{2}$ to $\frac{3\theta}{2}$ is one–one.)

In the second case, as the estimate of $\frac{\theta}{2}$ is approximately normally distributed, it can be divided by 3 to obtain an estimate of $\frac{\theta}{2}$ with RMSE = $\frac{1}{3\sqrt{N_A}}$, constituting a factor of $\frac{1}{\sqrt{3}}$ reduction in RMSE compared to the first, 'classical', alternative. This reveals that the source of quantum advantage in quantum amplitude estimation is the ability to coherently rotate a state in constant increments rather than being restricted to 'incoherently' averaging independent sample values (as in the classical case), and so amounts to the same essential feature of quantum mechanics that facilitates the quadratic advantage in Grover search.

Algorithm 9.1 Amplitude estimation without phase estimation

Require: N_{shots}; A, $A^\dagger$
1: $Q = AS_0A^\dagger S_\chi$
2: **for** $m = 0, 1, 2, 4, 8, 16, \ldots$ **do**
3:     Prepare and measure N_{shots} of the state $Q^m A|0\rangle$
4: **end for**
5: In classical post-processing estimate θ and hence a.

Turning to quantum amplitude estimation without phase estimation more generally, the original proposal for such an algorithm is given in Algorithm 9.1. Stepping through Algorithm 9.1, the first iteration of the for loop with $m = 0$ amounts to classical Monte Carlo integration with N_{shots} samples, which gives a probability distribution $\Pr\left(\frac{\theta}{2}|\mathcal{D}_0\right)$ (where $\mathcal{D}_m$ is the measurement data for the N_{shots} of the circuit $Q^m A$).

Moving onto the second iteration with $m = 1$, measurements of the state $Q^1 A|0^n\rangle = \cos\frac{3\theta}{2}|\Phi_0\rangle|0\rangle + \sin\frac{3\theta}{2}|\Phi_1\rangle|1\rangle$ provide an estimate of $\sin^2\frac{3\theta}{2}$. However, unlike in the simple example where it was promised that $\theta < \pi/3$, which in turn led to a one–one conversion from an estimate of $\sin^2\frac{3\theta}{2}$ to one of $\frac{3\theta}{2}$, in general, θ will be in the range $0 \leq \theta \leq \pi$ (as defined below Eqn. (9.27)), and so there will be three values of $\frac{3\theta}{2}$ consistent with any estimate of $\sin^2\frac{3\theta}{2}$. As a result, the probability distribution of the data obtained from the N_{shots} measurements of $Q^1A|0^n\rangle$ will have three peaks. Generally speaking, for each value of m, the number of peaks in the corresponding probability distribution is $2m + 1$, as depicted:

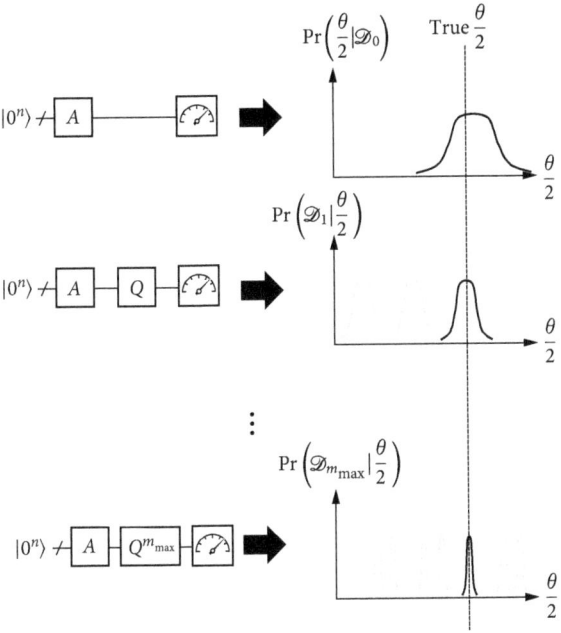

Figure 9.10

Notice the probability distributions in Fig. 9.10: for all but $m = 0$, $\Pr\left(\mathcal{D}_m|\frac{\theta}{2}\right)$ is shown, and so Bayes' law can be used at each iteration to update the probability of $\frac{\theta}{2}$ given *all* of the measurement data:

$$\Pr\left(\frac{\theta}{2}|\mathcal{D}_0\ldots\mathcal{D}_m\right) \propto \Pr\left(\mathcal{D}_m|\frac{\theta}{2}\right)\Pr\left(\frac{\theta}{2}|\mathcal{D}_0\ldots\mathcal{D}_{m-1}\right) \quad (9.51)$$

where the constant of proportionality is a normalising constant to ensure that the total probability sums to 1. The effect of Bayes' law multiplying together these distributions is to 'knock out' all but one of the peaks, as emphasised in Fig. 9.10: at each step, the peaks – shown as greyed out – that are outside of the region where the probability is concentrated in the prior distribution (for previous values of m) are suppressed to close to zero by the Bayesian update.

Jumping ahead to the final (largest) value of m ($m_{\max}$) which provides an estimate of $\sin^2\left((2m_{\max} + 1)\frac{\theta}{2}\right)$, in turn, giving an estimate of $(2m_{\max}+1)\frac{\theta}{2}$ with RMSE = $\frac{k}{\sqrt{N_{\text{shots}}}}$ (where k remains the constant of proportionality defined above). Even though there are $2m_{\max} + 1$ values of $\frac{\theta}{2}$ that are consistent with any value of $\sin^2\left((2m_{\max} + 1)\frac{\theta}{2}\right)$, the effect of the aforementioned 'knocking out' means that just a single peak remains.

Therefore, the conversion from an estimate of $(2m_{max}+1)\frac{\theta}{2}$ to one of $\frac{\theta}{2}$ can be treated as one–one, and as such, it is possible to obtain an estimate of $\frac{\theta}{2}$ with

$$\text{RMSE}_{quantum} = \frac{k}{(2m_{max}+1)\sqrt{N_{shots}}} \qquad (9.52)$$

In amplitude estimation without phase estimation, to obtain a quadratic quantum advantage compared to the classical strategy of simply measuring $A|0^n\rangle$ for the same total number of uses of A, the exponentially increasing sequence (i.e., $m = 0, 1, 2, 4, 8, 16, \ldots$) is crucial. The exponentially increasing sequence is the fastest growing sequence that enables the process of 'knocking out' the incorrect peaks to work: sequences that increase more rapidly cannot achieve the required disambiguation. Conversely, sequences that increase more gradually are inefficient because too few of the total number of uses of A and $A^\dagger$ are deployed in the final iteration with $m = m_{max}$. The total number of uses A and $A^\dagger$ to prepare N_{shots} of $Q^m A$ is equal to $(2m + 1)N_{shots}$, and for the exponentially increasing sequence, about half of the total number of uses of A and $A^\dagger$ are deployed in the final iteration, so the total number of uses of A and $A^\dagger$ is $2(2m_{max}+1)N_{shots}$. If the same number of uses was instead used in the classical strategy, such that $2(2m_{max} + 1)N_{shots}$ samples are obtained by measuring the final qubit of $A|0^n\rangle = \cos\frac{\theta}{2}|\Phi_0\rangle|0\rangle + \sin\frac{\theta}{2}|\Phi_1\rangle|1\rangle$, then an estimate of $\frac{\theta}{2}$ with

$$\text{RMSE}_{classical} = \frac{k}{\sqrt{2(2m_{max}+1)N_{shots}}} \qquad (9.53)$$

would be obtained – which is quadratically worse than the quantum case in Eqn. (9.52) when N_{shots} is constant (and so the total number of uses is proportional to m_{max}).

9.5 Quantum counting

Quantum amplitude estimation can be used to approximately count the number of marked solutions and hence determine the required number of Grover iterates required in Grover's algorithm. The Grover iterate, $(W \otimes I)V$, can be expressed as a rotation in the subspace spanned by $|y\rangle|-\rangle, |z\rangle|-\rangle$ (where $|y\rangle$ and $|z\rangle$ are defined in Eqn. (8.3)),

$$(W \otimes I)V = \begin{bmatrix} \cos\theta & -\sin\theta \\ \sin\theta & \cos\theta \end{bmatrix} \qquad (9.54)$$

and so $(W \otimes I)V$ can be substituted directly for Q in quantum amplitude estimation to obtain an estimate of θ; also using the fact that Eqn. (8.3) shows that the uniform superposition $|\psi\rangle$ is a linear combination of $|y\rangle$ and $|z\rangle$ and, hence, is a suitable initial state for the second register.

The objective of approximate counting is not to determine θ, but rather the number of marked solutions, M, which is related to θ by $\sin\frac{\theta}{2} = \sqrt{\frac{M}{N}}$ (see Eqn. (8.12)).

This means that for quantum counting, a slightly different error analysis is required: whereas quantum amplitude estimation estimates $a = \sin^2 \frac{\theta}{2} = \frac{M}{N}$ to accuracy ϵ_a, in quantum counting, the relevant error is that for an estimate of M, which we define ϵ_M, and is therefore related to ϵ_a by $\frac{|\epsilon_M|}{N} = |\epsilon_a|$. For this purpose, a suitable error analysis starts with Eqn. (9.45), neglecting the square of the error (which is appropriate for small error) but no longer using the bound $\sin \frac{\theta}{2} \leq 1$:

$$|\epsilon_a| \leq \left(2\sin\frac{\theta}{2} + \frac{|\epsilon_\theta|}{2}\right)\frac{|\epsilon_\theta|}{2}$$

$$\approx \sin\frac{\theta}{2}|\epsilon_\theta|$$

$$\implies \frac{|\epsilon_M|}{N} \leq \sin\frac{\theta}{2}|\epsilon_\theta|$$

$$= \sqrt{M/N}\,|\epsilon_\theta|$$

$$\implies |\epsilon_M| \leq \sqrt{MN}\,|\epsilon_\theta| \qquad (9.55)$$

Using $\lceil n/2 \rceil$ qubits (plus a further constant amount to suppress the failure probability) in the first quantum phase estimation register gives $|\epsilon_\theta| \leq 2^{-\lceil n/2 \rceil} \leq 1/\sqrt{N}$ and thus, substituting into Eqn. (9.55):

$$|\epsilon_M| \leq \sqrt{M} \qquad (9.56)$$

This is an acceptable relative error for quantum counting, as it is proportional to a sub-linear function of M. With $\lceil n/2 \rceil$ qubits (plus a further constant number to suppress the failure probability) in the first quantum phase estimation register, the required number of Grover iterates is $\Theta(2^{n/2}) = \Theta(\sqrt{N})$. As this is the same complexity as Grover search itself, it follows that it is possible to use quantum amplitude estimation to approximately count the number of solutions prior to running Grover's search algorithm, without increasing the overall complexity.

Chapter problems

1. Consider two-bit phase estimation of a unitary, U, with the circuit:

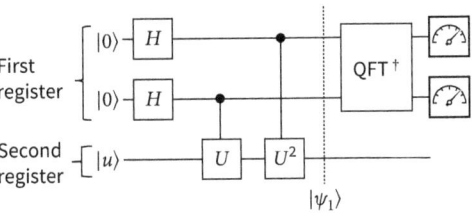

Figure 9.11

(a) Show that $U = \frac{1}{\sqrt{5}}\begin{bmatrix} 1 & 2 \\ 2 & -1 \end{bmatrix}$ is unitary; also that $\begin{bmatrix} 1+\sqrt{5} \\ 2 \end{bmatrix}$ and $\begin{bmatrix} 1-\sqrt{5} \\ 2 \end{bmatrix}$ are (un-normalised) eigenvectors of U, and give the eigenvalues.

(b) Find the quantum state, $|\psi_1\rangle$, before the inverse quantum Fourier transform is executed.

(c) What possible measurement outcomes can occur (i.e., after the inverse quantum Fourier transform, with measurement in the computational basis)? Give probabilities for each possible outcome.

2. Quantum phase estimation is to be performed for the two-qubit unitary:

$$U = T \otimes H = \begin{bmatrix} 1/\sqrt{2} & 1/\sqrt{2} & 0 & 0 \\ 1/\sqrt{2} & -1/\sqrt{2} & 0 & 0 \\ 0 & 0 & 1/2 + i/2 & 1/2 + i/2 \\ 0 & 0 & 1/2 + i/2 & -1/2 - i/2 \end{bmatrix} \quad (9.57)$$

(a) Derive the eigenvectors, eigenvalues and eigenvalue phases of U.

(b) If the second register is initialised in the state $|00\rangle$, what are the possible outcomes of running quantum phase estimation, and what is the probability with which each occurs? How many bits of precision are required for quantum phase estimation to correctly estimate the phase for any initial state?

3. Figure 9.12 shows the circuit for quantum phase estimation of a Hadamard gate.

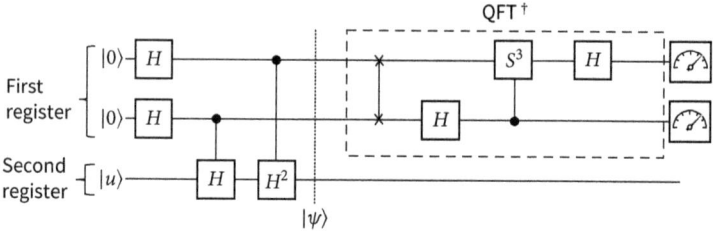

Figure 9.12

(a) To how many bits of precision is the estimate of the phase given?

(b) The Hadamard gate has matrix $\begin{bmatrix} 1/\sqrt{2} & 1/\sqrt{2} \\ 1/\sqrt{2} & -1/\sqrt{2} \end{bmatrix}$. What are its eigenvectors and corresponding eigenvalues? Express each eigenvector as a quantum state (that is, as superposition of computational basis states).

(c) Simplify the circuit in the figure such that when the initial state of the first register is $|00\rangle$ as specified, the only gate occurring on one of the wires is a swap.

(d) Quantum phase estimation is performed using the circuit given in the figure with $|u\rangle = a|0\rangle + b|1\rangle$. Express the three-qubit state $|\psi\rangle$ in terms of a and b. Verify that if $|u\rangle$ is a correctly normalised quantum state, then so is $|\psi\rangle$.

(e) Let $a = 0$ and $b = 1$, i.e., $|u\rangle = |1\rangle$. Knowing that the intended function of the circuit is to perform quantum phase estimation, what possible measurement outcomes do you expect, and with what probabilities do they occur? Does this match the actual measurement outcomes and the probabilities with which they occur?
4. What would happen if the amplitude amplification operator were incorrectly implemented as $Q = AS_0^2 A^\dagger S_\chi$ in quantum amplitude estimation?
5. In quantum amplitude estimation, the objective is to estimate $a = \sin^2 \frac{\theta}{2}$ for a general quantum state expressed in the form:

$$|\psi\rangle = \cos\frac{\theta}{2}|\Phi_0\rangle|0\rangle + \sin\frac{\theta}{2}|\Phi_1\rangle|1\rangle \qquad (9.58)$$

How can the algorithm be adapted to estimate instead $\cos^2 \frac{\theta}{2}$?
6. Consider the state

$$|\psi\rangle = 0.6725\,|00\rangle + 0.6725\,|10\rangle + 0.3090\,|11\rangle \qquad (9.59)$$

(a) Express $|\psi\rangle$ in the form $\cos\frac{\theta}{2}|\Phi_0\rangle|0\rangle + \sin\frac{\theta}{2}|\Phi_1\rangle|1\rangle$ and find θ, $|\Phi_0\rangle$ and $|\Phi_1\rangle$.
(b) Design a quantum circuit using amplitude amplification that prepares the state $|\Phi_1\rangle|1\rangle$.

Further reading

Forms of the FFT can be traced back as far as the works of Carl Friedrich Gauss (unpublished), although its modern form is usually credited to Cooley and Tukey [1965]. The quantum Fourier transform is due to Peter Shor [1994]; quantum phase estimation was proposed by Alexei Kitaev [1995]. Quantum counting was outlined by Boyer et al [1998] and fully developed by Brassard et al [1998]; an expanded group of authors then proposed the closely related quantum amplitude amplification and estimation [2002]. Performing amplitude estimation without the phase estimation subroutine was originally suggested by Suzuki et al [2020], and the presentation here closely resembles this original proposal – indeed, the illustration of coinciding peaks in Fig. 9.10 is based on Fig. 1 therein – although it is worth noting that Grinko et al [2021] proposed a more rigorous algorithm to achieve the same. Quantum Monte Carlo integration in its simplest form is little more than a corollary of quantum amplitude estimation, but it is worth noting that Montanaro [2015] wrote an excellent piece on quantum Monte Carlo methods in general.

10
Order finding, period finding, and quantum factoring

In 1994, Peter Shor discovered a quantum algorithm to factor numbers in polynomial time. Without doubt, Shor's algorithm remains one of the crowning achievements of the field of quantum computation (indeed, few in the field would raise an eyebrow if it were declared *the* crowning achievement). Classically, factoring is widely believed to be a *one-way* function: one-way functions are those that are easy to perform 'forward', but hard to invert – that is, on average, and not just in the worst case. In the case of factoring, this is because multiplying two prime numbers p and q to give a semi-prime, N, is easy, but the inverse, factoring a semi-prime N into factors p and q, is mathematically hard. The discovery of Shor's algorithm has both theoretical and practical implications. After decades of research, no polynomial time classical algorithm for factoring has been found – the state of the art is the *number field sieve*, which requires $\exp(\Theta(n^{1/3}\log^{2/3} n))$ operations, where $n = \lceil \log_2 N \rceil$, i.e., the number of bits required to express N – leading experts to believe that it is a problem outside of bounded-error probabilistic polynomial time (BPP). However the existence of Shor's factoring algorithm shows that factoring *is* inside of bounded-error quantum polynomial time (BQP), and therefore, Shor's algorithm provides strong evidence of a separation between BQP and BPP. On a practical note, public-key crypotosystems rely on one-way functions, and in particular, the very widely used Rivest–Shamir–Adleman cryptosystem (RSA) public-key cryptosystem uses factoring as a one-way function; therefore, Shor's algorithm means that once sufficiently large-scale quantum computers exist, a significant amount of the world's encrypted data and communications will no longer be secure.

10.1 Order finding

Shor's algorithm relies on the reduction of factoring to *order finding*. For co-prime positive integers x and N, such that $x < N$, the *order of x modulo N* is defined to be the least positive integer r such that $x^r = 1 \mod N$. To find the order of x modulo N quantumly, we show that it suffices to apply quantum phase estimation to the unitary:

$$U|y\rangle = |(xy) \mod N\rangle \quad (10.1)$$

where y is an integer such that $0 \le y < N$.

First, we show that U is a permutation matrix and hence *is* unitary, when x and N are co-prime. To see this, consider the following argument (in which $y_1 > y_2 \geq 0$):

- If $U|y_1\rangle = U|y_2\rangle$, then $xy_1 = xy_2 + kN$ for some integer $k \neq 0$.
- Therefore $y_1 - y_2 = \frac{kN}{x}$, i.e., $\frac{kN}{x}$ is an integer.
- However, as N and x are co-prime, the smallest integer k, which satisfies this is $k = x$, i.e., $y_1 - y_2 = N$, and so it cannot be the case that $0 \leq y_1, y_2 < N$.

It follows that, for each integer y such that $0 \leq y < N$, $U|y\rangle$ gives a different integer between 0 and $N - 1$ inclusive, so the upper left $N \times N$ block of U is a permutation matrix. As the remainder of U is irrelevant, for $N \leq y < 2^n$, we define $U|y\rangle \to y$, such that (the entire) U is a permutation matrix.

10.1.1 Order finding with quantum phase estimation

Recall from Chapter 9 that the quantum phase estimation circuit is:

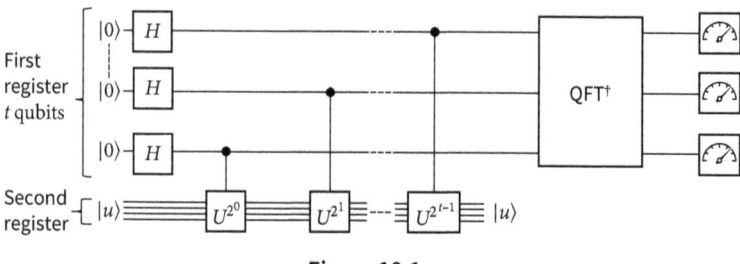

Figure 10.1

10.1.2 Preparing the second register in a suitable initial state

Quantum phase estimation requires that the initial state, $|u\rangle$, of the second register is a suitable eigenstate (or superposition of suitable eigenstates) of U. States of the form

$$|u_s\rangle = \frac{1}{\sqrt{r}} \sum_{k=0}^{r-1} e^{-2\pi i s k/r} |x^k \mod N\rangle \quad (10.2)$$

for integer s, $0 \leq s \leq r - 1$ are eigenstates of U, because:

$$U|u_s\rangle = \frac{1}{\sqrt{r}} \sum_{k=0}^{r-1} e^{-2\pi i s k/r} |x^{k+1} \mod N\rangle$$

$$= \frac{1}{\sqrt{r}} \sum_{k=1}^{r} e^{-2\pi i s (k-1)/r} |x^k \mod N\rangle$$

$$= e^{2\pi i s/r} \frac{1}{\sqrt{r}} \sum_{k=1}^{r} e^{-2\pi i s k/r} |x^k \mod N\rangle$$

$$= e^{2\pi i s/r} \frac{1}{\sqrt{r}} \sum_{k=0}^{r-1} e^{-2\pi i s k/r} |x^k \mod N\rangle$$

$$= e^{2\pi i s/r} |u_s\rangle \tag{10.3}$$

where the penultimate line follows by the definition $x^r = 1 = x^0 \mod N$, and also $e^{-2\pi i s(r/r)} = 1 = e^{-2\pi i s(0/r)}$. The phase of $|u_s\rangle$ is s/r, which is what enables quantum phase estimation to be able to find r; however, $|u_s\rangle$ is not a state that can usually be efficiently prepared. Indeed, to do so would ostensibly seem to require already knowing r – the very value that order finding is trying to compute! The trick is to prepare not a single eigenstate, but a superposition of eigenstates for $0 \le s \le r-1$:

$$\frac{1}{\sqrt{r}} \sum_{s=0}^{r-1} |u_s\rangle = \frac{1}{\sqrt{r}} \sum_{s=0}^{r-1} \frac{1}{\sqrt{r}} \sum_{k=0}^{r-1} e^{-2\pi i s k/r} |x^k \mod N\rangle$$

$$= \frac{1}{r} \left(\sum_{s=0}^{r-1} e^0 |x^0 \mod N\rangle + \sum_{k=1}^{r-1} \underbrace{\left(\sum_{s=0}^{r-1} e^{-2\pi i s k/r} \right)}_{=0 \text{ when } k>0} |x^k \mod N\rangle \right)$$

$$= \frac{1}{r} r |1\rangle$$

$$= |1\rangle \tag{10.4}$$

The parenthesised term in the second line is equal to zero, as the terms within the sum are equally spaced around the unit circle (on the Argand diagram) and, hence, sum to zero.

10.1.3 Extracting the order from the phase

As $|u_s\rangle$ has eigenvalue $e^{2\pi i s/r}$, i.e., its phase is s/r, running quantum phase estimation, with the second register in the state $|1\rangle = \frac{1}{\sqrt{r}} \sum_{s=0}^{r-1} |u_s\rangle$, has the following effect:

$$\frac{1}{\sqrt{r}} \sum_{s=0}^{r-1} |0\rangle^{\otimes n} |u_s\rangle \xrightarrow{\text{QPE}} \frac{1}{\sqrt{r}} \sum_{s=0}^{r-1} |2^t \widetilde{s/r}\rangle |u_s\rangle \tag{10.5}$$

The measurement then collapses the state such that quantum phase estimation returns an estimate of the phase $\widetilde{s/r}$ for some (unknown) integer s. The key remaining questions are: (i) how can the order r be extracted from $\widetilde{s/r}$, and (ii) how many qubits are required in the first register to do so.

The answers to these two questions are intrinsically connected, and we now show that there is a way to extract the phase from $\widetilde{s/r}$ when the first register contains $2n + 1 + \lceil \log(2 + 1/2\epsilon) \rceil$ qubits, where $n = \lceil \log_2 N \rceil$ is the size of the second register and $\log(2 + 1/2\epsilon)$ guarantees that quantum phase estimation is successful with probability at least $1 - \epsilon$.

Even though quantum phase estimation returns an estimate of s/r, as s is an unknown integer, there are an infinite number of such (integer) ratios that equal the same value, and so it may seem that finding r without knowledge of s is a hopeless task. However, perhaps surprisingly, there *is* an algorithm that returns r. In particular, the algorithm finds a pair of integers s' and r' such that $s'/r' \approx \widetilde{s/r}$, and the fraction s'/r' is irreducible. s'/r' being irreducible means that s' and r' are co-prime, and this turns out to be crucial as there is a good chance that s and r are themselves co-prime. More rigorously, if co-prime r and s is taken as a necessary condition for the extraction of r from $\widetilde{s/r}$, then the number or random retries can be upper-bounded as follows.

As the initial state of the second register, $|1\rangle$, is the *equal* superposition of $|u_s\rangle$ for all $s = 0, \ldots, r-1$, this means that there is an equal chance of collapsing to a state with phase s/r for each value of $0 \le s < r$. The number of these values of s that are co-prime to r is lower-bounded by the number of primes less than r, which is at least $\frac{r}{2\log r}$. Therefore, after $2 \log r \le 2 \log N$ retries, one can expect to have collapsed to a phase corresponding to prime s. This expected number of retries is, therefore, proportional to the number of bits, $\lceil \log N \rceil$, required to express the number being factored and, hence, represents an acceptable increase in the overall complexity. Moreover, sampling until a prime s is obtained is actually unnecessarily strong, and a tighter bound can be derived such that a constant number of retries is sufficient.

Proceeding on the assumption that quantum phase estimation has returned an estimate of s/r for co-prime s and r, the fact that the set-up is such that the estimate, $\widetilde{s/r}$, is accurate to $2n + 1$ bits guarantees that the following holds:

$$\left|\frac{s}{r} - \widetilde{s/r}\right| \le 2^{-(2n+1)} = \frac{1}{2(2^n)^2} \le \frac{1}{2r^2} \tag{10.6}$$

since $r \le N \le 2^n$. This, in turn, enables the *continued fractions algorithm* to be used to obtain an estimate of s and r.

A continued fraction is a number of the form:

$$[a_0, a_1, a_2, a_3 \ldots] = a_0 + \cfrac{1}{a_1 + \cfrac{1}{a_2 + \cfrac{1}{a_3 \ldots}}} \tag{10.7}$$

For example:

$$\frac{17}{15} = 1 + \frac{2}{15}$$

$$= 1 + \cfrac{1}{\left(\cfrac{15}{2}\right)}$$

$$= 1 + \cfrac{1}{7 + \cfrac{1}{2}} \tag{10.8}$$

so here $[a_0, a_1, a_2] = [1, 7, 2]$ and the algorithm for finding continued fractions has been defined implicitly (alternately separating out the integer part and then taking a double reciprocal, so a top heavy fraction appears in the denominator of the next part of the continuation). Every rational number has a (finite) continued fraction representation and if, as above, the final value thereof, a_M, is not allowed to be equal to 1, then the representation is unique. From each such continued fraction representation, an equivalent representation can be obtained with final term $a_{M+1} = 1$ by noticing $a_M = (a_M - 1) + \frac{1}{1}$ and so setting $a_M \leftarrow a_M - 1$; $a_{M+1} \leftarrow 1$. In the above example, this gives:

$$\frac{17}{15} = 1 + \cfrac{1}{7 + \cfrac{1}{1 + \cfrac{1}{1}}} \tag{10.9}$$

such that $[a_0, a_1, a_2, a_3] = [1, 7, 1, 1]$. This means that the continued fraction representation of any rational number can be chosen as having either an even *or* odd number of terms – a fact that is later used.

The key to using continued fractions to extract s and r is that when Eqn. (10.6) holds and $\widetilde{s/r}$ is rational (this will always be the case, as it has been obtained from a finite bit string), then s/r will appear at some known point in the continued fraction expansion of $\widetilde{s/r}$. To see this, first consider that for a fraction p/q with continued fraction representation $[a_0, \ldots, a_M]$, it is possible to define p_m and q_m recursively, such that $p_M/q_M = p/q$. Specifically, set $p_0 = a_0$, $q_0 = 1$, $p_1 = 1 + a_0 a_1$, and $q_1 = a_1$ and for $m \geq 2$:

$$p_m = a_m p_{m-1} + p_{m-2} \tag{10.10}$$

$$q_m = a_m q_{m-1} + q_{m-2} \tag{10.11}$$

The claim $p_M/q_M = p/q$ can easily be checked for $M = 0, 1, 2$:

$$M = 0: \quad \frac{p_0}{q_0} = \frac{a_0}{1} = a_0 = \frac{p}{q} \tag{10.12}$$

$$M = 1: \quad \frac{p_1}{q_1} = \frac{1 + a_0 a_1}{a_1} = a_0 + \frac{1}{a_1} = \frac{p}{q} \tag{10.13}$$

$$M = 2: \quad \frac{p_2}{q_2} = \frac{a_2 p_1 + p_0}{a_2 q_1 + q_0}$$

$$= \frac{a_2(1 + a_0 a_1) + a_0}{a_2 a_1 + 1} = \frac{a_0(a_2 a_1 + 1) + a_2}{a_2 a_1 + 1} = a_0 + \cfrac{1}{a_1 + \cfrac{1}{a_2}} = \frac{p}{q} \tag{10.14}$$

These now provide the base case to prove by induction that this holds also for $M \geq 3$. In particular, assuming that $p_M/q_M = p/q$ holds for every p/q that has a continued fraction representation $[a_0, \ldots, a_M]$, we show that this implies that the ratio p_{M+1}/q_{M+1} is indeed equal to p/q for every p/q with a continued fraction representation $[a_0, \ldots, a_{M+1}]$. To do so, we use the fact that a continued fraction representation, which goes up to some a_{M+1}, can be expressed as a continued fraction representation, $[a'_0, \ldots, a'_M]$, i.e., only up to an Mth term, by setting $a'_m = a_m$ for $m = 0, \ldots, M-1$ and $a'_M = a_M + 1/a_{M+1}$:

$$[a_0, \ldots, a_{M+1}] = [a_0, \ldots, a_M + 1/a_{M+1}] \tag{10.15}$$

Starting with the inductive hypothesis that for every continued fraction representation with terms up to a_M, $p/q = p_M/q_M$, and substituting in the continued fraction representation in Eqn. (10.15) along with the definitions of p_m and q_m in Eqns. (10.10) and (10.11) gives:

$$\begin{aligned}
\frac{p}{q} &= \frac{p_M}{q_M} \\
&= \frac{(a_M + 1/a_{M+1})p_{M-1} + p_{M-2}}{(a_M + 1/a_{M+1})q_{M-1} + q_{M-2}} \\
&= \frac{(a_{M+1}a_M + 1)p_{M-1} + a_{M+1}p_{M-2}}{(a_{M+1}a_M + 1)q_{M-1} + a_{M+1}q_{M-2}} \\
&= \frac{a_{M+1}(a_M p_{M-1} + p_{M-2}) + p_{M-1}}{a_{M+1}(a_M q_{M-1} + q_{M-2}) + q_{M-1}} \\
&= \frac{a_{M+1}p_M + p_{M-1}}{a_{M+1}q_M + q_{M-1}} \\
&= \frac{p_{M+1}}{q_{M+1}} \tag{10.16}
\end{aligned}$$

thus completing the inductive proof.

For the following analysis, it is also necessary to show $q_m p_{m-1} - p_m q_{m-1} = (-1)^m$ for $m \geq 1$. This can be proven by induction, with the base case of $m = 1$:

$$q_m p_{m-1} - p_m q_{m-1} = q_1 p_0 - p_1 q_0 = a_1 a_0 - (1 + a_0 a_1) \times 1 = -1 = (-1)^1 \tag{10.17}$$

and for the inductive step, assuming $q_m p_{m-1} - p_m q_{m-1} = (-1)^m$:

$$q_{m+1}p_m - p_{m+1}q_m = (a_{m+1}q_m + q_{m-1})p_m - (a_{m+1}p_m + p_{m-1})q_m \tag{10.18}$$

$$= -(q_m p_{m-1} - p_m q_{m-1}) + a_{m+1}q_m p_m - a_{m+1}q_m p_m$$

$$= (-1)(q_m p_{m-1} - p_m q_{m-1})$$

$$= (-1)(-1)^m$$

$$= (-1)^{m+1} \tag{10.19}$$

thus completing the inductive proof.

This result can also be used to prove that the fraction $\frac{p_m}{q_m}$ is always irreducible (i.e., p_m and q_m are co-prime). Let $p_m = wp'_m$ and $q_m = wq'_m$, where all terms are integers and p'_m and q'_m are co-prime, i.e., w is the greatest common divisor (gcd) of p_m and q_m; then, substituting these definitions into Eqn. (10.19) gives:

$$(-1)^{m+1} = q_{m+1}p_m - p_{m+1}q_m$$
$$= w(q_{m+1}p'_m - p_{m+1}q'_m) \tag{10.20}$$

which can only hold if $w = 1$ (as $(q_{m+1}p'_m - p_{m+1}q'_m)$ is an integer); hence, p_m and q_m are co-prime and $\frac{p_m}{q_m}$ is irreducible.

Having shown these general results about continued fractions, we can now return to the question of how this enables the phase to be extracted from the estimate $\widetilde{s/r}$. From Eqn. (10.6), it is possible to write

$$\widetilde{s/r} = \frac{s}{r} + \frac{\delta}{2r^2} \tag{10.21}$$

for some $|\delta| < 1$. In the following, we actually assume $0 \leq \delta < 1$, i.e., quantum phase estimation always provides an overestimate of $\widetilde{s/r}$, which can be achieved in practice by increasing the accuracy a little and adding a suitable constant to the estimate. As $\frac{s}{r}$ is rational, it has a continued fraction representation $[a_0, \ldots, a_M]$ such that $\frac{s}{r} = \frac{p_M}{q_M}$, and furthermore, as both $\frac{s}{r}$ and $\frac{p_M}{q_M}$ are irreducible (the former by the assumption that the order-finding algorithm has been repeated until co-prime s and r are obtained; the latter by the proof, above), we have that $s = p_M$ and $r = q_M$. This gives:

$$\widetilde{s/r} = \frac{p_M}{q_M} + \frac{\delta}{2q_M^2} \tag{10.22}$$

from which we define

$$\lambda = 2\left(\frac{q_M p_{M-1} - p_M q_{M-1}}{\delta}\right) - \frac{q_{M-1}}{q_M} \tag{10.23}$$

Using (i) $q_M p_{M-1} - p_M q_{M-1} = 1$ for even M (recalling that the continued fraction representation can always be chosen thus); (ii) $0 \leq \delta < 1$; and (iii) q_m is increasing so $\frac{q_{M-1}}{q_M} < 1$, gives:

$$\lambda = \frac{2}{\delta} - \frac{q_{M-1}}{q_M} > 1 \tag{10.24}$$

Note that λ is rational. Substituting Eqn. (10.23) into Eqn. (10.22) gives (see Chapter problems):

$$\widetilde{s/r} = \frac{\lambda p_M + p_{M-1}}{\lambda q_M + q_{M-1}} \tag{10.25}$$

which is in the standard recursive form of the continued fraction expression shown above. If λ were not just rational but an integer, we would immediately have $\widetilde{s/r} = [a_0, \ldots, a_M, \lambda]$. As λ is not (necessarily) an integer, but is rational and greater than 1, it has a continued fraction representation $\lambda = [\tilde{a}_0, \ldots, \tilde{a}_{\tilde{M}}]$, and so $\widetilde{s/r} = [a_0, \ldots, a_M, \tilde{a}_0, \ldots, \tilde{a}_{\tilde{M}}]$. This result guarantees that the continued fraction representation, $[a_0, \ldots, a_M]$, for $\frac{s}{r}$ appears within the continued fraction representation of $\widetilde{s/r}$. To obtain r, the continued fraction representation of $\widetilde{s/r}$ can be computed, and for each a_m that is obtained, it can be checked whether the corresponding $\frac{p_m}{q_m} = \frac{s}{r}$ is such that $q_m = r$ (as is guaranteed for some value of m). Each returned value can be checked to see if it is indeed the order, and this is computationally inexpensive, as is the process of computing a continued fraction representation (i.e., as $\widetilde{s/r}$ is a $t \in \Theta(n)$ bit number).

10.1.4 Efficiently implementing the controlled unitaries

Having shown how the order can be extracted from an estimate of the phase of U, we return to the quantum phase estimation circuit to show how the controlled-U gates may be efficiently implemented. For this, it is helpful to think of the entire sequence as composing a single reversible operation on a t-bit input register. If the first register is initially a t qubit computational basis state $|j\rangle$, such that $j = j_1 \ldots j_t = \sum_{i=1}^{t} j_i 2^{t-i}$, then the overall operation of the sequence of controlled-U gates can be expressed

$$\prod_{i=1}^{t} x^{j_i 2^{t-i}} y \mod N = x^{\sum_{i=1}^{t} j_i 2^{t-i}} y \mod N = x^j y \mod N \quad (10.26)$$

Here, y is the value that is in the second register prior to the controlled unitaries. In order finding, as established in Section 10.1.2, this is set to $|1\rangle$, so $y = 1$. This means that an identical state (i.e., prior to the inverse quantum Fourier transform) is prepared if, instead, the reversible circuit

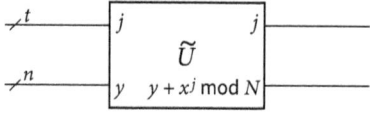

Figure 10.2

is used with the initial state of second register set as $|0\rangle$, so $y = 0$. It follows that the state after the inverse quantum Fourier transform is also identical, and so the analysis of how the order can be extracted from the measurement outcome continues to hold.

In this way, the problem of how to construct a suitable sequence of controlled unitaries, as shown in Fig. 10.1, can be replaced with the problem of how to construct a reversible circuit $\tilde{U}|j\rangle|y\rangle = |j\rangle|y + x^j \mod N\rangle$ for a pre-known, fixed, value of $x < N$. For this, we can use the usual result that a reversible circuit of this form can always be constructed directly from the corresponding classical algorithm, with the same asymptotic complexity. The classical algorithm for x^j mod N is termed *modular exponentiation* and is given in Algorithm 10.1.

Algorithm 10.1 Modular exponentiation to compute x^j mod N for any fixed $x < N$

Require: t-bit unsigned integer, $j = j_1 j_2 \ldots j_t$
1: $z \leftarrow 1$
2: **for** $i = 0 : t - 1$ **do**
3:    **if** $j_{t-i} = 1$ **then**
4:       $z \leftarrow xz$ mod N
5:    **end if**
6:    $x \leftarrow x^2$ mod N
7: **end for**
8: Return z

Algorithm 10.1 works because ab mod $c = (a$ mod $c)(b$ mod $c)$ for all a, b, c. This is crucial as every multiplication is performed modulo N, and so the values being multiplied never become too large. Multiplication takes $\mathcal{O}(m^2)$ operations, where m is the number of bits needed to specify the numbers being multiplied (using textbook 'long multiplication' – slightly more efficient variations do exist). Algorithm 10.1 consists of a loop of t iterations, in which there are (up to) two modular multiplications of n-bit integers (namely xz and x^2), so the overall computational complexity is $\Theta(n^2 t)$. As shown in Section 10.1.3, to extract the order from the phase, it suffices that $t \in \mathcal{O}(n)$, and putting these together, we have that the number of operations required to perform the series of controlled-U gates is $\mathcal{O}(n^3)$.

10.2 Shor's algorithm

Shor's algorithm uses quantum order finding to factor an n-bit composite integer N and is given in Algorithm 10.2. (In some presentations of Shor's algorithm, a couple of computationally efficient classical steps – omitted here – are also included at the start to check (i) if N is even, in which case 2 is a factor that can be returned, and (ii) if N is the power of some prime, in which case this prime can be computed and returned.)

The order-finding subroutine in Algorithm 10.2 requires co-prime x and N. Co-primality can be efficiently classically checked using Euclid's algorithm, as in Line 2, to find the gcd of x and N. If this returns the value 1, then x and N are co-prime as required, and if not, then the gcd is itself a non-trivial factor of N that, therefore, constitutes a solution to be returned. To show that Shor's algorithm indeed factors

> **Algorithm 10.2** Shor's algorithm
>
> **Require:** N
> 1: Randomly choose some x, such that $1 < x < N$
> 2: Run Euclid's algorithm to compute $\gcd(x; N)$
> 3: **if** $\gcd(x; N) \neq 1$ **then**
> 4:   Return $\gcd(x; N)$
> 5: **else**
> 6:   Use quantum order finding as a subroutine to find the order, r, of x mod N.
> 7:   **if** r is odd **then**
> 8:   Goto Line 1
> 9:   **else**
> 10:   Apply efficient classical post processing to extract a factor of N from r.
> 11:   **if** post processing fails **then**
> 12:   Goto Line 1
> 13:   **else**
> 14:   Return the factor
> 15:   **end if**
> 16:   **end if**
> 17: **end if**

N when x and N are co-prime, it thus remains to reduce factoring to order finding – which only holds for even r, hence the penultimate `if` statement.

10.2.1 Reduction of factoring to order finding

The order-finding subroutine produces r such that $x^r \mod N = 1$, i.e.,

$$(x^r - 1) \mod N = 0 \tag{10.27}$$

if r is even this factorises:

$$(x^{r/2} - 1)(x^{r/2} + 1) \mod N = 0 \tag{10.28}$$

for which there are four possibilities:

1. $x^{r/2} = 1 \mod N$: but we know this cannot actually occur, as r is the *least* integer satisfying $x^r = 1 \mod N$.
2. $x^{r/2} = -1 \mod N$: in which case the classical post-processing fails.
3. $(x^{r/2} - 1)(x^{r/2} + 1) = N$, in which case $(x^{r/2} - 1)$ and $(x^{r/2} + 1)$ are factors which can be returned directly.

4. $(x^{r/2} - 1)(x^{r/2} + 1) = kN$ for some $k \geq 2$, in which case one of $\gcd(x^{r/2} + 1 \mod N; N)$ or $\gcd(x^{r/2} - 1 \mod N; N)$ is a non-trivial factor of N, which can be found by Euclid's algorithm and then returned. (See chapter problems.)

The failure mode in the second item is covered by the final `if` statement in Algorithm 10.2. To ascertain the overall computational complexity of Shor's algorithm, it is necessary to quantify all of the possible failure modes, as well as the running time for a successful execution.

10.2.2 Shor's algorithm: computational complexity

A single run of Shor's algorithm only returns a factor with a certain probability. In particular:

1. The continued fractions algorithm may fail to extract the order on any given run (for instance if the collapse is such that s and r are not co-prime). As it is quick to check whether the returned value is indeed the order, Line 6 of Shor's algorithm should implicitly be taken as repeating the quantum subroutine until success.
2. The order-finding subroutine could return odd r (as explicitly noted in Line 8).
3. The order-finding subroutine could return r such that $x^{r/2} = -1 \mod N$ (which constitutes a failure in post-processing, as in Line 11).

All of these failure modes occur with a probability that does not grow with the problem size, and so $\mathcal{O}(1)$ repeats suffice. Therefore, the overall computational complexity is proportional to that of a single run of Shor's algorithm in which failure does not occur.

The quantum circuit for Shor's algorithm consists of:

1. Initial state preparation: the first register requires $t \in \mathcal{O}(n)$ Hadamard gates to prepare the equal superposition of all t-bit strings; the second register is initially the computational basis state, $|0\rangle$ (when the modular exponentiation unitary is used) and, hence, incurs no gates to prepare.
2. The modular exponentiation block, which has been shown to correspond to a reversible circuit implementation of a $\Theta(n^3)$ classical algorithm and, therefore, requires $\Theta(n^3)$ quantum gates.
3. The inverse quantum Fourier transform, which requires $\Theta(n^2)$ gates.

All of the classical subroutines called within Shor's algorithm are also efficient. Therefore, as claimed, Shor's algorithm indeed gives a way to factor numbers in time that is only polynomial in the problem size. By contrast, recall that the best classical algorithm for factoring requires $\exp(\Theta(n^{1/3}\log^{2/3} n))$ operations, and so for the problem of factoring Shor's algorithm provides a super-polynomial speed-up compared to the best classical algorithm.

10.2.3 Shor's algorithm explained in terms of period finding

Presenting the sequence of controlled-U operations as a single reversible operation acting on a register of input bits allows (the quantum part of) Shor's algorithm to be cast as an instance of *period finding*. A function, $f : \mathbb{R} \to \mathbb{R}$, is periodic with period $T \in \mathbb{R}^+$ if $f(x + T) = f(x)$ for all $x \in \mathbb{R}$. The reversible circuit, $\tilde{U}$, in Shor's algorithm is periodic with period r:

$$\begin{aligned}
\tilde{U}|j+r\rangle|y\rangle &= |j+r\rangle|y + x^{j+r} \mod N\rangle \\
&= |j+r\rangle|y + (x^{j+r} \mod N) \mod N\rangle \\
&= |j+r\rangle|y + (x^{j} \mod N)(x^{r} \mod N) \mod N\rangle \\
&= |j+r\rangle|y + (x^{j} \mod N) \times 1) \mod N\rangle \\
&= |j+r\rangle|y + x^{j} \mod N\rangle
\end{aligned}
\qquad (10.29)$$

In Shor's algorithm, using period finding to obtain the period of the modular exponentiation unitary, $\tilde{U}$, is therefore equivalent to finding the order of U.

The quantum circuit for period finding is as follows, where U_f is a reversible encoding of any periodic function, $f(x)$, with period T, that is taken as a black box:

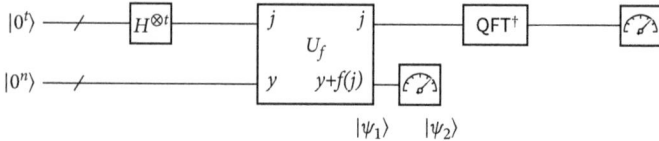

Figure 10.3

Note that when $U_f = \tilde{U}$, i.e., the modular exponentiation unitary, then this is identical to circuit for order finding. Stepping through the circuit, the state, $|\psi_1\rangle$, after U_f is:

$$\begin{aligned}
|\psi_1\rangle &= \frac{1}{\sqrt{2^t}} \sum_{j=0}^{2^t-1} |j\rangle|f(j)\rangle \\
&= \frac{1}{\sqrt{2^t}} \sum_{f(j)} \left(\sum_{l \in L_{CZ}} |j + lT\rangle \right) |f(j)\rangle
\end{aligned}
\qquad (10.30)$$

In the second line, the first summation is over each value of $f(j)$ that appears in the superposition, and the subset of integers L included as the range of the second summation is implicitly defined by $0 \le j + lT < 2^t$. This expression is used to emphasise the fact that computational basis states in the first register that differ by an integer multiple of the period are entangled with the same state in the second register.

Measurement of the second register is not actually required in period finding; however, it simplifies the analysis to assume (by the principle of implicit measurement) that it is measured, and furthermore to treat this as occurring *before* the inverse quantum Fourier transform. This collapses the state to:

$$|\psi_2\rangle = \frac{1}{C} \sum_{l \in LCZ} |j + lT\rangle |f(j)\rangle \qquad (10.31)$$

for some normalising constant, C, and some particular (arbitrary) value of j. The second register is omitted in the remaining analysis.

Eqn. (10.31) shows that the state prior to the inverse quantum Fourier transform is a superposition of computational basis states such that only those separated by the period, T, have non-zero coefficients (and the non-zero terms all have equal coefficients). Using the (inverse quantum) Fourier transform to extract this period, T, resonates strongly with a classical understanding of the function of the Fourier transform in digital signal processing. To see the detail of how this works, it is necessary to approximately express $|\psi_2\rangle$ as:

$$|\psi_2\rangle \approx |\psi_2'\rangle = \frac{1}{C'} \sum_{x=-j}^{2^t-1-j} e^{2\pi i s x / T} |j + x\rangle$$

$$= \frac{1}{C'} e^{-2\pi i s j / T} \sum_{x'=0}^{2^t-1} e^{2\pi i s x' / T} |x'\rangle \qquad (10.32)$$

for any constant $0 < s < T$. This approximation holds as for each x which is a multiple of T, $e^{2\pi i s x / T} = 1$, and for all other values of x, the corresponding values of $e^{2\pi i s x / T}$ are approximately equally spaced around the unit circle, and so approximately sum to zero. (When analysing the same circuit as an instance of order finding – i.e., for $U_f = \tilde{U}$ – this approximation is subsumed in the quantum phase estimation subroutine, which provides an approximation of the phase if it is not exactly representable in t bits.) It is interesting to note that the particular value of s is actually related to the state collapsed to by the measurement of the second register; however, this is immaterial in the remaining analysis. The final step is the inverse quantum Fourier transform and measurement of the first register:

$$|\psi_2'\rangle \xrightarrow{QFT^\dagger} |\widetilde{2^t s/T}\rangle \qquad (10.33)$$

ignoring the global phase $e^{-2\pi i s j / T}$. As in order finding, the period T can be extracted from $\widetilde{s/T}$ using with the continued fractions algorithm.

As the function U_f is a black box, the problem of period finding is a query problem – one which we have shown can be solved with $\mathcal{O}(1)$ quantum queries. Why this is so can be explained in terms of the principles outlined in Section 7.5: the query problem concerns a family of black box functions that are *promised* to have a certain *structure*, namely periodicity. This structure can be leveraged by a quantum algorithm to extract a certain property, in this case the period itself. Conversely, any classical strategy would require querying the black box a large number of times. This, in turn, reveals something of the source of quantum advantage in Shor's algorithm: as the problem of finding the period of U_f is one that has enough structure for an efficient *quantum* algorithm (namely Shor's), but there is not (believed to be) sufficient structure to facilitate an efficient *classical* algorithm.

Chapter problems

1. Show that permutation matrices are unitary.
2. Write down the following as continued fractions:
   (a) $\dfrac{25}{17}$
   (b) $\dfrac{63}{4}$
3. Derive the expression for $\widetilde{s/r}$ in Eqn. (10.25) from the definition of λ in Eqn. (10.23). **Hint**: it is sufficient just to prove this for even M.
4. Show that if $(x^{r/2} + 1)(x^{r/2} - 1) = 0 \mod N$ and $x^{r/2} \neq \pm 1$, then the gcd of N and at least one of $(x^{r/2} + 1) \mod N$ or $(x^{r/2} - 1) \mod N$ is a non-trivial factor of N. **Hint**: begin by reducing each of $x^{r/2} + 1$ and $x^{r/2} - 1$ modulo N and use the fundamental theorem of arithmetic to express N in terms of its prime factors.
5. This question concerns using Shor's algorithm to factor the number 21 with $x = 10$.
   (a) Find the order of 10 mod 21.
   (b) Say we were to run Shor's algorithm in full with $x = 10$, and were to measure the phase corresponding to the eigenvector u_1 (i.e., as a special case of u_s in the chapter), express this eigenvector (in full, not as abbreviated by a sum) and its eigenvalue.

Further reading

Peter Shor [1994; 1997] proposed Shor's algorithm; RSA was proposed by Rivest et al [1978].

11
Hamiltonian simulation and ground state energy estimation

Richard Feynman famously asserted that in order to simulate a quantum mechanical system, a computer itself operating on the principles of quantum mechanics would be required. In this chapter, we see that gate-based quantum computers indeed fulfil this vision for many quantum systems of interest. The ground state energy problem is also introduced as one of the most important real-world applications of Hamiltonian simulation – for example in quantum chemistry and condensed matter physics.

Hamiltonian simulation is the problem of determining the state $|\psi_t\rangle$ of a quantum system at some time, t, in the future, given its (time-invariant) Hamiltonian, H, and its current state, $|\psi_0\rangle$. This entails solving or *simulating* the Schrödinger equation:

$$|\psi_t\rangle = e^{-iHt}|\psi_0\rangle \qquad (11.1)$$

For convenience, rather than explicitly stating the normalisation by Planck's constant, here it has implicitly been incorporated it into the Hamiltonian itself (in contrast to the introduction of the Schrödinger equation in Chapter 3).

For the analysis in this chapter, it is taken as a starting point that the Hamiltonian of interest is already in a form that applies to a system of qubits. However, this is not always the case, and some preparatory work is often needed to express the physics/chemistry of the quantum mechanical system in an appropriate form for processing on a qubit-based quantum computer. Two particularly interesting and important types of systems are *spin systems*, such as the Ising model, and *Fermionic systems*, such as the electrical interactions in chemistry. The former is essentially already in qubit form – indeed, some quantum computing researchers of a particularly physics-oriented mindset speak and write of 'up' and 'down' spin states $|\uparrow\rangle$ and $|\downarrow\rangle$ rather than the more familiar computational basis states $|0\rangle$ and $|1\rangle$. It follows that spin systems do not require translation into qubit form, and indeed, the Ising model is an important starting point for understanding quantum optimisation, as is the topic of Chapter 12. However, the same cannot be said for Fermionic systems, where a transformation, such as the Jordan–Wigner transformation, is needed to express the Fermionic system in the form of a spin (qubit) system. Such transformations are a rich and fascinating subject in their own right; however, they are beyond the scope of this book.

11.1 Hamiltonian decomposition

Every quantum circuit (notably those we have studied in the preceding chapters) is nothing more than an instance of e^{-iHt}, and so it is important to explain what, precisely, is the objective we are aiming at in this chapter. One way to motivate the problem we are tackling is that, whereas in the previous chapters we were concerned with Hamiltonian evolutions (quantum circuits) that have been built up from elementary quantum gates to achieve a computational effect, here we flip the problem on its head and *start* with a Hamiltonian whose evolution we wish to decompose into a product of elementary quantum gates. To this end, we are aided by the fact that, though in general e^{-iHt} may not be a unitary that is trivial to decompose into a product of standard quantum gates (at least not using the techniques introduced in earlier chapters), often for real quantum systems of interest, the Hamiltonian can be decomposed as a sum of terms,

$$H = \sum_{l=0}^{L-1} H_l \tag{11.2}$$

where L is small (in a sense that will become clear), and the physical nature of the system is such that for all l, $e^{-iH_l t}$ is a unitary evolution that *can* be efficiently synthesised on a gate-based quantum computer.

This decomposition takes advantage of the 'locality' of the physical system, where the qubits represent a spatially extended system in which, for example, interactions between distant 'bodies' can safely be neglected. This is significant, and computationally advantageous, as it means that the Hamiltonian can be decomposed into a sum of terms such that each term is a tensor product of a (non-trivial) Hermitian operator acting on a constant number of qubits with the identity acting on all the other qubits.

As a simple example, consider a four-qubit system, where the qubits are physically arranged in a line (or 'linear array') and the system is such that all interactions apart from those between neighbouring qubits can be neglected. That is, there are three 'nearest-neighbour' interactions that contribute to the overall Hamiltonian, which are denoted $\tilde{H}_0$, $\tilde{H}_1$, and $\tilde{H}_2$, respectively, and can be depicted:

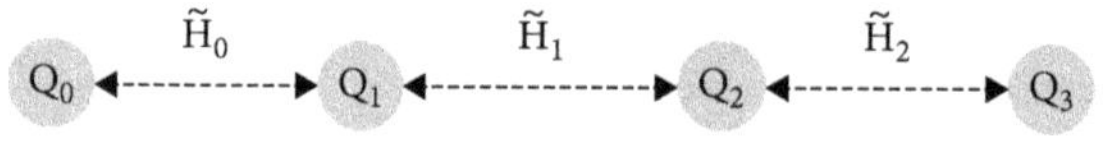

Figure 11.1

The system Hamiltonian is the sum of these three components

$$H = \tilde{H}_0 \otimes I_4 + I_2 \otimes \tilde{H}_1 \otimes I_2 + I_4 \otimes \tilde{H}_2 \tag{11.3}$$

If one extends the linear array to include any number of qubits, it remains the case that each qubit will only interact with (at most) two others. This, in turn, manifests as each of the component Hamiltonians being row (and column) sparse: the number of non-zero elements in each row (and column) is a constant and does not grow with the system size (number of qubits). Simulation of Hamiltonians that are sparse but not necessarily locally structured (in the sense described here) is an important topic that is developed further in Section 11.2.3; however, for now we continue our focus on systems that do have local structure. In particular, continuing with the example of the Hamiltonian of a four-qubit linear array with nearest-neighbour interactions, without making further assumptions about the physical system, each of $\tilde{H}_1$, $\tilde{H}_2$, and $\tilde{H}_3$ could be any 4×4 Hermitian matrix – however, they can always be further decomposed into a sum of tensor products of Pauli matrices. That is, any Hermitian matrix, H, can be written as:

$$H = \sum_i s_i P_i^{(1)} \otimes P_i^{(2)} \otimes \cdots \otimes P_i^{(n)} \tag{11.4}$$

where each $P_i^{(\cdot)}$ (for 'Pauli') is one of $\{I, X, Y, Z\}$, and s_i are numbers. This decomposition is possible because linear combinations of $\{I, X, Y, Z\}$ can be used to represent any 2 × 2 matrix. That is,

$$\frac{I+Z}{2} = \begin{bmatrix} 1 & 0 \\ 0 & 0 \end{bmatrix} \quad \frac{I-Z}{2} = \begin{bmatrix} 0 & 0 \\ 0 & 1 \end{bmatrix} \quad \frac{X+iY}{2} = \begin{bmatrix} 0 & 1 \\ 0 & 0 \end{bmatrix} \quad \frac{X-iY}{2} = \begin{bmatrix} 0 & 0 \\ 1 & 0 \end{bmatrix} \tag{11.5}$$

and so an appropriately weighted sum of these four terms can be made to equal any 2×2 matrix. Indeed, by taking an appropriate tensor product of these four operators, any single element of a matrix of any size can be 'picked out', and so every matrix can be decomposed in this way. It is further the case that every $s_i \in \{s_*\}$ is real for Hermitian matrices – for instance, a general 2 × 2 Hermitian matrix may be written (for real a, b, c, d)

$$\begin{bmatrix} a & b-ci \\ b+ci & d \end{bmatrix} = a\frac{I+Z}{2} + (b-ci)\frac{X+iY}{2} + (b+ci)\frac{X-iY}{2} + d\frac{I-Z}{2}$$

$$= \frac{a+d}{2}I + bX + cY + \frac{a-d}{2}Z \tag{11.6}$$

and this 'cancelling out' of imaginary terms by Hermitianity persists for any matrix size.

Of course, the linear array with only nearest-neighbour interactions is one of the simplest systems, and in practice even for Hamiltonians that pertain to local physical and chemical systems, there is usually less restriction on which qubits interact. For instance, in a two-dimensional grid, with nearest-neighbour interactions, each qubit interacts with four others, and similarly in a three-dimensional grid with nearest-neighbour interactions, each qubit interacts with six others. Another possibility is that a nearest-neighbour model is too restrictive to accurately capture the physics or chemistry of the system, and instead nearest-but-one neighbour interactions (in each direction) must be treated as non-negligible. More generally still, though useful for motivation, grid models (and lattice models in general) only apply to certain quantum mechanical systems (such as those in condensed matter physics) and for areas such as molecular chemistry, a more sophisticated representation is required. Hence, to keep things sufficiently general, we stick to the case where we only insist on a decomposition of the form Eqn. (11.2) such that each component Hamiltonian acts non-trivially on a constant number of qubits.

By way of two decompositions: one being the physical locality manifesting as a decomposition of the form given in Eqn. (11.2), and the other a decomposition into a sum of Pauli operators, as in Eqn. (11.4), Hamiltonians for a wide range of physical and chemical systems of real-world interest can be expressed as a sum of (an acceptably small number of) terms which are each a tensor product of Pauli operators. Moreover, for one form of Hamiltonian simulation in particular, namely Trotterisation, it is significant that any exponentiated tensor product of Pauli operators can be (efficiently) implemented as a quantum circuit. To demonstrate this, first consider a term of the form $e^{-iZ^{\otimes n}s}$ – that is, the Hamiltonian is a tensor product of Pauli Z operators, scaled by the real number s – which can be executed with the following 'Pauli-gadget' circuit:

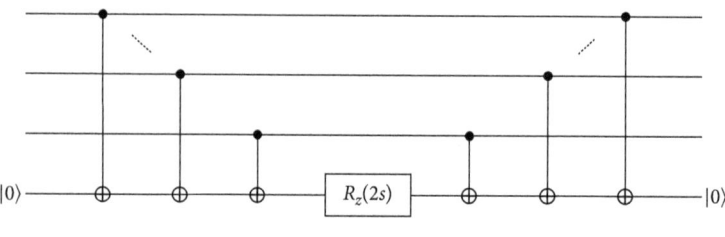

Figure 11.2

This has the claimed operation as $e^{-iZ^{\otimes n}s}$ is a diagonal matrix with e^{-is} on leading diagonal elements corresponding to computational basis states with an *even* number of ones when written in binary, and e^{is} on leading diagonal elements corresponding to computational basis states with an *odd* number of ones when written in binary. In the circuit, the bank of CNOTs has the effect of 'picking out' the computational basis states with an odd number of ones and applying the phase e^{is}, and applying the phase e^{-is} otherwise. The final bank of CNOTs performs the uncomputing to put the ancilla back to $|0\rangle$.

Turning to general tensor products of Pauli operators, note that: (i) when $P_i^{(\cdot)}$ is the identity, then the corresponding qubit can just be left alone; and (ii) Pauli-X and Y operators can easily be constructed from Pauli-Z operations, by the relations:

$$X = HZH; \quad Y = (HSH)^\dagger Z(HSH) = (HS^\dagger H)Z(HSH) \qquad (11.7)$$

from which the general Pauli-gadget circuit follows, by using the corresponding quantum gates to swap the qubit bases before the 'picking out' CNOTs, and also after the uncomputing CNOTs to return the state to the correct form. For example, the operator $e^{-i(X\otimes I\otimes Y\otimes Z)s}$ can be implemented by the circuit:

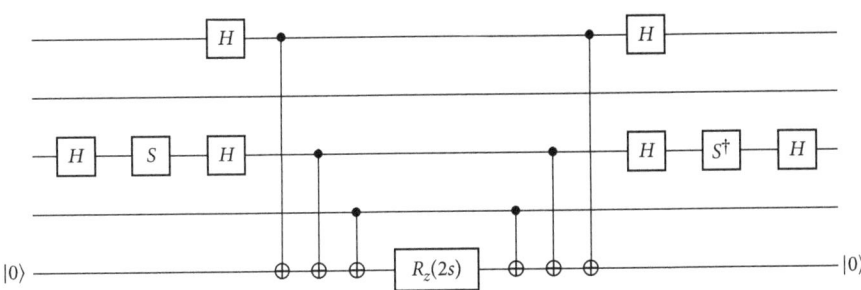

Figure 11.3

In this way, any Hamiltonian that has been decomposed into a sum of terms, each of which is a tensor product of Paulis (and the identity), can be exponentiated. Returning to the Schrödinger equation, this, with an appropriately decomposed Hamiltonian, gives

$$|\psi_t\rangle = e^{-iHt}|\psi_0\rangle = e^{-i\sum_l H_l t}|\psi_0\rangle \tag{11.8}$$

where each $e^{-iH_l t}$ can be implemented by way of a Pauli gadget. However, there is still a problem, as in general:

$$e^{-i\sum_l H_l t} \neq \prod_l e^{-iH_l t} \tag{11.9}$$

and so it is still not obvious how to *directly* take advantage of the fact that each $e^{-iH_l t}$ can implemented as a simple quantum circuit.

11.2 Simulating Hamiltonian evolution

In this section, we introduce two widely-used forms of Hamiltonian simulation, and for the first of these, Trotterisation, it turns out that the decomposition of H as $\sum_l H_l$ is the crucial property that enables Hamiltonian simulation on gate-based quantum computers.

11.2.1 Trotterisation

The *Trotter formula* is:

$$\lim_{m\to\infty} \left(\prod_l e^{-iH_l t/m}\right)^m = e^{-i\sum_l H_l t} \tag{11.10}$$

which can be shown by starting with the Taylor series definition of an exponentiated matrix:

$$e^{-iH_l t/m} = I - \frac{1}{m} iH_l t + \mathcal{O}\left(\frac{1}{m^2}\right)$$

$$\Rightarrow \prod_l e^{-iH_l t/m} = I - \frac{1}{m} i \sum_l H_l t + \mathcal{O}\left(\frac{1}{m^2}\right) \tag{11.11}$$

$$\Rightarrow \left(\prod_l e^{-iH_l t/m}\right)^m = I + \sum_{j=1}^m \binom{m}{j} \frac{1}{m^j} \left(-i \sum_l H_l t\right)^j + \mathcal{O}\left(\frac{1}{m}\right) \tag{11.12}$$

$\mathcal{O}\left(\frac{1}{m^2}\right)$ and $\mathcal{O}\left(\frac{1}{m}\right)$ have been used slightly informally here, and the terms gathered up in these are not simply numbers whose values vary with m, but rather matrices where the size of the coefficient multiplying the matrix has the specified asymptotic behaviour. For this to be meaningful, it is necessary that $||H_l t|| \leq 1$, for all l. The right-hand side of Eqn. (11.12) contains the term $\binom{m}{j}\frac{1}{m^j}$, which can be re-expressed

$$\binom{m}{j}\frac{1}{m^j} = \frac{m!}{j!(m-j)!}\frac{1}{m^j} = \frac{m(m-1)(m-2)\cdots(m-j+1)}{m^j}\frac{1}{j!} = \frac{1}{j!}\left(1 + \mathcal{O}\left(\frac{1}{m}\right)\right) \tag{11.13}$$

from which it follows that

$$\left(\prod_l e^{-iH_l t/m}\right)^m = I + \sum_{j=1}^m \frac{(-i\sum_l H_l t)^j}{j!}\left(1 + \mathcal{O}\left(\frac{1}{m}\right)\right) + \mathcal{O}\left(\frac{1}{m}\right)$$

$$= \sum_{j=0}^m \frac{(-i\sum_l H_l t)^j}{j!}\left(1 + \mathcal{O}\left(\frac{1}{m}\right)\right) + \mathcal{O}\left(\frac{1}{m}\right) \tag{11.14}$$

$$\Rightarrow \lim_{m \to \infty} \left(\prod_l e^{-iH_l t/m}\right)^m = \sum_{j=0}^\infty \frac{(-i\sum_l H_l t)^j}{j!}$$

$$= e^{-i \sum_l H_l t} \tag{11.15}$$

thus proving the Trotter formula (using the Taylor series of an exponentiated matrix).

In the case of simulation, it is always the case that the Hamiltonian evolution will continue only for a finite duration, say Δt, and so Eqn. (11.11) can be re-stated:

$$\prod_l e^{-iH_l \Delta t} = I - i \sum_l H_l \Delta t + \mathcal{O}((\Delta t)^2) \tag{11.16}$$

However, directly exponentiating the sum of Hermitian matrices, and expressing the Taylor series also gives:

$$e^{-i \sum_l H_l \Delta t} = I - i \sum_l H_l \Delta t + \mathcal{O}((\Delta t)^2) \tag{11.17}$$

and therefore:

$$\left| e^{-i\sum_l H_l \Delta t} - \prod_l e^{-iH_l\Delta t} \right| = \mathcal{O}((\Delta t)^2) \tag{11.18}$$

which is significant for proposing a Hamiltonian simulation algorithm.

An algorithm for Hamiltonian simulation

From the above approximation of the Hamiltonian evolution, an algorithm, most commonly known as *Trotterisation* but also as the *product formula* method of Hamiltonian simulation, immediately follows. Starting from a Hamiltonian, that can be decomposed into a sufficiently succinct sum of terms each of which can be exponentiated, and whose evolution is to be simulated, the following unitary operation (which can be executed on the quantum computer) can be defined:

$$U_{\Delta t} = e^{-iH_1\Delta t} e^{-iH_2\Delta t} \cdots e^{-iH_L\Delta t} \tag{11.19}$$

The idea is to simulate $|\tilde{\psi}_t\rangle \approx |\psi_t\rangle = e^{-iHt}|\psi_0\rangle$, by simulating many intervals of duration Δt in succession, with Δt chosen to be sufficiently small that the overall desired accuracy holds. The algorithm for this is given in Algorithm 11.1. (Note that by allowing Δt to be set in this way, it is always possible to enforce $||H\Delta t|| \leq 1$.)

Algorithm 11.1 Hamiltonian simulation by Trotterisation

Require: $U_{\Delta t}, t, |\psi_0\rangle$
1: **Initialise** $|\tilde{\psi}_0\rangle = |\psi_0\rangle; j = 0$
2: **while** $j\Delta t < t$ **do**
3:     $|\tilde{\psi}_{j+1}\rangle \leftarrow U_{\Delta t}|\tilde{\psi}_j\rangle$
4:     $j \leftarrow j + 1$
5: **end while**
6: **Return** $|\tilde{\psi}_t\rangle = |\tilde{\psi}_j\rangle$

As the transformation from $|\psi_0\rangle$ to $|\tilde{\psi}_t\rangle$ amounts to the coherent evolution of a quantum state, Algorithm 11.1 is essentially a description for how to build the quantum circuit:

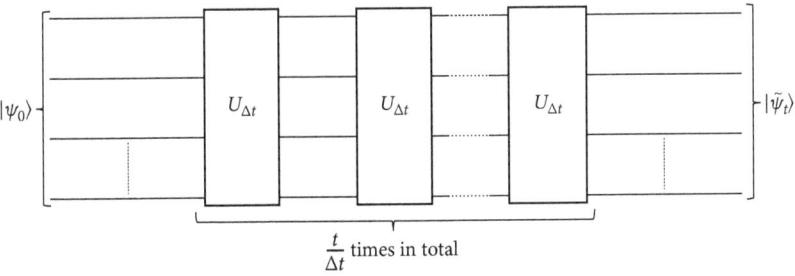

Figure 11.4

Error and run-time analysis

Different means of Hamiltonian simulation can be compared in terms of the total run-time required to simulate a Hamiltonian evolution of duration t, with error at most ϵ. According to Eqn. (11.18), the error grows as $\mathcal{O}((\Delta t)^2)$, or to be more precise, the coefficient of any erroneous unitary matrix term grows as $\mathcal{O}(\Delta t)^2$. Taking the operator norm as a suitable measure of the error (as in Section 2.4.9), which is sub-additive, and noting that the overall Hamiltonian evolution is simulated by a total of $t/\Delta t$ of these unitary blocks, the total error is therefore $\mathcal{O}\left(\frac{t}{\Delta t}(\Delta t)^2\right) = \mathcal{O}(t\Delta t)$. To express the run-time in the standard form, it is necessary to eliminate Δt, using the fact that the run-time is proportional to $t/\Delta t$, giving the error $\epsilon = \mathcal{O}(t^2/\text{run-time})$. This can be rearranged to give an expression of the run-time required to simulate the Hamiltonian evolution for time t, within maximum error ϵ:

$$\text{run-time} = \mathcal{O}\left(\frac{t^2}{\epsilon}\right) \tag{11.20}$$

A simple one-qubit example of Trotterisation

To see an example Trotterisation in action, consider a Hamiltonian evolved for a certain time such that $Ht = X + Z$. For a single-qubit example, it is easy enough to calculate the corresponding unitary operation directly:

$$e^{-i(X+Z)} = \begin{bmatrix} 0.156 - 0.699i & 0.699i \\ 0.699i & 0.156 + 0.699i \end{bmatrix} \tag{11.21}$$

which gives the exact solution to which a Trotterised approximation, $e^{-i(X+Z)} \approx (e^{-iX/m}e^{iZ/m})^m$, can be compared for various values of m (i.e., numbers of Trotter steps), as shown in Table 11.1. These tabulated results illustrate that increasing the number of Trotter steps suppresses the discrepancy between the exact unitary, and that synthesised with the simulated Hamiltonian evolution, as the theoretical results predict.

An even simpler two-qubit example of Trotterisation

As another example, consider a two-qubit Hamiltonian, evolved for a certain time such that $Ht = Z \otimes I + I \otimes X$, which means that the first qubit has Hamiltonian Pauli-Z and the second qubit has Hamiltonian Pauli-X. Again, two qubits is small enough to explicitly calculate and express the exponentiated matrix:

$$e^{-i(Z\otimes I + I\otimes X)} = \begin{bmatrix} 0.292 - 0.455i & -0.708 - 0.455i & 0 & 0 \\ -0.708 - 0.455i & 0.292 - 0.455i & 0 & 0 \\ 0 & 0 & 0.292 + 0.455i & 0.708 - 0.455i \\ 0 & 0 & 0.708 - 0.455i & 0.292 + 0.455i \end{bmatrix} \tag{11.22}$$

Table 11.1 An illustration of the increased accuracy of Trotterisation with number of Trotter steps, m, for the Hamiltonian evolution $e^{-i(X+Z)}$. 'Discrepancy' here is the absolute value of $e^{-i(X+Z)} - (e^{-iX/m}e^{iZ/m})^m$ on an element by element basis; even though this is not the formal way that 'closeness of matrices' is measured, presenting it in this way allows a nice visualisation of how the discrepancy shrinks with the number of Trotter steps.

Trotter steps (m)	Trotterised Unitary $((e^{-iX/m}e^{iZ/m})^m)$	Discrepancy from $e^{-i(X+Z)}$
1	$\begin{bmatrix} 0.292 - 0.455i & 0.708 - 0.455i \\ -0.708 - 0.455i & 0.292 + 0.455i \end{bmatrix}$	$\begin{bmatrix} 0.279 & 0.749 \\ 0.749 & 0.279 \end{bmatrix}$
2	$\begin{bmatrix} 0.186 - 0.648i & 0.345 - 0.648i \\ -0.345 - 0.648i & 0.186 + 0.648i \end{bmatrix}$	$\begin{bmatrix} 0.0588 & 0.358 \\ 0.358 & 0.0588 \end{bmatrix}$
3	$\begin{bmatrix} 0.169 - 0.677i & 0.234 - 0.677i \\ -0.234 - 0.677i & 0.169 + 0.677i \end{bmatrix}$	$\begin{bmatrix} 0.0252 & 0.235 \\ 0.235 & 0.0252 \end{bmatrix}$
⋮	⋮	⋮
10	$\begin{bmatrix} 0.157 - 0.697i & 0.0699 - 0.697i \\ -0.0699 - 0.697i & 0.157 + 0.697i \end{bmatrix}$	$\begin{bmatrix} 0.0022 & 0.0699 \\ 0.0699 & 0.0022 \end{bmatrix}$

It is interesting to observe that in this case just a single Trotter step suffices to exactly implement the Hamiltonian evolution. That is,

$$e^{-i(Z \otimes I + I \otimes X)} = e^{-iZ \otimes I} e^{-iI \otimes X} \tag{11.23}$$

(where the left-hand side is the Hamiltonian evolution itself, and the right-hand side can be thought of as an 'approximation' with one Trotter step). We are, therefore, prompted to ask, is this just a lucky break? In fact, the key property at work here is that $Z \otimes I$ and $I \otimes X$ commute (which can easily be verified), and this is an instance of the general fact that if matrices A and B commute, then $e^{A+B} = e^A e^B$. This can be seen using the Taylor series expansion:

$$e^{A+B} = \sum_{i=0}^{\infty} \frac{1}{i!}(A+B)^i$$

$$= \sum_{i=0}^{\infty} \frac{1}{i!} \sum_{j=0}^{i} \binom{i}{j} A^j B^{i-j}$$

$$= \sum_{i=0}^{\infty} \sum_{j=0}^{i} \frac{1}{(i-j)!j!} A^j B^{i-j}$$

$$= \sum_{j=0}^{\infty} \sum_{i=j}^{\infty} \frac{1}{(i-j)!j!} A^j B^{i-j}$$

$$= \sum_{j=0}^{\infty} \sum_{j'=0}^{\infty} \frac{1}{j'!j!} A^j B^{j'}$$

$$= \sum_{j=0}^{\infty} \frac{A^j}{j!} \sum_{j'=0}^{\infty} \frac{B^{j'}}{j'!}$$

$$= e^A e^B \qquad (11.24)$$

using the substitution $j' = i - j$. The requirement of commutation occurs in the second step, where all products of j lots of A with $i - j$ lots of B are grouped as $A^j B^{i-j}$.

11.2.2 Hamiltonian simulation with the truncated Taylor series

An alternative to Trotterisation is to use the Taylor series of the exponentiated Hamiltonian directly. The Taylor series of a Hermitian matrix is a sum of *Hermitian* matrices; however, in order to simulate the Hamiltonian evolution as follows, it is necessary to have a sum – or more properly linear combination – of *unitary* matrices. The result in Eqn. (11.4) provides one way to decompose any Hermitian matrix into a linear combination of unitary matrices, and this is once again used to achieve Hamiltonian simulation. If each term in the Taylor series of some Hermitian matrix can be decomposed into (at most) some L unitary matrices, this gives:

$$e^{-iHt} = \sum_{k=0}^{\infty} \frac{(-iHt)^k}{k!} = \sum_{k=0}^{\infty} \sum_{l=0}^{L-1} \tilde{b}_{k,l} \tilde{U}_{k,l} \qquad (11.25)$$

where $\tilde{U}_{k,l}$ are unitary.

The *infinite* summation in Eqn. (11.25) causes a problem, and to simulate the Hamiltonian evolution in this way, it must be truncated (to include only the terms $0, \ldots, K-1$). Once the summation has been truncated, it is convenient to re-write Eqn. (11.25) as a single summation:

$$e^{-iHt} \approx \sum_{k'=0}^{KL-1} b_{k'} U_{k'} \qquad (11.26)$$

To gain an intuitive grasp of how Hamiltonian simulation with a linear combination of unitaries (LCU) works, we consider a very simple example with two LCU terms:

$$e^{-iHt} \approx A = b_0 U_0 + b_1 U_1 \qquad (11.27)$$

In this case, the LCU can (up to a scale factor) be executed by the following circuit, with the measurement outcome 0 being post-selected:

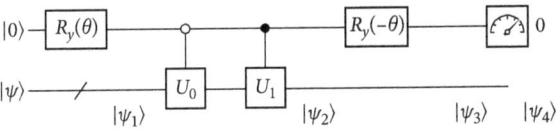

Figure 11.5

The value of θ is chosen such that $\cos^2(\theta/2) = \alpha b_0$ and $\sin^2(\theta/2) = \alpha b_1$, for some constant α. Stepping through the circuit, the $R_y(\theta)$ rotation transforms the state to:

$$|\psi_1\rangle = (\cos(\theta/2)|0\rangle + \sin(\theta/2)|1\rangle)|\psi\rangle \tag{11.28}$$

and subsequently, following the two controlled unitaries:

$$|\psi_2\rangle = \cos(\theta/2)|0\rangle(U_0|\psi\rangle) + \sin(\theta/2)|1\rangle(U_1|\psi\rangle) \tag{11.29}$$

which contains the two unitary terms to be linearly combined, but in the incorrect proportions and entangled with different states of the controlling qubit, which the second R_y gate remedies:

$$\begin{aligned}|\psi_3\rangle &= \cos(\theta/2)(\cos(\theta/2)|0\rangle - \sin(\theta/2)|1\rangle)(U_0|\psi\rangle) \\ &\quad + \sin(\theta/2)(\sin(\theta/2)|0\rangle + \cos(\theta/2)|1\rangle)(U_1|\psi\rangle) \\ &= |0\rangle\left(\cos^2(\theta/2)(U_0|\psi\rangle) + \sin^2(\theta/2)(U_1|\psi\rangle)\right) \\ &\quad + |1\rangle(\cos(\theta/2)\sin(\theta/2)(-(U_0|\psi\rangle) + (U_1|\psi\rangle)))\end{aligned} \tag{11.30}$$

Post-selecting on the measurement outcome 0 (that is, adopting a 'retry until success' strategy, where success is the measurement outcome zero) indeed yields the desired state (now omitting the first qubit):

$$|\psi_4\rangle = \alpha'(b_0 U_0|\psi\rangle + b_1 U_1|\psi\rangle) = \alpha'(b_0 U_0 + b_1 U_1)|\psi\rangle \tag{11.31}$$

for some constant, α', accounting for the renormalisation. As the operation $\alpha' A$ is applied to $|\psi\rangle$ when the first qubit is initialised as $|0\rangle$ and the measurement outcome 0 on the first qubit is post-selected, this means that the overall unitary operation of the circuit is

$$\begin{bmatrix} \alpha' A & \cdot \\ \cdot & \cdot \end{bmatrix} \tag{11.32}$$

(the elements denoted '$\cdot$' represent block matrices that are unimportant for our purposes). For this reason, we say that the (unitary matrix of the) circuit *block encodes* $\alpha' A$ – and block encoding is a central concept in Chapter 14. Its specific relation to LCU is further developed in the chapter problems.

Turning to the general case, where the LCU contains some KL terms, the essential principle illustrated in the simple example remains the same, but it is first necessary to introduce the notion of a *state preparation circuit*, denoted PREP (short for 'prepare'), that acts on $\lceil \log_2(KL) \rceil$ qubits and achieves the following:

$$\text{PREP} \left| 0^{\lceil \log_2(KL) \rceil} \right\rangle = \alpha \sum_{k'} \sqrt{b_{k'}} |k'\rangle \tag{11.33}$$

for some (new) constant α. The prepared state is then to used to 'pick out' each of the unitaries to be linearly combined, by way of multiply-controlled unitaries (variously controlled on one and zero). For example, for a three-term LCU, $\lceil \log_2 3 \rceil = 2$ qubits are needed (with one spare bit string), and an appropriate circuit is:

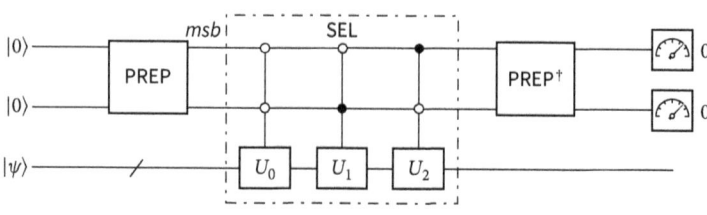

Figure 11.6

where '*msb*' means most significant bit and, as shown, the central block of the circuit is termed 'select' and denoted SEL. This has the desired effect

$$|00\rangle |\psi\rangle \xrightarrow{\text{PREP} \otimes I} \left(\alpha \sum_{k'} \sqrt{b_{k'}} |k'\rangle \right) |\psi\rangle$$

$$\xrightarrow{\text{SEL}} \alpha \sum_{k'} \sqrt{b_{k'}} |k'\rangle (U_{k'} |\psi\rangle)$$

$$\xrightarrow{\text{PREP}^\dagger \otimes I} \alpha |00\rangle \sum_{k'} b_{k'} (U_{k'} |\psi\rangle) + \ldots \tag{11.34}$$

which thus correctly prepares a state proportional to $\sum_{k'} b_{k'} U_{k'} |\psi\rangle$ when the measurement outcome zero is post-selected on each of the first two qubits. This approach generalises to any finite number of unitaries in the linear combination.

There are a few variations on exactly how the LCU circuit is constructed, for instance one alternative is to employ a state preparation circuit that uses *unary* encoding, PREP_U – in which case a total of KL qubits are required therein:

$$\text{PREP}_U |0\rangle = \alpha \sum_{k'=0}^{KL-1} \sqrt{b_{k'}} |0^{KL-k'-1} 1 0^{k'}\rangle \tag{11.35}$$

that is, the bit string corresponding to the k'th term is all zeros except for a single one at the $(k'+1)$th element. (Recall here that the exponent in the ket is read as 'how many times the symbol being exponentiated is repeated', so, for example, when $k' = 0$, then there are no trailing zeros.)

Fig. 11.6 can be adjusted to use PREP$_U$:

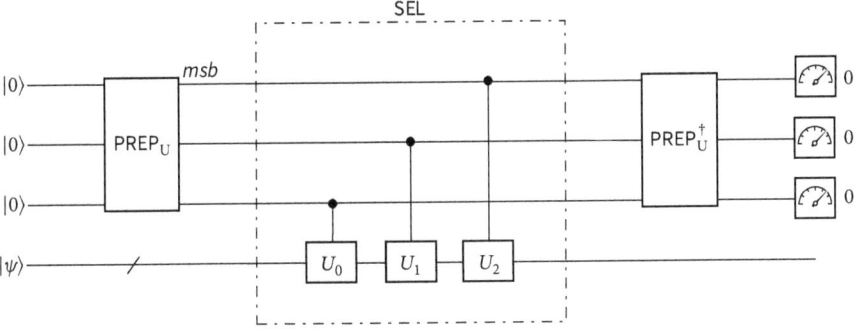

Figure 11.7

Stepping through the circuit in this case gives:

$$|000\rangle|\psi\rangle \xrightarrow{\text{PREP}_U \otimes I} \left(\alpha \sum_{k'} \sqrt{b_{k'}} |0^{KL-k'-1} 1 0^{k'}\rangle\right)|\psi\rangle$$

$$\xrightarrow{\text{SEL}} \alpha \sum_{k'} \sqrt{b_{k'}} |0^{KL-k'-1} 1 0^{k'}\rangle (U_{k'}|\psi\rangle)$$

$$\xrightarrow{\text{PREP}_U^\dagger \otimes I} \alpha |000\rangle \sum_{k'} b_{k'} (U_{k'}|\psi\rangle) + \ldots \quad (11.36)$$

and so, once again, appropriate post-selection yields the desired LCU applied to the input state, $|\psi\rangle$.

Error and run-time analysis

The run-time for simulating Hamiltonian evolution by a truncated Taylor series can be evaluated (or more precisely, bounded) for the case where the Hamiltonian is decomposed into a sum of tensor products of Paulis and unary encoding is used. This setting simplifies the analysis without losing anything essential. In this case, the PREP and PREP† blocks consist of $\mathcal{O}(KL)$ gates, and the SEL block consists of KL controlled Pauli operators. Recalling that L is the number of terms in the decomposition of the Hermitian matrix (i.e., a property of the physics of the system in question) and K is the number of terms retained in the truncated Taylor series, L can be taken as a constant (for a given system), but K depends on the required accuracy. In order to keep the error below a desired bound, it is necessary that the LCU method is not used to evolve the Hamiltonian for the entire time, but rather for a short interval Δt, chosen such that $\|H\|\Delta t < 1$. The error from truncating the Taylor series can be derived:

$$\left\|\sum_{k=0}^{K-1} \frac{(iH\Delta t)^k}{k!} - e^{iH\Delta t}\right\| = \left\|\sum_{k=K}^{\infty} \frac{(iH\Delta t)^k}{k!}\right\|$$

$$\leq \sum_{k=K}^{\infty} \frac{\|(iH\Delta t)^k\|}{k!}$$

$$= \sum_{k=K}^{\infty} \frac{\|H\Delta t\|^k}{k!}$$

$$= \sum_{k=K}^{\infty} \frac{(\|H\|\Delta t)^k}{k!}$$

$$= \frac{(\|H\|\Delta t)^K}{K!} \sum_{k'=0}^{\infty} \frac{(\|H\|\Delta t)^{k'}}{(K+k')!/K!}$$

$$\leq \frac{(\|H\|\Delta t)^K}{K!} \sum_{k'=0}^{\infty} \frac{(\|H\|\Delta t)^{k'}}{k'!}$$

$$= \frac{(\|H\|\Delta t)^K e^{\|H\|\Delta t}}{K!} \quad (11.37)$$

This shows why it is necessary that $\|H\|\Delta t < 1$, as this guarantees that the error decreases rapidly in K, the length of the truncated Taylor series. Fixing $\|H\|\Delta t$ as some constant, it suffices to choose some K, such that:

$$K = \mathcal{O}\left(\frac{\log(1/\epsilon)}{\log\log(1/\epsilon)}\right) \quad (11.38)$$

to approximate the Hamiltonian evolution up to a maximum error ϵ (see chapter problems).

To simulate the Hamiltonian evolution for a total time, t, a number of intervals of duration Δt are required. To keep Δt such that $\|H\|\Delta t < 1$ as is always required, the number of intervals is an integer that is upper-bounded by a term that grows proportionally with t (i.e., Δt only varies to ensure that the $t/\Delta t$ is an integer). We let t' be the number of intervals (as rationalised above $t' = \Theta(t)$), and furthermore, as the error is sub-additive (at worst the total error is the sum of those of the component intervals, as discussed below Eqn. (4.44)), to guarantee a *total* error no worse that ϵ (i.e., for the entire duration, t, of the simulated Hamiltonian evolution), it suffices to substitute ϵ/t' for ϵ in Eqn. (11.38). As justified in the opening paragraph of this run-time analysis, the number of gates and therefore run-time of a single interval of Hamiltonian simulation is K, giving a total run-time:

$$\text{run-time} = \mathcal{O}(t'K) = \mathcal{O}\left(t'\frac{\log(1/(\epsilon/t'))}{\log\log(1/(\epsilon/t'))}\right) = \mathcal{O}\left(t\frac{\log(t/\epsilon)}{\log\log(t/\epsilon)}\right) \quad (11.39)$$

One aspect of Hamiltonian simulation by truncated Taylor series that has been conspicuous by its absence in this run-time analysis is the fact that the post-selection condition must be achieved in practice by retrying until success. Ostensibly, this would appear to be particularly problematic as the number of qubits that have a post-selection condition is KL, and so if there is a *fixed* probability of getting the desired outcome, it would seem that the (expected) number of retries grows exponentially with KL.

Thankfully, this is not actually the case – as can be seen by returning to the block-encoded picture of the quantum circuit. The fact that there are multiple PREP qubits that must be prepared and post-selected in the $|0\rangle$ state does not alter the fact that the matrix A appears in the top-left block of the unitary pertaining to the entire circuit. In the case of Hamiltonian simulation, the goal is for A to approximate some exponentiated Hamiltonian, such that the matrix of the circuit is approximately $\begin{bmatrix} e^{-iHt} & \cdot \\ \cdot & \cdot \end{bmatrix}$. However, e^{-iHt} is itself unitary, and so if the top-left block is *exactly* unitary, the only way for the entire matrix to be unitary is if it is of the form $\begin{bmatrix} e^{-iHt} & 0 \\ 0 & \cdot \end{bmatrix}$. This means that the post-selection condition is always met (when the prepared state of the PREP qubits is $|0\rangle$). More generally, the more closely that the encoded block approximates a unitary, the higher the probability that the post-selection condition is met.

11.2.3 Simulation of sparse Hamiltonian evolution

The principal focus of this chapter is on the use of gate-based quantum computers to simulate physical and chemical quantum mechanical systems, and so it is natural that the local structure of the Hamiltonians of such systems has been leveraged in the simulation techniques. However, Hamiltonian simulation is of wider interest than just this application – in particular for the evolution of *sparse* Hermitian operators (for instance in quantum linear system solving, as in Chapter 13) – and so it is important to understand how well the Hamiltonian simulation methods perform more generally.

The distinction between locality and sparsity is somewhat subtle: Hamiltonians that are local, in the sense described in Section 11.1, can be decomposed into a sufficiently succinct summation of terms that are tensor products of Paulis and identities. The fact that each term acts non-trivially on only a constant number of qubits means that each matrix in the summation is sparse. However, the reverse implication does not hold: general sparse matrices do not necessarily have a decomposition as a succinct summation of terms that are tensor products of Paulis and identities. This is because the strategy of 'picking out' the non-zero elements individually relies on the fact that most of the qubits in a local system are not involved at all, and if the same technique of picking out the non-zero elements is applied to a general sparse $N \times N$ matrix, the each term will be of the form:

$$\bigotimes_{i=1}^{n} \frac{A_i \pm B_i}{2} \tag{11.40}$$

where $\bigotimes_{i=1}^{n}$ signifies an n-fold tensor product, $A_i \in \{I, X\}$, and $B_i \in \{Z, iY\}$. Once the tensor product is multiplied out, there will be a sum with a total of $2^n = N$ terms that are tensor products of Paulis and the identity. Naturally, 2^n constitutes a prohibitively large number of terms for *efficient* simulation, and so alternatives to the Pauli product decomposition – used in both Trotterisation and Hamiltonian simulation by the truncated Taylor series– must be sought.

In neither case, however, does this turn out to be fatal for general sparse Hamiltonian simulation. Indeed, in the case of Trotterisation, the condition of sparsity *is*

Table 11.2 Comparison of the run-time (complexity) of different methods of sparse Hamiltonian simulation. Note that for higher order product formulas, d must be even.

Simulation Method	Complexity
Trotterisation (first order product formula)	$\mathcal{O}(t^2/\epsilon)$
dth-order product formula	$\mathcal{O}\left(5^d t^{1+1/d}/\epsilon^{1/d}\right)$
LCU of Truncated Taylor series	$\mathcal{O}\left(\dfrac{t\log^2(t/\epsilon)}{\log\log(t/\epsilon)}\right)$
Quantum signal processing	$\mathcal{O}(t + \log(1/\epsilon))$

sufficient: even though the analysis herein has focused on Pauli gadgets for exponentiating local Hamiltonians (appropriately decomposed), other techniques exist to efficiently exponentiate 1-sparse Hamiltonians (that is, when each row and column has at most one non-zero entry), and so the same overall run-time can be achieved by different means. The same, however, in not true in the case of simulating the Hamiltonian by a truncated Taylor series, where the upshot is that the term, L, previously justified as a constant for a given Hamiltonian now varies with the evolution time, t, and maximum permissible error, ϵ. In this case, it turns out that the complexity can be improved to $\mathcal{O}(K \log L)$ and furthermore that it suffices to set $L \in \mathcal{O}(t/\epsilon)$, giving an overall complexity $\mathcal{O}\left(\frac{t\log^2(t/\epsilon)}{\log\log(t/\epsilon)}\right)$.

Table 11.2 summarises the run-time (complexity) of various means of simulating the evolution of sparse Hamiltonians. As well as Trotterisation and simulation by the truncated Taylor series, quantum signal processing and higher-order product formulas are also included. The former is analysed in detail in Chapter 14, whilst the latter is worthy of inclusion as it improves upon Trotterisation. Notably, one of the most unattractive features of Trotterisation is that the simulation time grows quadratically with the evolution time (by contrast, LCU of truncated Taylor series grows quasi-linearly, and quantum signal processing grows linearly). The *no fast forwarding principle* states that, even when restricted to sparse Hamiltonians, it is impossible to simulate Hamiltonian evolution faster than linearly in general, and so there is a premium on finding techniques that do achieve linear complexity in the evolution time.

11.3 Estimating the ground state energy of a Hamiltonian

Hamiltonian simulation has a prominent role in many algorithms for finding the *ground state energy* of a system. Let H be any Hamiltonian with spectral decomposition $\sum_i E_i |\lambda_i\rangle\langle\lambda_i|$, where $\{E_*\}$ are ordered from smallest to largest. Physically, the Hamiltonian is a representation of the energy of a quantum system, and its ground state energy is $\min_{|\psi\rangle} \langle \psi|H|\psi \rangle$ (this is an observable measurement, as introduced in Section 3.2.5).

$\{\lambda_*\}$ form an orthonormal basis (see Chapter 2 for full details) so any state $|\psi\rangle$ (of the appropriate dimension) can be written as a superposition in this basis.

$$|\psi\rangle = \sum_i a_i |\lambda_i\rangle \qquad (11.41)$$

Therefore:

$$\langle \psi|H|\psi\rangle = \left(\sum_i a_i^* \langle\lambda_i|\right)\left(\sum_i E_i|\lambda_i\rangle\langle\lambda_i|\right)\left(\sum_i a_i|\lambda_i\rangle\right) = \sum_i |a_i|^2 E_i \qquad (11.42)$$

For every quantum state, the normalisation is such that $\sum_i |a_i|^2 = 1$, and clearly $\forall i, |a_i|^2 \geq 0$, so Eqn. (11.42) is minimised when $a_0 = 1$ (recalling that E_i is ordered from smallest to largest). Therefore the ground state is obtained when $|\psi\rangle = |\lambda_0\rangle$, and the ground state energy itself is equal to E_0, the smallest eigenvalue of H.

11.3.1 Using quantum phase estimation to find the ground state energy

Finding the ground state energy of a Hamiltonian is a computationally hard problem in general on a classical computer; however, on a quantum computer, quantum phase estimation can be used. To do this, recall from Section 2.4.7 that for a Hamiltonian H with spectral decomposition $H = \sum_i E_i |\lambda_i\rangle\langle\lambda_i|$, the unitary $U = e^{2\pi i H}$ (which can be implemented using a suitable Hamiltonian simulation technique) also has eigenvectors $\{|\lambda_*\rangle\}$ with eigenvalues $\{e^{2\pi i E_*}\}$. Letting the second quantum phase estimation register be any quantum state, $|\psi\rangle$, expressed as a superposition of the eigenvectors of H,

$$|\psi\rangle = \sum_i a_i|\lambda_i\rangle \qquad (11.43)$$

and applying the quantum phase estimation circuit of Fig. 9.7 prepares the state (before the measurement):

$$\sum_i a_i |\tilde{E}_i\rangle |\lambda_i\rangle \qquad (11.44)$$

where $\tilde{E}_i$ is a t-bit approximation of the eigenvalue E_i.

Leaving the input $|\psi\rangle$ as an *arbitrary* quantum state means that the ground state energy problem is still not quite solved. Once the final measurements are made, for a general input state in the second register as expressed in Eqn. (11.43), the probability of collapsing into the desired state $|\tilde{E}_0\rangle$ is $|a_0|^2$. To remedy this, the second register should be initialised in a state that is dominated by $|\lambda_0\rangle$, i.e., such that $|a_0|$ is sufficiently large that the t-bit approximation of the ground state energy is obtained (with high probability) with a just few repeats of quantum phase estimation.

Such a requirement is perhaps unsurprising when put like this: in order to find the ground state *energy*, first a reasonable approximation of the ground state itself is required. The search for efficient methods of ground state preparation is a rich and very active research topic in its own right; one way is to adiabatically evolve from an easy to prepare ground state to the ground state of the Hamiltonian of interest (see Chapter 12 for details of adiabatic quantum evolution).

11.3.2 Ground state energy estimation in quantum chemistry

Calculating the ground state energy of a Hamiltonian is something that is done in many scientific disciplines; however, there is a particular focus on doing so for chemical Hamiltonians. Chemistry is the study of how atoms 'bond' to form molecules and crystals. A molecule or crystal will form when the motion of the electrons relative to each other and also relative to the positions of (positively-charged) nuclei of the constituent atoms is such that the electrical forces there between yield a stable (or meta-stable) equilibrium. As sub-atomic particles, electrons are poorly described by classical physics and should be treated as entities governed by the principles of quantum mechanics in any computational model that is fit for purpose. In other words, a molecule or crystal can only be adequately described by a quantum mechanical Hamiltonian.

Quantum chemistry is a wide-ranging and multi-faceted subject that seeks to probe and quantify myriad different properties of chemicals and materials. The ground state energy is just one of the properties one may wish to quantify regarding some specific chemical Hamiltonian, but it is a particularly important property because ground state energies can be used to predict which chemical reactions are likely to occur. For even relatively modest sized molecules, finding the ground state energy is a hard problem, for the simple reasons that classical computers run out of memory before the ground state and Hamiltonian can even be encoded in full therein. This has catalysed substantial interest in quantum computing for quantum chemistry. Indeed, as innately quantum systems, simulation of chemicals fits perfectly with Feynman's picture of the principal purpose a quantum computer (see Chapter 1).

11.4 The variational quantum eigensolver

Using quantum phase estimation to calculate ground state energies in quantum chemistry is expected to require a *full-scale error-corrected quantum computer*. However, there has been a great deal of effort expended attempting to instead use resource-constrained quantum computers to find ground state energies of Hamiltonians, and these efforts have typically centred on the idea that only shallow-depth quantum circuits, in which only a small of error is expected, may be used. This has, in turn, led to 'hybrid' quantum-classical approaches, the most famous of which is the *variational quantum eigensolver*. The variational quantum eigensolver tackles the problem of computing the ground state energy by using the Rayleigh–Ritz variational principle:

$$\langle \psi(\theta)|H|\psi(\theta)\rangle \geq E_0 \qquad (11.45)$$

where $|\psi(\theta)\rangle$ is a quantum state parameterised by θ. This bound can be shown to hold by expressing $|\psi(\theta)\rangle = \sum_i a_i |\lambda_i\rangle$, i.e., as a superposition of the eigenstates of H:

$$\begin{aligned}
\langle \psi(\theta)|H|\psi(\theta)\rangle &= \sum_{i,j} a_i^* \langle \lambda_i | H \, a_j | \lambda_j \rangle \\
&= \sum_{i,j} a_i^* \langle \lambda_i | E_j a_j | \lambda_j \rangle \\
&= \sum_i |a_i|^2 E_i \langle \lambda_i | \lambda_i \rangle \\
&\geq E_0 \sum_i |a_i|^2 \\
&= E_0 \qquad (11.46)
\end{aligned}$$

which uses the fact that any orthonormal basis is such that $\langle \lambda_i | \lambda_j \rangle = 0$ when $i \neq j$. The parameterised definition of the state $|\psi(\theta)\rangle$ is deliberately suggestive: the variational quantum eigensolver aims to converge on the ground state energy by finding the value of parameters that minimise $\langle \psi(\theta)|H|\psi(\theta)\rangle$. Specifically, the variational quantum eigensolver is a hybrid quantum–classical algorithm that uses a *parameterised quantum circuit* (sometimes known as a *variational quantum circuit* or *ansatz* – the latter especially in quantum chemistry) as a subroutine called by a classical optimisation loop. Algorithmically, the variational quantum eigensolver is an iterating loop, as shown in Algorithm 11.2 and Fig. 11.8.

Algorithm 11.2 Variational Quantum Eigensolver

Require: PQC $U(\theta)$ (with suitably initialised θ), H
1: **while** Convergence condition not satisfied **do**
2:     Run a shallow-depth quantum circuit $U(\theta) : |0\rangle \xrightarrow{U(\theta)} |\psi(\theta)\rangle$ to prepare $|\psi(\theta)\rangle$
3:     Measure to estimate the observable $E(\theta) = \langle \psi(\theta)|H|\psi(\theta)\rangle$.
4:     Perform classical optimisation to update θ aiming to reduce $\langle \psi(\theta)|H|\psi(\theta)\rangle$.
5: **end while**
6: **Return** $E(\theta) \approx E_0$

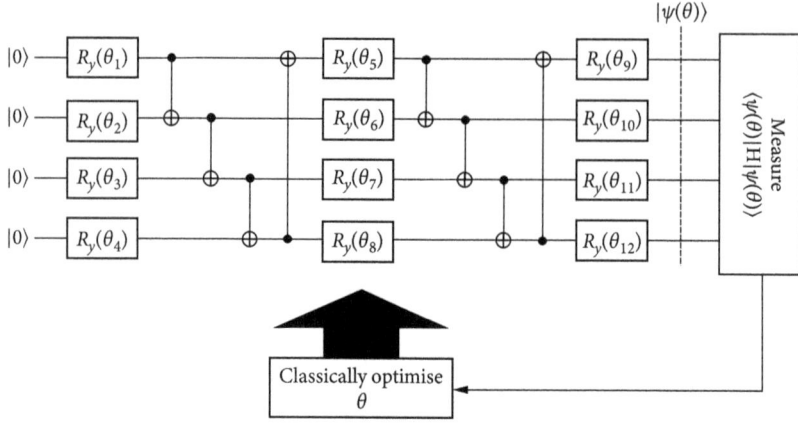

Figure 11.8

Unlike for the case of using quantum phase estimation for ground state energy calculation, the variational quantum eigensolver does not need an input which is the ground state itself – the variational quantum eigensolver finds the ground state *and* the ground state energy. However, whilst this is an ostensibly attractive feature, the diagram of the operation of the variational quantum eigensolver (Fig. 11.8) also prompts a number of important questions about its practicality: is there any guarantee that the parameterised quantum circuit is such that there even exists parameter values that prepare the desired state; what does it actually mean to measure an observable in practice; and is there an overall quantum advantage? Taking these questions in turn: for applications in quantum chemistry, it is generally important that the ansatz circuit is appropriately 'matched' to the physical system from which the (qubit) Hamiltonian derives. There is a growing body of literature on how to achieve this, whilst maintaining the necessary shallow circuit depth. On a tangential note, the circuit structure in Fig. 11.8 is unlikely to be used for chemical applications. Rather this is the sort of *hardware efficient ansatz* structure, widely-used in quantum machine learning, (see Box 11.1), with alternating blocks of parameterised rotation gates and entangling gates – where the latter respects the layout of the physical device, so swap gates are not needed. (For example, in Fig. 11.8, it can be seen that the blocks of CNOTs can be implemented directly without swap gates if the four qubits are joined in a ring.)

The second question, above, amounts to how one can obtain the measured observable value using computational basis measurement (which may conservatively be assumed to be the unique measurement available 'natively' on the quantum hardware). One way to achieve this again uses the property that every $2^n \times 2^n$ Hermitian matrix can be expressed as a sum of terms, each of which is an n-fold tensor product of the Pauli matrices and the 2×2 identity.

Box 11.1 Parameterised quantum circuits as machine learning models

Perhaps more than any other sub-field of quantum computing, quantum machine learning defies pigeon-holing as a 'special-case' or 'variation' of any other established branch of the subject. However, there are broadly two approaches: one uses quantum linear algebra to enhance machine learning, and the other stems from the functional similarity between parameterised quantum circuits and artificial neural networks (ANNs). Indeed, viewing the diagram of the VQE in Fig. 11.8 without knowing that context, one may be struck by the similarity between the training of ANNs and the variational tuning of parameterised quantum circuits (PQCs). Although PQCs have been variously proposed as machine learning (ML) models for use in most ML tasks, the analogy is strongest when thinking about generative models, for the simple reason that quantum measurement is intrinsically a form of sampling. In a classical generative model, an ANN provides a map from the latent space (a uniformly random bit string, or possible a sample from a Gaussian distribution) to a sample from a target distribution that has been trained. In particular, training involves automated tuning of the ANN neuron weights such that the distribution sampled from converges to the target distribution. A quantum generative model is a near exact analogue, where the parameterised gate values replace the neuron weights, and there is no need for latent space, as the PQC prepares a state whose squared amplitudes provide the distribution that is sampled from.

Notably, there is strong theoretical evidence that there exist certain probability distributions from which samples can be prepared in polynomial time quantumly but not classically (see Chapter 6). Or to put it another way, it is widely believed that a modest size PQC can prepare certain samples that cannot possibly be prepared by any reasonably-sized ANN. However, questions remain about whether such distributions are relevant to any real-world application, and in general the barren plateau problem (see Section 11.5) still persists when PQCs are trained as machine learning models.

Using the sum of n-fold tensor products of Paulis in place of the Hermitian matrix (i.e., as given in Eqn. (11.4)) in the observable measurement:

$$\langle\psi(\theta)|H|\psi(\theta)\rangle = \langle\psi(\theta)| \left(\sum_i s_i P_i^{(1)} \otimes P_i^{(2)} \otimes \cdots \otimes P_i^{(n)} \right) |\psi(\theta)\rangle$$

$$= \sum_i s_i \langle\psi(\theta)| \left(P_i^{(1)} \otimes P_i^{(2)} \otimes \cdots \otimes P_i^{(n)} \right) |\psi(\theta)\rangle \quad (11.47)$$

each term of which can be estimated individually, and then summed to get the desired observable measurement outcome. Once again, physical locality can be used to argue that for many practical situations of interest, the number of terms in this summation will be small enough to perform the decomposed measurement. Measurement of each term can be achieved using computational basis measurements, as whenever $P_i^{(\cdot)}$ is X or Y, a unitary transform can be applied to transform the state such that the term in the observable is Z: using the relations given in Eqn. (11.7), Eqn. (11.47) can be re-written:

$$\sum_i s_i \langle \psi(\theta) | \left(P_i^{(1)} \otimes P_i^{(2)} \otimes \cdots \otimes P_i^{(n)} \right) | \psi(\theta) \rangle$$

$$= \sum_i s_i \langle \psi(\theta) | U_i^\dagger \left(\tilde{P}_i^{(1)} \otimes \tilde{P}_i^{(2)} \otimes \cdots \otimes \tilde{P}_i^{(n)} \right) U_i | \psi(\theta) \rangle$$

$$= \sum_i s_i \langle \tilde{\psi}_i(\theta) | \left(\tilde{P}_i^{(1)} \otimes \tilde{P}_i^{(2)} \otimes \cdots \otimes \tilde{P}_i^{(n)} \right) | \tilde{\psi}_i(\theta) \rangle \qquad (11.48)$$

where for each i, U_i is a tensor product of terms which are either the identity, H or HSH and $|\tilde{\psi}_i(\theta)\rangle = U_i |\psi(\theta)\rangle$ such that $\{\tilde{P}_i^{(*)}\}$ are now all either the identity or Z.

Focusing on a single term of the sum in Eqn. (11.48), i.e., a single value of i, the observable $\tilde{P}_i^{(1)} \otimes \tilde{P}_i^{(2)} \otimes \cdots \otimes \tilde{P}_i^{(n)}$ is a tensor product of terms which are either the identity or Pauli-Z, and so is diagonal in the computational basis. This means that its eigenvectors are the computational basis states, and each eigenvalue is either 1 or −1. In particular, for computational basis states that have an odd number of ones corresponding to qubits where $\tilde{P}_i^{(\cdot)} = Z$, the eigenvalue is −1, and the eigenvalue is 1 otherwise. (In the chapter problems, you are asked to verify this for a simple example.) This means that the observable expectation can be numerically estimated by averaging over a large number of computational basis measurements of the qubits where $\tilde{P}_i^{(\cdot)} = Z$; interpreting the outcome as −1 when a bit string with an odd number of ones is obtained, and 1 otherwise. To summarise, it is possible to measure any state $|\psi\rangle$ for any observable H using only computational basis measurements, by performing the steps given in Algorithm 11.3.

Algorithm 11.3 Observable measurement

1: Decompose H as a sum of tensor products of Paulis and the identity as in Eqn. (11.47).
2: **for** Nshots **do**
3: Apply a layer of single-qubit gates from $\{I, H, HSH\}$ to the prepared state such that the observable measurement amount to a sum of observables, each of which is a tensor product of Pauli Zs and identities, as in Eqn. (11.48).
4: For each qubit corresponding to a Pauli Z in the observable, measure in the computational basis. If a bit string with an odd number of ones is obtained, interpret the outcome as −1; and 1 otherwise.
5: **end for**
6: Average the outcomes to get the observable measurement outcome.

The final question raised by the discussion of the variational quantum eigensolver is whether there is an overall quantum advantage, and it is worth broadening this question to cover the other techniques in this chapter too.

11.5 Is there a quantum advantage available in Hamiltonian simulation and ground state energy estimation?

The question of whether there is a quantum advantage available in Hamiltonian simulation amounts to the question of whether there are Hamiltonians that can be hard (that is, *exponentially* hard) to simulate on classical computers, but easy to simulate on quantum computers. As every quantum circuit is, in a sense, nothing more than a certain Hamiltonian evolution, applied to a suitable initial state, it is an uncontroversial and widely-held view that the answer is 'yes' – there is a quantum advantage available. Moreover, the fact that there is no efficient classical algorithm available for (for example) ground state energy estimation is taken as good reason to investigate quantum computational approaches; and the fact that the quantum algorithm involves the transformation of highly-entangled quantum states is taken as reason to believe that no efficient classical algorithm *can* exist. This idea deserves to be emphasised. Computer science (in very rough terms) tends to deal with 'classical problems' (such as factoring) where quantum computation merely provides (or is conjectured to provide) a clever shortcut, thus meaning that an elaborate system of computational complexity classes is needed to establish beliefs around their classical tractability. By contrast, in scientific computing, the 'natively quantum' nature of some problems (such as Hamiltonian simulation) is often taken as sufficient to assert that quantum computation is necessary unless shown otherwise. To put it another way, Feynman's original assertion that the simulation of quantum mechanical systems necessarily requires some sort of quantum mechanical computer remains as relevant today as ever.

This means that describing the 'source of quantum advantage' in Hamiltonian simulation is not as cumbersome as for other algorithms and subroutines. Quantum states are described by vectors in Hilbert space – which, therefore, grow exponentially with the problem (system) size – and quantum computers are uniquely placed to manipulate these states. Moreover, some of these manipulations amount to finding useful quantities, such as the ground state energy. That is not to say that there has been no work to investigate how well classical computers can do at such problems. Classical approaches to quantum chemistry are very mature, and in some cases the approximations that must necessarily be made still enable useful solutions to be discovered. Moreover, an important question in general is when, exactly, does the vague assertion that 'manipulating highly-entangled quantum states must require a quantum computer' actually apply? Advances in this direction have brought the region of quantum advantage in Hamiltonian simulation into better focus.

Nevertheless, it remains the case that the problem of finding the ground state energy for real-world systems of interest is often hard for classical computers, and this was a crucial factor in the widespread early enthusiasm for the variational quantum eigensolver. That the state prepared by the variational quantum eigensolver is out of reach classically – even if finding it requires an iterative training loop – was taken as sufficient reason that the variational quantum eigensolver would provide a quantum advantage. However, more detailed analysis since that early enthusiasm has revealed that the training complexity should be neglected at one's peril. In the worst case, the training is NP-hard, and perhaps even more significantly, it has been theoretically established and experimentally verified that the cost function landscape in most variational quantum eigensolver problems is ridden with *barren plateaus* – regions where the cost function is almost entirely flat, punctuated by *narrow gorges* in which the optimal solution is found – rendering optimisation thereof practically impossible. For this reason, even though the variational quantum eigensolver will always have the distinction of being the first genuine quantum algorithm that was widely used by experimentalists on real hardware (albeit only for small problem sizes), many practitioners are doubtful that it will scale to provide useful quantum advantage for real-world problems.

Chapter problems

1. Verify that the circuit in Figure 11.3 correctly executes $e^{-i(X \otimes I \otimes Y \otimes Z)s}$
2. (a) Show that:

$$e^{-i(H_1+H_2)\Delta t} = e^{-iH_1 \Delta t} e^{-iH_2 \Delta t} + \mathcal{O}(\Delta t^2) \qquad (11.49)$$

  (b) Show that a more accurate simulation is obtained if instead

$$e^{-i(H_1+H_2)\Delta t} \approx e^{-iH_1 \Delta t/2} e^{-iH_2 \Delta t} e^{-iH_1 \Delta t/2} \qquad (11.50)$$

  is used, and give the error.
3. If quantum phase estimation is performed on a n-qubit Hamiltonian to estimate its ground state energy, and the second register is prepared as a uniform superposition of the eigenvectors of the Hamiltonian, what is the probability that a single run of quantum phase estimation returns the ground state?
4. Let A of Eqn. (11.27) be such that:

$$A = X + Z \qquad (11.51)$$

  i.e., $b_0 = b_1 = 1$ and $U_0 = X$, $U_1 = Z$.
  (a) Suggest an alternative to the $R_y(.)$ gate to do the prepare circuit in this case.
  (b) Show that the circuit in Figure 11.5 *block encodes* A, that is $\tilde{b}A = (\langle 0| \otimes I_2) U(|0\rangle \otimes I_2)$ and find the value of $\tilde{b}$.

(c) What is the probability that the post-selection condition is met for any single run of the circuit?

(d) Design a circuit using unary encoding to perform this LCU, and confirm that it has the correct operation.

5. As $\frac{(\|H\|\Delta t)^K e^{\|H\|\Delta t}}{K!}$ is an upper-bound on the error, as in Eqn. (11.37), and $\|H\|\Delta t < 1$ is a constant, show that it suffices to set

$$K = \mathcal{O}\left(\frac{\log(1/\epsilon)}{\log\log(1/\epsilon)}\right) \quad (11.52)$$

to upper-bound the error by ϵ.

6. Consider the observable $H = Z \otimes I \otimes Z$.

(a) Express the matrix for H.

(b) Give the eigenvalues and eigenvectors of H and verify that the claimed (below Fig. 11.2) connection between the eigenvectors and eigenvalues holds in this case.

Further reading

The notion that universal quantum computers may be capable of quantum simulation can be traced back to Feynman [1982], and it was Seth Lloyd [1996] and Daniel Abrams [1997] who fleshed out the idea in full and showed it to be realisable. The Trotter formula is due to Hale Trotter [1959] and was also proven by Chernoff [1968]. Hamiltonian simulation by a truncated Taylor series was proposed by Berry et al [2015], and the complexity analysis herein is a simplified version of that in this paper; Childs and Wiebe [2012] had earlier proposed the use of an LCU for Hamiltonian simulation. The no fast-forwarding principle is well-known within the quantum computing community and appears, for example, in Berry [2006]. More generally, this chapter has only scratched the surface of the vast fields of Hamiltonian simulation and quantum chemistry and Childs et al [2018] provide a nice overview of the former (indeed, Table 11.2 is a simplified version of [Childs et al 2018, Table 1]), whilst McArdle et al [2020] provide a comprehensive overview of the latter. The Jordan–Wigner transform is due to Jordan and Wigner [1928], and Michael Nielsen [2005] provides a nice explanation thereof – particularly for computer scientists.

The variational quantum eigensolver was proposed by Peruzzo et al [2014]; however, the origins of the notion of using PQCs as machine learning models is somewhat obscure. Benedetti et al [2019b] wrote one of the earliest papers and provided a comprehensive set of citations of all earlier works hinting at the idea, whilst Benedetti et al [2019a] and Liu and Wang [2018] pioneered the use of PQCs as generative models. That PQC training landscapes suffer from barren plateaus was discovered by McClean et al [2018]; and Bittel and Kliesch [2021] proved that training variational quantum algorithms is NP-hard.

12
Quantum optimisation

The preceding chapters have focused almost exclusively on quantum circuits as the model of computation, as gate-based quantum computing has emerged as the prevailing approach to building real-life quantum computers, and the quantum circuit model provides a way of reasoning about quantum algorithms that is both intuitive and amenable to formal analysis. However, for one particular application – namely optimisation – it is more natural to start from the adiabatic model of quantum computation.

12.1 The quantum adiabatic theorem

Adiabatic quantum computation depends on the *quantum adiabatic theorem*, which, in turn, requires the introduction of the *time-dependent* Schrödinger equation:

$$i\frac{d|\psi(t)\rangle}{dt} = H(t)|\psi(t)\rangle \qquad (12.1)$$

The quantum adiabatic theorem. *Let $H(t)$ be a time-varying Hamiltonian, which is initially H_0 at $t = 0$ and subsequently H_f at some later time, $t = T$; then, if the system is initially in the ground state of H_0, and as long as the time evolution of the Hamiltonian is sufficiently slow, the state is likely to remain in the ground state throughout the evolution, therefore being in the ground state of H_f at $t = T$.*

The quantum adiabatic theorem can be proven using the property that at all times, the *instantaneous* Hamiltonian is still Hermitian and, therefore, has a corresponding decomposition into instantaneous eigenstates $|\lambda_0(t)\rangle, |\lambda_1(t)\rangle, \ldots$, with corresponding (real) eigenvalues, $\lambda_0(t), \lambda_1(t), \ldots$. In the following, it is important that the eigenvalues are always ordered in increasing value, so λ_0 is the smallest eigenvalue (i.e., the ground state energy). At any time, the state can, therefore, be written in terms of the instantaneous eigenstates,

$$|\psi(t)\rangle = \sum_i c_i(t) e^{i\theta_i(t)} |\lambda_i(t)\rangle \qquad (12.2)$$

where $\theta_i = -\int_0^t \lambda_i(\tau) d\tau$. Note that this expression is without loss of generality as $c_i(t)$ has not yet been specified and is defined this way to make the subsequent analysis a little more straightforward. Eqn. (12.2) can be substituted into the time-dependent

Schrödinger equation (12.1) also taking the inner product of each side with some eigenstate, $|\lambda_j(t)\rangle$. Starting with the left-hand side of Eqn. (12.1):

$$\langle\lambda_j(t)|\left(i\frac{d\left(\sum_i c_i(t)e^{i\theta_i(t)}|\lambda_i(t)\rangle\right)}{dt}\right)$$

$$= i\langle\lambda_j(t)|\left(\sum_i \frac{dc_i(t)}{dt}e^{i\theta_i(t)}|\lambda_i(t)\rangle + ic_i(t)\frac{d\theta_i(t)}{dt}e^{i\theta_i(t)}|\lambda_i(t)\rangle + c_i(t)e^{i\theta_i(t)}\frac{d|\lambda_i(t)\rangle}{dt}\right)$$

$$= i\langle\lambda_j(t)|\left(\sum_i \frac{dc_i(t)}{dt}e^{i\theta_i(t)}|\lambda_i(t)\rangle - ic_i(t)\lambda_i(t)e^{i\theta_i(t)}|\lambda_i(t)\rangle + c_i(t)e^{i\theta_i(t)}\frac{d|\lambda_i(t)\rangle}{dt}\right)$$

(12.3)

where the final equality following from the definition of $\theta_i(t)$ being such that $\frac{d\theta_i(t)}{dt} = -\lambda_i(t)$. For the right-hand side of Eqn. (12.1):

$$\langle\lambda_j(t)|H(t)|\psi(t)\rangle = \langle\lambda_j(t)|\left(\sum_i \lambda_i(t)c_i(t)e^{i\theta_i(t)}|\lambda_i(t)\rangle\right) \quad (12.4)$$

which is such that the right-hand side is exactly the central term of Eqn. (12.3). Equating the left- and right-hand sides (i.e., Eqns. (12.3) and (12.4)), therefore, gives:

$$i\langle\lambda_j(t)|\left(\sum_i \frac{dc_i(t)}{dt}e^{i\theta_i(t)}|\lambda_i(t)\rangle + c_i(t)e^{i\theta_i(t)}\frac{d|\lambda_i(t)\rangle}{dt}\right) = 0 \quad (12.5)$$

$$\implies \frac{dc_j(t)}{dt} = -\sum_i c_i(t)e^{i(\theta_i(t)-\theta_j(t))}\langle\lambda_j(t)|\frac{d|\lambda_i(t)\rangle}{dt} \quad (12.6)$$

the right-hand side of which can be expressed in terms of the original Hamiltonian by differentiating both sides of the equality $H(t)|\lambda_i(t)\rangle = \lambda_i(t)|\lambda_i(t)\rangle$ with respect to time, and again taking the inner product with $|\lambda_j(t)\rangle$:

$$\langle\lambda_j(t)|\left(\frac{dH(t)}{dt}|\lambda_i(t)\rangle + H(t)\frac{d|\lambda_i(t)\rangle}{dt}\right) = \langle\lambda_j(t)|\left(\frac{d\lambda_i(t)}{dt}|\lambda_i(t)\rangle + \lambda_i(t)\frac{d|\lambda_i(t)\rangle}{dt}\right) \quad (12.7)$$

which, when $i \neq j$, gives:

$$\langle\lambda_j(t)|\frac{dH(t)}{dt}|\lambda_i(t)\rangle + \lambda_j(t)\langle\lambda_j(t)|\frac{d|\lambda_i\rangle}{dt} = \lambda_i(t)\langle\lambda_j(t)|\frac{d|\lambda_i(t)\rangle}{dt} \quad (12.8)$$

$$\implies \langle\lambda_j(t)|\frac{d|\lambda_i(t)\rangle}{dt} = \frac{\langle\lambda_j(t)|\frac{dH(t)}{dt}|\lambda_i(t)\rangle}{\lambda_i(t) - \lambda_j(t)} \quad (12.9)$$

The adiabatic condition can now be stated as follows: when the Hamiltonian varies slowly relative to the minimum difference between any pair of eigenvalues, then the right-hand side of Eqn. (12.9) is approximately equal to zero. Recalling that this holds for all $i \neq j$ and substituting into Eqn. (12.6) gives:

The quantum adiabatic theorem

$$\frac{dc_j(t)}{dt} \approx -c_j(t)e^{i(\theta_j(t)-\theta_j(t))}\langle\lambda_j(t)|\frac{d|\lambda_j(t)\rangle}{dt} = -c_j(t)\langle\lambda_j(t)|\frac{d|\lambda_j(t)\rangle}{dt} \quad (12.10)$$

It can be shown that $\langle\lambda_j(t)|\frac{d|\lambda_j(t)\rangle}{dt}$ must be imaginary by the following argument, which takes as a starting point that even though the instantaneous eigenstates vary with time, the fact that they are unit vectors means that their moduli do not vary with time:

$$0 = \frac{d(\langle\lambda_j(t)|\lambda_j(t)\rangle)}{d(t)}$$

$$= \frac{d(\langle\lambda_j(t)|)}{d(t)}|\lambda_j(t)\rangle + \langle\lambda_j(t)|\frac{d(|\lambda_j(t)\rangle)}{d(t)}$$

$$= \left(\langle\lambda_j(t)|\frac{d(|\lambda_j(t)\rangle)}{d(t)}\right)^* + \langle\lambda_j(t)|\frac{d(|\lambda_j(t)\rangle)}{d(t)}$$

$$= 2\mathrm{Re}\left(\langle\lambda_j(t)|\frac{d(|\lambda_j(t)\rangle)}{d(t)}\right) \quad (12.11)$$

Using the fact that $\langle\lambda_j(t)|\frac{d|\lambda_j(t)\rangle}{dt}$ is imaginary, Eqn. (12.10) can be integrated to give

$$c_j(t) \approx c_j(0)e^{i\gamma_j(t)} \quad (12.12)$$

where $\gamma_j(t) = i\int_0^t \langle\lambda_j(\tau)|\frac{d|\lambda_j(\tau)\rangle}{d\tau}d\tau$. When the adiabatic condition holds, and so the approximation in Eqn. (12.12) is valid, we can see that the magnitude of each coefficient, c_j, is approximately constant throughout the (adiabatic) Hamiltonian evolution. This actually constitutes a stronger result than is required for adiabatic quantum computation, in which the initial state is by design the ground state. In this case, the adiabatic condition can be correspondingly weakened to just requiring that the difference between the smallest two eigenvalues – the spectral gap – is large relative to the time variation of the Hamiltonian.

To see how large the spectral gap – or put more usefully, how slow the time evolution – must be, we return to Eqn. (12.6) and consider the case where $c_0(0) = 1$, i.e., the initial state is a ground state, which gives:

$$\frac{dc_j(t)}{dt} = -c_0(t)e^{i(\theta_0(t)-\theta_j(t))}\langle\lambda_j(t)|\frac{d|\lambda_0(t)\rangle}{dt}$$

$$= c_0(t)e^{i(\theta_0(t)-\theta_j(t))}\frac{\langle\lambda_j(t)|\frac{dH(t)}{dt}|\lambda_0(t)\rangle}{\lambda_j(t) - \lambda_0(t)} \quad (12.13)$$

The next step is to integrate Eqn. (12.13) for a short time from $t = 0$ to $t = T$, during which the eigenvalues remain approximately constant as does $\frac{dH(t)}{dt}$. By the premise, $c_0(0) = 1$, and the goal is to quantify how much 'leakage' there is into the first 'excited' state (i.e., with that coefficient c_1). For this, we use

$$\theta_0(t) - \theta_1(t) = -\int_0^t \lambda_0(\tau)d\tau + \int_0^t \lambda_1(\tau)d\tau = \int_0^t \lambda_1(\tau) - \lambda_0(\tau)d\tau = \Delta t \quad (12.14)$$

where $\Delta = \lambda_1 - \lambda_0$. Substituting this into Eqn. (12.13) for the case $j = 1$ gives

$$c_1(T) = \int_0^T c_0(t) e^{i\Delta t} \frac{\langle \lambda_1(t)| \frac{dH(t)}{dt} |\lambda_0(t)\rangle}{\Delta} dt$$

$$= i \left(c_0(0) - c_0(T) e^{i\Delta T} \right) \frac{\langle \lambda_1(t)| \frac{dH(t)}{dt} |\lambda_0(t)\rangle}{\Delta^2}$$

$$\approx i \left(1 - e^{i(\gamma_0(t) + \Delta T)} \right) \frac{\langle \lambda_1(t)| \frac{dH(t)}{dt} |\lambda_0(t)\rangle}{\Delta^2} \quad (12.15)$$

using $c_0(T) \approx e^{i\gamma_0(T)} c_0(0) = e^{i\gamma_0(T)}$, as in Eqn. (12.12).

$\langle \lambda_1(t)| \frac{dH(t)}{dt} |\lambda_0(t)\rangle$ may be thought of as a matrix element of the Hamiltonian differentiated with respect to time (if represented as diagonal in the instantaneous eigenbasis of the time-varying Hamiltonian). In the literature, it is common to upper-bound this value by the *largest* matrix element thereof, which we denote h_{max}. In order for there to be little leakage into the first excited state – such that the adiabatic condition holds – the value of Eqn. (12.15) must remain small, which is satisfied when $dh_{max}(t)/dt \propto \Delta^2$, and the constant of proportionality is small.

Finally, the spectral gap can vary throughout the evolution, and allowing the evolution to speed up and slow down accordingly is crucial for obtaining quantum advantage. Therefore, to give the adiabatic condition in a form that is suitable for analysing adiabatic quantum algorithms, let the evolution be fixed as a function of some variable, s:

$$H(s) = (1-s)H_0 + sH_f \quad (12.16)$$

for $s = [0, 1]$. From this, we define a function $s(t)$ to get a time evolution that satisfies the adiabatic condition:

$$\frac{dh_{max}(s)}{dt} \propto \Delta^2$$

$$\implies \frac{dh_{max}(s)}{ds} \frac{ds}{dt} \propto \Delta^2$$

$$\implies \frac{dt}{ds} \propto \frac{1}{\Delta^2} \quad (12.17)$$

where Δ is (now taken as) the instantaneous spectral gap.

12.2 Adiabatic quantum computation

Adiabatic quantum computation follows directly from the quantum adiabatic theorem and requires:

1. An initial Hamiltonian, H_0, whose ground state is easy to prepare.

2. A final Hamiltonian, H_f, whose ground state encodes the solution to the problem of interest. A measurement is performed to extract the solution from the state.
3. An adiabatic evolution path, $s(t)$.

This constitutes a universal model of computation that is *polynomially equivalent* to gate-based quantum computing. Any adiabatic quantum algorithm can be converted into a gate-based algorithm with a factor increase in the number of qubits and running time that is only a polynomial of the number of qubits of the original adiabatic quantum algorithm. This can be achieved by performing Hamiltonian simulation (now for a time-varying Hamiltonian) using standard techniques. In the other direction, any gate-based algorithm can likewise be converted into an adiabatic quantum algorithm (again the factor increase in running time and number of qubits is a polynomial of the original number of qubits). This is achieved using the ostensibly rather odd approach of constructing a Hamiltonian that encodes the *entire* quantum circuit in its ground state. For a polynomial depth quantum circuit, this strategy, therefore, requires a number of qubits that is only a polynomial of the number of qubits in the original algorithm. Moreover, it has been proven that an adiabatic evolution path whose duration amounts to a polynomial overhead relative to the execution time of the original circuit always suffices.

However, whilst the polynomial equivalence of gate-based and adiabatic quantum computation has been formally proven, in some ways, this gives little indication about how adiabatic quantum computers may be used in practice. For one thing, the model of computation requires initial and final Hamiltonian matrices, which are exponentially large in the number of qubits, to be defined. Without further qualification, this cannot, therefore, be presented as a *scalable* programming paradigm. The problem here is in some ways analogous to the role of unitary matrices in gate-based quantum computing, which are exponentially large (in the number of qubits) but never need to be explicitly expressed, as the quantum circuit model automatically provides a representation in which they are decomposed into a product of one- and two-qubit gates. In the case of adiabatic quantum computation, the most serious work at practical realisation has focused on hardware that is specialised to optimisation tasks. This is best introduced by first showing how adiabatic quantum computation can be used to solve NP-complete problems.

12.2.1 Using adiabatic quantum computation for unstructured search

A suitable starting point for showing how adiabatic quantum algorithms for NP-complete decision problems can be designed is to develop an algorithm – in some ways analogous to Grover's algorithm in gate-based quantum computing – that solves the problem of unstructured search. In Grover's algorithm, there is an oracle that can be queried to a reveal 'marked' element. Clearly, this set-up does not transfer to adiabatic quantum computation directly; however, a similar set-up is that there is some Hamiltonian H_f, which has a single unknown computational basis state as its ground state (that is, the analogue of a marked element in the oracle). The subscript of H_f is deliberately suggestive as the goal is to find an adiabatic evolution that arrives at H_f in a manner that is quantumly advantageous.

More specifically, let H_f be an n-qubit Hamiltonian such that

$$H_f = \sum_{x \in \{0,1\}^n \setminus x=z} |x\rangle\langle x| \tag{12.18}$$

for some n-qubit marked computational bases state z. Thus, H_f is diagonal in the computational basis, with all computational basis states having eigenvalue equal to one, apart from z which has eigenvalue 0.

It is also necessary to define a suitable starting state, bearing in mind that z is unknown. A convenient choice is:

$$H_0 = \sum_{\tilde{x} \in \{+,-\}^n \setminus +^n} |\tilde{x}\rangle\langle \tilde{x}| \tag{12.19}$$

which is thus diagonal in the $\{|+\rangle, |-\rangle\}$ basis, with ground state, $|+^n\rangle$, having eigenvalue equal to zero and all other eigenstates having eigenvalue 1. Thus, the adiabatic evolution commences in the state $|+^n\rangle$, which is easy to prepare.

These initial and final states must be connected by a suitable adiabatic evolution path, the design of which is aided by the fact that only a two-dimensional subspace of the N-dimensional space is significant for the spectral gap. To see this, let $|z\rangle$ be the marked state, as above, and (following the same convention for Grover search), let

$$|y\rangle = \frac{1}{\sqrt{N-1}} \sum_{x \in \{0,1\}^n \setminus z} |x\rangle \tag{12.20}$$

It follows that $|+^n\rangle = \frac{1}{\sqrt{N}}|z\rangle + \frac{\sqrt{N-1}}{\sqrt{N}}|y\rangle$, and therefore, if $|\psi\rangle$ is any vector in the subspace that is orthogonal to that spanned by $|z\rangle, |y\rangle$, then it is certainly the case that:

$$((1-s)H_0 + sH_f)|\psi\rangle = (1-s)|\psi\rangle + s|\psi\rangle = |\psi\rangle \tag{12.21}$$

as the entire subspace has eigenvalue 1 for both H_0 and H_f. With regard to the eigenvalues and eigenvectors in the subspace spanned by $|z\rangle$ and $|y\rangle$, it can be shown that the eigenvectors are:

$$|\lambda_0\rangle = \cos\frac{\theta(s)}{2}|z\rangle + \sin\frac{\theta(s)}{2}|y\rangle \tag{12.22}$$

$$|\lambda_1\rangle = -\sin\frac{\theta(s)}{2}|z\rangle + \cos\frac{\theta(s)}{2}|y\rangle \tag{12.23}$$

with eigenvalues, respectively $\lambda_0 = \frac{1}{2}(1 - \Delta(s))$ and $\lambda_1 = \frac{1}{2}(1 + \Delta(s))$, where

$$\cos\theta(s) = \frac{1}{\Delta(s)}\left(1 - 2(1-s)\left(1 - \frac{1}{N}\right)\right) \tag{12.24}$$

$$\sin\theta(s) = \frac{2}{\Delta(s)}(1-s)\frac{1}{\sqrt{N}}\sqrt{1 - \frac{1}{N}} \tag{12.25}$$

which have already been defined in terms of the spectral gap,

$$\Delta(s) = \sqrt{\frac{N + 4(N-1)(s^2 - s)}{N}} \tag{12.26}$$

(In the chapter problems, you are asked to prove these.)

The shortest duration adiabatic evolution path requires $s(t)$ to be set such that $dt/ds = k/(\Delta(s))^2$, for some small constant, k. From this, the total duration T of the adiabatic evolution can be computed:

$$T = \int_0^1 \frac{dt}{ds}ds = \int_0^1 \frac{k}{(\Delta(s))^2}ds = k\frac{N\arctan(\sqrt{N-1})}{\sqrt{N-1}} \in \mathcal{O}(\sqrt{N}) \tag{12.27}$$

thus matching the asymptotic run-time of Grover's algorithm.

12.2.2 Using adiabatic quantum computation to solve NP-complete decision problems

There is a natural way of constructing a Hamiltonian such that its ground state coincides with a satisfying assignment for *1-in-3-SAT* problems. 1-in-3-SAT is a variation of 3-SAT, which concerns the question of satisfiability when a logical expression is expressed in conjunctive normal form (CNF). A CNF statement is a logical expression in the form of a 'product of sums', $\bigwedge_i C_i$, where each C_i is of the form of a sum of Boolean variables, or negations of Boolean variables (together referred to as 'literals') and is referred to as a 'clause'. In particular, 3-SAT concerns CNF expressions where each clause contains at most three literals. 1-in-3-SAT concerns a further restricted case where each clause has exactly three literals, of which exactly one must be satisfied – however, there is a reduction from 3-SAT to 1-in-3-SAT, and both are NP-complete.

How 1-in-3 SAT can be encoded into a Hamiltonian is best shown through an explicit example. Consider the problem of finding x_1, x_2, and x_3 such that y is satisfied (is true) with an assignment such that *exactly* one literal in each clause is true:

$$y = (x_1 \text{ OR } x_2 \text{ OR } x_3) \text{ AND } (\neg x_1 \text{ OR } \neg x_2 \text{ OR } x_3) \text{ AND } (\neg x_1 \text{ OR } x_2 \text{ OR } \neg x_3) \tag{12.28}$$

where ¬ denotes negation. By inspection, $x_1 = 1$; $x_2 = x_3 = 0$ is the only satisfying assignment such that exactly one literal in each clause is true. For *any* 1-in-3-SAT, a Hamiltonian whose ground state corresponds to the satisfying assignment can be constructed, as can be illustrated by continuing with the same example. The first step towards achieving this is to write each clause as one minus the sum of the literals, and to then take the sum of the squares all of the clauses written in this way (also writing out negation as one minus the corresponding literal). For Eqn. (12.28), this gives:

$$\tilde{y} = (1 - x_1 - x_2 - x_3)^2 + (1 - (1 - x_1) - (1 - x_2) - x_3)^2 + (1 - (1 - x_1) - x_2 - (1 - x_3))^2 \tag{12.29}$$

which is equal to zero if and only if the 1-in-3-SAT is satisfied, and a positive number otherwise. Eqn. (12.29) is a quadratic in x_1, x_2, x_3, which can thus be written (using the fact that for $x_i \in \{0, 1\}$, $x_i^2 = x_i$):

$$\tilde{y} = 3 - 3x_1 + x_2 + x_3 + 2x_1x_2 + 2x_1x_3 - 2x_2x_3$$
$$= g + \sum_i h_i x_i + \sum_{i,j<i} J_{i,j} x_i x_j \tag{12.30}$$

where

$$g = 3; \quad \mathbf{h} = \begin{bmatrix} -3 \\ 1 \\ 1 \end{bmatrix}; \quad J = \begin{bmatrix} 0 & 0 & 0 \\ 2 & 0 & 0 \\ 2 & -2 & 0 \end{bmatrix} \tag{12.31}$$

The next step is to map the binary values x_1, x_2, and x_3 to s_1, s_2, and s_3, respectively, which take values $\{-1, +1\}$ according to:

$$x_i = \frac{1 - s_i}{2} \tag{12.32}$$

which gives:

$$\tilde{y} = g + \sum_i h_i \frac{1 - s_i}{2} + \sum_{i,j<i} J_{i,j} \frac{(1 - s_i)(1 - s_j)}{4} \tag{12.33}$$

from which the Hamiltonian can be constructed according to:

$$H_y = gI + \sum_i h_i \frac{I - Z_i \otimes I}{2} + \sum_{i,j<i} J_{i,j} \frac{(I - Z_i \otimes I - Z_j \otimes I + Z_i \otimes Z_j \otimes I)}{4} \tag{12.34}$$

where $Z_i \otimes I$ means a Pauli-Z on the ith qubit, tensor multiplied by identities on the remaining qubits, and similarly $Z_i \otimes Z_j \otimes I$ represents Pauli-Z matrices on the ith and jth qubits, tensor multiplied by identities on the remaining qubits. As identities and Pauli-Z unitaries are all diagonal (in the computational basis), it follows that all of the terms in Eqn. (12.34) are diagonal, and hence, the final Hamiltonian is diagonal. Continuing with the same example, this gives

$$H_y = \begin{bmatrix} 3 & 0 & 0 & 0 & 0 & 0 & 0 & 0 \\ 0 & 4 & 0 & 0 & 0 & 0 & 0 & 0 \\ 0 & 0 & 4 & 0 & 0 & 0 & 0 & 0 \\ 0 & 0 & 0 & 3 & 0 & 0 & 0 & 0 \\ 0 & 0 & 0 & 0 & 0 & 0 & 0 & 0 \\ 0 & 0 & 0 & 0 & 0 & 3 & 0 & 0 \\ 0 & 0 & 0 & 0 & 0 & 0 & 3 & 0 \\ 0 & 0 & 0 & 0 & 0 & 0 & 0 & 4 \end{bmatrix} \qquad (12.35)$$

As this is a diagonal matrix, the values on the leading diagonal are the eigenvalues, with the corresponding computational basis states as eigenvectors. Thus, if the state is indeed adiabatically evolved such that $H_f = H_y$, then it can be seen directly that H_y has smallest eigenvalue of 0 corresponding to the vector $[00001000]^T$, i.e., the state $= |100\rangle$ – which corresponds to the satisfying assignment $x_1 = 1$, $x_2 = 0$, $x_3 = 0$ as expected (and so can be extracted by measurement of the final ground state.)

The above mapping works in general for 1-in-3-SAT expressions: when a candidate input bit string is expressed as a computational basis state, then (for each input bit) the Pauli-Z operation has exactly the desired effect of mapping 0 to 1 and 1 to −1, thus meaning that H_y constructed in this way will have zero on the leading diagonal if and only if the corresponding input bit string (computational basis state) is a satisfying assignment and a positive number otherwise. Diagonal Hamiltonians constructed from quadratic expressions in this way are known as *Ising Hamiltonians* as they relate to physical systems known as Ising models. Ising models are of great interest in statistical physics, and they play an important role in quantum optimisation. Moreover, the final Hamiltonian (in Eqn. (12.34)), which defines the problem instance to be solved, is now decomposed into a sum of (simple) terms. This can be thought of as the 'programme input', and whilst the exact adiabatic quantum computation procedure is still a little opaque, this is clearly closer to a realistic programming paradigm than the original situation, in which it appeared that the (exponentially large) final Hamiltonian would need to be expressed in full and taken as an input.

By NP-completeness, any instance of a problem in NP can be efficiently reduced to an instance of 1-in-3 SAT, and so it follows that NP problems can (in general) be encoded such that the solution is the ground state of a Hamiltonian. Thus, the problem is expressed in a suitable form to be solved on an adiabatic quantum computer, and the quadratic quantum advantage for the instance of unstructured search (in Section 12.2.1) suggests that a significant speed-up may be available. However, this still constitutes an exponentially long running time and, furthermore, as for Grover search, it is significant that (i) classical algorithms may solve NP-complete problems much faster than exhaustive search; and (ii) exponential running time is often prohibitive. On the second of these, it is notable that adiabatic quantum computation can also be used as a starting point for designing metaheuristics, and to see how this is so, we first introduce adiabatic quantum *optimisation* properly.

12.3 Adiabatic quantum optimisation

The previous section shows that *deciding* if there exists a satisfying assignment to a 1-in-3-SAT problem can be mapped to the problem of deciding whether the ground state of an Ising Hamiltonian is equal to zero. However, the adiabatic quantum algorithm for this task delivers a great deal more: it returns the input bit string, which obtains the *minimum* value of the corresponding quadratic expression. As such, the adiabatic quantum algorithm for finding the ground state of an Ising Hamiltonian is an *optimisation* algorithm. Mathematical optimisation is the process of finding x such that $f(x)$ is minimised, for some $f : \mathbb{R}^n \to \mathbb{R}$ of interest. f is known as the *objective* or *cost* function, and optimisation problems can optionally be constrained by inequality and/or equality constraints, for example, finding x in the range $0 \le x \le 10$ such that some $f(x)$ is minimised. (Note that for problems that are naturally posed as the *maximisation* of a cost function, it is trivial to define a corresponding minimisation by inverting or negating the cost function.) From a practical point of view, the term 'optimisation' tends to be used for both the computational problem of returning the assignment of variables that minimises the objective function (e.g., the choice of $\{x_i\}$ that minimises $\tilde{y}$ in Eqn. (12.30)) and returning the optimal cost function value itself.

12.3.1 NP-hardness and optimisation

In order to further explain how adiabatic quantum computation can be applied to hard optimisation problems, it is necessary to give a bit more detail regarding the connection between decision and optimisation problems. In particular, every optimisation algorithm also solves the problem of *deciding* whether there is a solution that is below a certain threshold. If the 'decision version' of an optimisation problem is NP-complete, then the optimisation problem itself is NP-hard – NP-hard being the class of problems at least as hard as any problem in NP. This classification is immediate from the fact that any algorithm that solves the optimisation problem also solves the corresponding NP-complete decision version, and it is important to appreciate that whilst P and NP are strictly concerned with decision problems, by the above rationale, optimisation (and search) problems may be classified as NP-hard. (Note that there is also a set of complexity classes specifically for optimisation problems, but that is actually of less relevance here.)

Conversely, it is also the case that bisection search can be used to obtain an optimal value by just solving the decision version of an optimisation problem. This requires that there are known upper- and lower-bounds on the optimal value, and the threshold can then be set halfway between these. Once the existence of a solution below the threshold has been decided (by the decision algorithm), only one of the two halves can contain the minimal value, and so the upper and lower-bounds can be adjusted accordingly. Iterating this process, the fact that the region that contains the optimal solution is halved each time means that the decision problem must be solved a number of times that is logarithmic in the original range. As any NP-complete problem can be reduced to any other, this means that any NP-hard optimisation problem whose decision version is NP-complete can also be converted to any other (with only the modest logarithmic overhead from the bisection search). This ability to translate between different hard optimisation problems is of central importance when considering adiabatic quantum computation as a generic tool for optimisation.

12.3.2 Ising models and quadratic unconstrained binary optimisation

A general term for minimising expressions of the form given in Eqn. (12.30) is *quadratic unconstrained binary optimisation* (QUBO): 'quadratic' because the cost function is quadratic in its variables (i.e., $\{x_i\}$); 'binary' as these are binary variables; and 'unconstrained' as there are no equality/inequality constraints in addition to the cost function. As illustrated in Section 12.2.2, QUBO problems are equivalent to Ising models and so constitute a natural starting point for quantum optimisation. In statistical physics, Ising models are often defined such that the non-zero terms of J are restricted to 'nearest neighbour interactions' when the qubits are laid out on some lattice. For example, a common Ising model is that defined on a two-dimensional square grid – where terms $J_{i,j}$ can only take a non-zero value when i and j are neighbours in the grid. For two-dimensional grids, the Ising model can be solved in polynomial time; however, for square grids of three dimensions and higher, solving the Ising model is known to be NP-complete. By the argument in the previous subsection, this gives algorithms that optimise the Ising model great power in principle, because of their ability to also solve other optimisation and decision problems.

12.3.3 Max-cut

Although the approach outlined in Sections 12.3.1 and 12.3.2 means that *every* NP-hard optimisation problem with a decision version that is NP-complete can be solved via an NP-completeness reduction into 1-in-3 SAT, it is important to note that some other known NP-complete problems also have a natural representation in QUBO form. One such example is *max-cut*. In the optimisation version of max-cut, a connected graph is presented, and the goal is to partition the graph into two components in such a way that the maximum number of edges is 'cut' to achieve this partition.

Formally, let $G = (V, E)$ be a graph with vertex set V and edge set E. Each edge is a pair corresponding to the two vertices it connects. Max-cut may thus be stated as the problem of finding a subset, $\tilde{V}$ of vertices, that obtains

$$\min_{\tilde{V}} \sum_{u,v \in E,\ u \in \tilde{V},\ v \in V \setminus \tilde{V}} -1 \qquad (12.36)$$

To convert this into a quadratic cost function, a variable x_i is defined for each vertex, such that

$$x_i = \begin{cases} 0 & \text{if } v_i \in \tilde{V} \\ 1 & \text{otherwise} \end{cases} \tag{12.37}$$

which can be used to translate the above optimisation into a QUBO:

$$\min \sum_{i,j \in E} 2x_i x_j - x_i - x_j \tag{12.38}$$

The minima of Eqns. (12.36) and (12.38) coincide because in the latter, if i and j are in the same partitions (i.e., either both in $\tilde{V}$ or both in $V\backslash\tilde{V}$), the summand is equal to zero, and if they are in different partitions, then the summand is equal to -1. Thus, max-cut indeed has a natural representation in QUBO, and so the corresponding Ising Hamiltonian can be constructed to solve adiabatically.

12.3.4 Adiabatic quantum algorithms for NP-hard optimisation problems

To fully specify an adiabatic quantum algorithm that solves some NP-hard optimisation problem (whose decision version is NP-complete) as well as showing how the solution can be encoded in the ground state of some Hamiltonian, it is further necessary to specify an initial Hamiltonian (whose ground state is easy to prepare) and a suitable adiabatic evolution path. Although solving the NP-complete problem as an unstructured search may offer a quantum advantage, this is often uncompetitive with state-of-the-art classical algorithms (as was the case for Grover). In the case of adiabatic quantum optimisation, it is possible to give up the cast-iron guarantee of correctness by selecting a different (shorter duration) evolution path, thus reducing the running time. This idea rests on the premise that, even if adiabicity is not attained, it is still more likely that the final state will be a low-energy state (i.e., one whose eigenvalue is close to the minimal value). As the promise of obtaining the global minimum has now been sacrificed in the hope that good solutions will be obtained much more quickly, this brings us into the domain on metaheuristics. As a metaheuristic, the closely related *quantum annealing* rather than adiabatic quantum computation is usually considered.

12.4 Quantum annealing

Even though the complexity-theoretic classification of optimisation problems is important for formal analysis, from a practitioners standpoint, optimisation problems broadly fall into three categories:

1. Analytically solvable optimisation, for example finding x which minimises $f(x) = x^2 - 2x - 5$.
2. Convex optimisation, in which the problems have a single global minimum, and for which efficient optimisation algorithms usually exist (at least for very good approximations).
3. General optimisation, which concerns optimisation of any function, which may therefore have multiple 'local' minima:

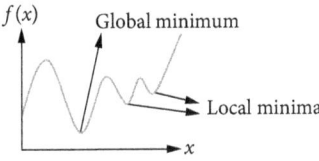

Figure 12.1

Optimisation problems may also be intrinsically discrete in nature and indeed on a digital computer even continuous optimisation problems are necessarily discretised – although it can still be helpful for visualisation purposes to think of the domain of the cost function as being some continuous range (or perhaps a grid overlaid thereon).

From a practical perspective, sufficiently large non-convex optimisation problems (especially those with many local minima) are typically treated as too hard to solve exactly within a reasonable time (indeed in some cases they may formally be shown to amount to NP-complete/hard problems). Thus, a more realistic goal is to find, within some restricted timescale, a 'good' rather than provably optimal solution. This is the purpose of metaheuristic optimisation algorithms, sometimes just termed 'metaheuristics'. In plain terms, a metaheuristic is a search policy that explores the optimisation function, $f(x)$, by evaluating it at certain values of x.

There are myriad metaheuristic algorithms that decide where next (at which value of x) to evaluate $f(x)$ given the history of function evaluations, but all are based on the same essential principle: that good solutions are likely to be near other good solutions, or in other words that the optimisation surface has some smoothness. This, in turn, reveals the *exploration versus exploitation* trade-off that all metaheuristics must make.

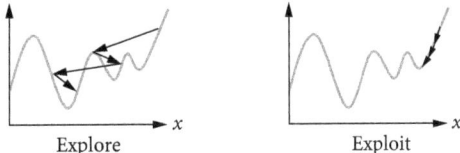

Figure 12.2

- A metaheuristic can *explore* the optimisation surface by making 'large movements' to discover whether another part of optimisation surface returns smaller values of $f(x)$. In this case, the vicinity of the global minimum may be found,

but the value of x returned may only be a fairly poor approximation of the global minimum itself.
- Alternatively, a metaheuristic can *exploit* its 'current' position, by descending incrementally. The risk is that this returns a (possibly not very good) local minimum.

Metaheuristic optimisation algorithms are typically set-up to search for a minimum in such a way that exploration is favoured at the start and exploitation at the end. Different metaheuristics do this in different ways and so are well-suited to different applications.

Simulated annealing is a metaheuristic inspired by the physical process of thermal annealing and applies to both combinatorial optimisation, and also to continuous optimisation problems whose cost function has a defined notion of neighbourhood. An initial x is chosen, and the algorithm proceeds as follows:

Algorithm 12.1 Simulated annealing (classical)

Require: Cost function $f(\cdot)$
1: Choose initial x
2: Evaluate $f(x)$
3: **while** unconverged **do**
4:     At random choose a neighbour, x' of x
5:     Evaluate $f(x')$.
6:     **if** $f(x') < f(x)$ **then**
7:         Set $x \leftarrow x'$
8:     **else**(i.e., $f(x') \geq f(x)$)
9:         Randomly decide whether to set $x \leftarrow x'$ or to leave x unchanged.
10:     **end if**
11: **end while**
12: Return $x, f(x)$.

Note that in Algorithm 12.1 the condition 'unconverged' should also be taken as running only up to a certain maximum number of iterations, even if convergence has not been obtained. Note also that, in practice, an implementation would save the smallest value in its entire history of function evaluations to output as the minimum, but this has been omitted for simplicity.

The random decision in Line 9 is such that the more that $f(x')$ exceeds $f(x)$ by the less likely it is that $x \leftarrow x'$. Additionally, this random decision varies throughout the repeated iterations of the algorithm, such that 'uphill' moves are more likely to be accepted at the start of the optimisation, and less likely towards the end. In this way, the exploration versus exploitation trade-off is made such that exploration dominates initially, and exploitation dominates at the end.

Taking inspiration from the physical process of thermal annealing, the acceptance probability (i.e., for accepting the update $x \leftarrow x'$) is usually determined by:

$$\Pr(\text{accept}) = \exp\left(-\frac{f(x') - f(x)}{T}\right) \tag{12.39}$$

where T is the 'temperature', which is 'cooled' as the algorithm progresses, such that uphill moves become less likely as the algorithm progresses.

Quantum annealing can be thought of as a *quantum* metaheuristic optimisation algorithm that is in some ways similar to simulating annealing. As for adiabatic quantum computation, in quantum annealing, the solution to an optimisation problem is encoded in the ground state of some 'final' Hamiltonian, H_f, and the goal of the quantum annealing is to converge on a ground state thereof. To perform quantum annealing, the following are also needed:

- A *transverse field* (or *mixing*) Hamiltonian, H_D, that does not commute with H_f.
- An *arbitrary initial state* (conventionally a uniform superposition).

In quantum annealing, the system is then evolved according to:

$$H(t) = H_f + \Gamma(t) H_D \tag{12.40}$$

where $\Gamma(t)$ is the *transverse field coefficient*, which is initially very high, and reduces to zero over time. H_D plays the role of 'exploring' the optimisation surface, which initially dominates. As the quantum annealing progresses, H_f comes to dominate, and the aim is that the final state will have a large overlap with the ground state or other low-energy states (which can be justified as being good, but not optimal solutions). If you squint a little, this essentially amounts to the 'exploitation' of the cost function.

This procedure of starting in some easy to prepare state and evolving the Hamiltonian such that the ground state of the final Hamiltonian encodes the solution is clearly reminiscent of adiabatic quantum computation, and indeed, in common parlance, the terms 'adiabatic quantum computation' and 'quantum annealing' are very often used interchangeably. The key distinction is that in adiabatic quantum computation, the system is initialised in an easy to prepare ground state, and the Hamiltonian is evolved slowly, so that the system remains in a ground state, whereas quantum annealing is a metaheuristic, which starts in an *arbitrary* initial state. It follows that in adiabatic quantum computation, the choice of initial state and Hamiltonian evolution path are intimately linked, whereas in quantum annealing, in principle, the transverse field Hamiltonian can be (and in practice *is*) chosen independently. This flexibility comes at the cost of sacrificing performance guarantees (as do all metaheuristics) and indeed the only analytic means of evaluating the performance of quantum annealing comes by selecting H_f, $\Gamma(t)$, and the initial state such that the quantum annealing exactly amounts to an adiabatic evolution.

However, quantum annealing *is* a metaheuristic and thus should be judged as such. This means that all one can really hope to do is to qualitatively pick out cost functions with features which make classical metaheuristics perform poorly, but that may be amenable to (better) optimisation with quantum annealing. For instance, simulated annealing involves 'walking over' the cost function surface with a single trajectory, with uphill movements being unlikely, and so the existence of local minima that are surrounded by steep gradients on all sides are likely to present 'traps' from which the algorithm fails to escape (and so cannot then find the global minimum). In time, empirical data are likely to reveal the types of optimisation problem to which quantum annealing is better suited than classical metaheuristics (if, indeed, any exist).

Real quantum annealing hardware has been built, and whilst the benefits thereof are still (at the time of writing) unclear, this does complete the picture of how programmable adiabatic quantum computers/quantum annealers appear to the user. The most prominent example is the series of quantum annealers from *D-wave*, which solve the Ising model for a certain physical structure. In particular, the qubits are connected according to a *Chimera* topology, which means that in the Ising model, certain quadratic terms (those for disconnected qubits) are fixed as zero; however, the non-zero terms suffice to define an NP-hard optimisation problem – in the same way that the Ising model on a three-dimensional grid, though also having many quadratic terms set to zero, is NP-complete. The terms that are not required to be zero represent the inputs that the user specifies to define the particular problem instance to be solved.

12.5 The quantum approximate optimisation algorithm

Optimisation is a natural application of adiabatic quantum computation and quantum annealing; however, in terms of actual available hardware, gate-based quantum computers dominate. As many real-world computational challenges do indeed amount to optimisation problems, it is therefore desirable to develop (quantumly advantageous) techniques that perform metaheuristic optimisation on gate-based quantum computers. One such technique, which has gained widespread attention, is the *quantum approximate optimisation algorithm* (QAOA), which can be explained in terms of quantum annealing (although it remains unclear whether QAOA will offer quantum advantage in practice).

As $H_f + \Gamma(t)H_D$ represents a continuously-varying Hamiltonian evolution, it is not directly executable on a gate-based quantum computer; however, one technique to approximately do so is Trotterisation, which yields the quantum circuit:

$$\prod_k e^{-i\alpha_k H_f} e^{-i\gamma_k H_D} \tag{12.41}$$

for some suitable α and γ, whose size is established in advance. Rather than analytically finding α and γ, which well-approximate some specific quantum annealing path, QAOA is a variational algorithm (similar to the variational quantum eigensolver, in Chapter 11), in which an outer optimisation loop seeks to converge on suitable parameters α and γ, such that the final state is indeed the ground state that encodes the solution of the optimisation problem.

12.6 Other forms of quantum optimisation

Using adiabatic quantum computation and related approaches to converge on Hamiltonian ground states that encode the solution of an optimisation problem is perhaps the most widely researched means of using quantum computers to enhance optimisation, but it is far from the only approach. There are too many other ideas to list them all; however, it is worth at least namechecking the most important alternative approaches to quantum optimisation. One approach is to use quantum computers to prepare *Gibbs states*, which can then be used to speed up some semidefinite programming (SDP) problems. However, SDP can be solved efficiently classically, and moreover, the cases where quantum SDP solvers offer the largest speed-up (when there is an exponential separation between the quantum and classical hardness of the corresponding Gibbs state preparation) are not known to correspond to any real-world optimisation problems.

Another important category of quantum optimisation algorithms use quantum amplitude amplification and/or quantum amplitude estimation to concentrate probability mass onto some state quadratically faster (in the former case) or to estimate some quantity quadratically faster (in the latter case). Typically, this will amount to speeding-up some subroutine in the overall computational workflow. One example is using quantum Monte Carlo integration for simulation-based optimisation, which applies when the cost function is itself an expectation – in which case there is a quadratic quantum speed-up available in the time taken to estimate the cost to a specified accuracy. In a similar vein, quantum algorithms exist to speed up the calculation of the gradient of optimisation surfaces – which can thus enhance any gradient-based classical optimiser.

Chapter problems

1. (a) Show that λ_0, λ_1, and $|\lambda_0\rangle$, $|\lambda_1\rangle$, as defined in Eqn. (12.24), Eqn. (12.25) and the surrounding text, are respectively the eigenvalues and corresponding eigenvectors of the 'unstructured search' Hamiltonian, as claimed.
   (b) Show that the spectral gap, $\Delta(s)$, given in Eqn. (12.26) is correct.
2. For the 1-in-3-SAT instance:

$$(x_1 \text{ OR } x_2 \text{ OR } x_3) \qquad (12.42)$$

find all satisfying assignments, and show that if a Ising Hamiltonian is constructed according to the procedure given in the chapter, then the Hamiltonian is diagonal with positive values on the leading diagonal apart from values equal to zero if and only if the corresponding bit string is a satisfying assignment.

3. Consider the 1-in-3-SAT instance

$$(x_1 \text{ OR } x_2 \text{ OR } x_3)$$
$$\text{AND}(\neg x_1 \text{ OR } x_2 \text{ OR } x_3)$$
$$\text{AND}(x_1 \text{ OR } \neg x_2 \text{ OR } x_3)$$
$$\text{AND}(x_1 \text{ OR } x_2 \text{ OR } \neg x_3)$$
$$\text{AND}(\neg x_1 \text{ OR } \neg x_2 \text{ OR } x_3)$$
$$\text{AND}(\neg x_1 \text{ OR } x_2 \text{ OR } \neg x_3)$$
$$\text{AND}(x_1 \text{ OR } \neg x_2 \text{ OR } \neg x_3) \tag{12.43}$$

What outcome do you expect if a state is evolved adiabatically to the ground state of an Ising Hamiltonian constructed according to the procedure given? Comment on your answer, and verify that the Ising Hamilotian has the predicted ground state.

4. Consider the graph:

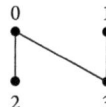

Figure 12.3

with vertices numbered as shown.
(a) How many ways are there to partition the vertices into two sets?
(b) By inspection, write down the max-cut.
(c) Write down the QUBO expression for each possible way of partitioning the vertices, and show that the max-cut found in (b) indeed corresponds to the minimum QUBO value of these.

Further reading

The idea of encoding an optimisation function into a Hamiltonian ground state can be traced back to Apolloni et al [1988; 1989], where the name 'quantum annealing' also first appeared. Adiabatic quantum computing itself (in an algorithmic sense) is due to Farhi et al [2000]; Albash and Lidar [2018] provide a nice survey of the field in general. Adiabatic unstructured search, as presented here, is closely based on that of Slutskii et al [2019]; and Farhi et al [2001] give more details on how adiabatic quantum computation can be applied to NP-complete problems more generally. Aharonov et al [2004] showed that adiabatic quantum computing is equivalent to gate-based.

D-Wave's quantum processor qubit connectivities are discussed by Boothby *et al* [2020]; QAOA is due to Farhi *et al* [2014]; using quantum computers for SDP has been suggested by various parties, a subject of which Brandao and Svore [2017] is an authorative source; using quantum computing to speed up optimisation problems where the cost function is itself an expectation is a trivial corollary of quantum Monte Carlo integration, itself a corollary of quantum amplitude estimation, as discussed in Chapter 9; however, Gacon *et al* [2020] also develop the idea explicitly.

13
A quantum linear system solver

In a sense, gate-based quantum computers are just machines that apply (extremely) large matrices to input states (vectors), and so it is natural to ask whether linear algebra applications can benefit from quantum computing. In the case of solving linear systems, this question has been answered with an emphatic 'yes!'. In 2009, Aram Harrow, Avinatan Hassidim, and Seth Lloyd proposed a quantum algorithm to solve linear systems exponentially faster than the best classical counterpart. The quantum linear systems solver is ubiquitously known by the initials of its inventors, 'HHL'.

Solving a linear system means inverting a matrix to solve the following problem: let x be an unknown N-element vector, b be a known N-element vector, and A be an $N \times N$ matrix such that:

$$b = Ax \qquad (13.1)$$

Then, the goal is to find x.

Classically, if one makes no assumptions about the structure of the matrix itself, then Gaussian elimination provides one way to solve the system and incurs complexity $\mathcal{O}(N^3)$. Any classical method of matrix inversion – even one that leverages some promised structure of A (such as sparsity) – requires $\Omega(N)$ operations.

13.1 Solving linear systems with quantum computers

When it comes to solving linear systems, it is sometimes possible to do much better quantumly than classically. In the first instance, it is convenient to consider cases, where A is Hermitian and invertible (it has all non-zero eigenvalues), and where N is power of 2, and also to define $n = \log_2 N$ (i.e., $N = 2^n$). If N is not already a power of 2, it is trivial to 'pad' the system with some null variables, and solve:

$$\begin{bmatrix} b \\ \tilde{b} \end{bmatrix} = \begin{bmatrix} A & 0 \\ 0 & kI \end{bmatrix} \begin{bmatrix} x \\ \tilde{x} \end{bmatrix} \qquad (13.2)$$

which, when inverted, correctly produces $x = A^{-1}b$ and also $\tilde{x} = (1/k)\tilde{b}$ as an irrelevant by-product. This process does not affect the complexity analysis in any significant way, as the size of the linear system at most (nearly) doubles and k can always be chosen such that the dependence on condition number (see Section 13.2) is not adversely impacted by the padding. Henceforth, we, therefore, assume that N is a power of 2 and solve an equation of the form in Eqn. (13.1).

238 A quantum linear system solver

HHL requires that **b** is encoded as an n-qubit quantum state, $|b\rangle$, which is achieved by *amplitude encoding*, that is, encoding **b** such that $|b\rangle$ is a superposition of N computational basis states, with the coefficient of each basis state proportional to the corresponding element of **b**:

$$|b\rangle = \frac{1}{\sum_i |b_i|^2} \sum_{i=0}^{N-1} b_i |i\rangle \tag{13.3}$$

where $|i\rangle$ is the ith computational basis state, as usual. Preparing such a state $|b\rangle$ to load **b** into the quantum computer will not always be efficiently possible; however, later in this chapter, promising candidates for applications where this is possible are discussed.

Assuming that the initial loading of **b** has indeed been accomplished in a sufficiently succinct manner, the next step is to treat A as a Hamiltonian and to simulate its evolution. If A is sparse, then using standard Hamiltonian simulation techniques, it is possible to efficiently construct a quantum circuit to apply the unitary:

$$U_A = e^{2\pi i A} \tag{13.4}$$

raised to any power.

This Hamiltonian evolution is then used in quantum phase estimation. For the applications of quantum phase estimation addressed in earlier chapters, the focus has been on extracting a quantity, namely an eigenvalues phase, of a *unitary* matrix, U_A. However, HHL inverts the *Hermitian* matrix, A, and so it is more convenient to analyse the algorithm in terms of the eigenvalues of A, which we denote $\{E_*\}$. Recalling from Section 2.4.7 that the eigenvectors (eigenstates) of U_A and A are the same, denoted $\{|\lambda_i\rangle\}_{i \in \{0,\ldots,N-1\}}$, any $|b\rangle$ can be expressed:

$$|b\rangle = \sum_{i=0}^{N-1} \beta_i |\lambda_i\rangle \tag{13.5}$$

where β_i are coefficients. This completes the set-up necessary to apply quantum phase estimation, omitting the final measurement:

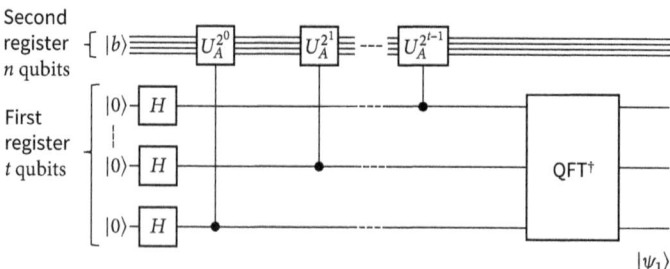

Figure 13.1

The 'second' register is shown above the 'first' register (according to the usual presentation of quantum phase estimation, i.e., in Chapter 9), as it makes the consequent qubit ordering a little easier for showing the effect of the subsequent operations. Labelling the entire circuit 'U_1', this gives:

$$|\psi_1\rangle = U_1(|b\rangle|0^t\rangle) = \sum_{i=0}^{N-1} \beta_i |\lambda_i\rangle|\tilde{E}_i\rangle \qquad (13.6)$$

which follows directly from the fact that the ith eigenvalue of U_A is $e^{2\pi E_i}$, and for each i, E_i is exactly the eigenvalue phase in the usual presentation of quantum phase estimation. There is an important subtlety here: when starting from the unitary, by construction, the phase is a number between 0 and 1; however, in the case where we start with A, the eigenvalues can be outside of this range, and hence, some information is lost unless the matrix is rescaled accordingly. Notably, this scaling should be such that positive and negative eigenvalues are disambiguated and it follows that it is therefore convenient to depart from the convention established in Chapter 9, where 2^t is explicitly stated in the phase estimate, and instead treat the phase estimation circuit as producing a value that is interpreted as E_i itself, as in Eqn. (13.6). Note also that such rescaling should be taken as implicit throughout this analysis, and when the run-time is analysed, it is crucial to note that the condition number is invariant under this rescaling.

A second unitary, U_2, is now required to perform a rotation that is inversely proportional to the value in a control register, that is:

$$|i\rangle|0\rangle \xrightarrow{U_2} \begin{cases} |i\rangle\left(\sqrt{1 - \frac{C^2}{i^2}}|0\rangle + \frac{C}{i}|1\rangle\right), & \text{if } i \neq 0 \\ |i\rangle|0\rangle, & \text{otherwise} \end{cases} \qquad (13.7)$$

for some computational basis state $|i\rangle$, which is interpreted as an encoding of the number i (in the manner described above for the eigenvalues, E_*), and some positive constant C. The split into two cases is required for the prepared state to be finite for any input. In the case of invertible matrices, i will always be non-zero, however when finding the pseudo-inverse of a non-invertible matrix – something that HHL naturally does too, as described in Section 13.3 – then doing nothing for the case where $i = 0$ turns out to be the correct thing to do.

The operation described in Eqn. (13.7) can be constructed as a reversible implementation of a (efficiently computable) classical function, and from this, the entire HHL circuit follows:

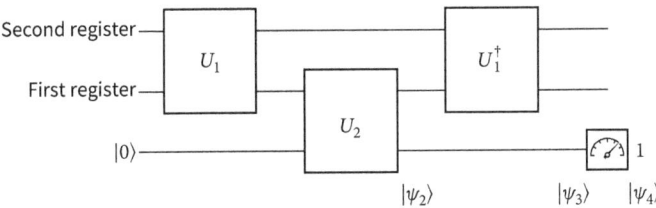

Figure 13.2

where the measurement outcome 1 is post-selected (in practice, this measurement outcome occurs with a sufficiently high probability, and is accounted for in the run-time/error analysis).

Stepping through the circuit (using the previous definition of $|\psi_1\rangle$ as the state prepared on the first and second registers by U_1), U_2 acts on a superposition of computational basis states, therefore:

$$|\psi_2\rangle = (I \otimes U_2)(|\psi_1\rangle|0\rangle) = \sum_{i=0}^{N-1} \beta_i |\lambda_i\rangle |\tilde{E}_i\rangle \left(\sqrt{1 - \frac{C^2}{\tilde{E}_i^2}} |0\rangle + \frac{C}{\tilde{E}_i} |1\rangle \right) \quad (13.8)$$

and the subsequent inverse of U_1 serves the purpose of uncomputing the first register, such that:

$$|\psi_3\rangle = (U_1^\dagger \otimes I)|\psi_2\rangle = \sum_{i=0}^{N-1} \beta_i |\lambda_i\rangle |0^t\rangle \left(\sqrt{1 - \frac{C^2}{\tilde{E}_i^2}} |0\rangle + \frac{C}{\tilde{E}_i} |1\rangle \right) \quad (13.9)$$

This relies on the fact that the eigenvalues are all real numbers, owing to the Hermitianity of A, and it is further necessary that $|C/\tilde{E}_i|$ is at most equal to 1, in order that a valid quantum state is prepared. This means that the circuit U_2 should be set up such that it does not 'over-rotate' even for the eigenvalue with smallest magnitude. For the following run-time considerations, the implication of this is that $C = \mathcal{O}(1/\kappa)$, where κ is the condition number, that is the ratio of the largest to the smallest (non-zero) singular value of the matrix. In the case of Hermitian matrices, the singular values have the same magnitudes as the eigenvalues, which are all real (but may be negative or positive). Finally, when the measurement outcome is one, as is post-selected,

$$|\psi_4\rangle = \frac{1}{C\sqrt{\sum_i |\beta_i|^2/|\tilde{E}_i|^2}} \sum_{i=0}^{N-1} \frac{C}{\tilde{E}_i} \beta_i |\lambda_i\rangle |0^t\rangle |1\rangle = C' A^{-1} |b\rangle |0^t\rangle |1\rangle = C' |x\rangle \otimes (|0^t\rangle |1\rangle) \quad (13.10)$$

where C' is the constant, $C' = \frac{1}{\sqrt{\sum_i |\beta_i|^2/|\tilde{E}_i|^2}}$, and the final '$\otimes$' is explicitly included to emphasise that, up to a constant of proportionality, $|x\rangle$ has been computed and is merely composed with a further $t + 1$ qubits.

HHL has, therefore, prepared $|x\rangle$, which is a quantum (amplitude) encoding of x, prompting us to ask whether this is actually a useful thing to have computed. If the goal is to obtain x in its entirety, then the answer is 'not really': because $|x\rangle$ is of length $\log N$ qubits, whereas x is an N-element vector. In general, to *read out* an N-element vector, any algorithm (classical or quantum) will be at least linear in N. If HHL were adapted to do this, then it would amount to repeatedly preparing and measuring $|x\rangle$ to obtain an estimate of x, and the quantum advantage would be lost.

Thus, the relevant question is not whether HHL computes x in a quantumly advantageous manner, but rather whether the output that HHL *does* produce, namely $|x\rangle$, is useful. Broadly speaking, the answer is that the amplitude encoding, $|x\rangle$, allows 'global' properties of x to be quickly extracted – and some possible applications are discussed in Sections 13.5 and 13.6.

13.2 Run-time and error analysis

To express the run-time (computational complexity) of HHL as a function of the matrix size N and the acceptable error ϵ, it is necessary to consider three factors:

- the required value of t, and what this means for the computational complexity of the Hamiltonian simulation blocks;
- the complexity of the entire quantum circuit;
- the probability of the measurement outcome being 1, as this, in turn, determines the expected number of repeats to achieve the post-selected state by a repeat until success strategy.

Taking these in turn, the quantum phase estimation subroutine is the dominant source of error and estimates any eigenvalue to t bits of precision, so the relative error is about 2^{-t}. However, this assumes that the t bits are used in an optimal manner, something that is possible when only a single eigenvalue is calculated. In HHL, quantum phase estimation is applied to an arbitrary superposition of eigenvalues, and so further consideration is required to find the error in this case. As A is Hermitian, it has real eigenvalues, and for simplicity, we consider the case where these are all positive, and ordered such that $E_0 \leq E_1 \leq \cdots \leq E_{N-1}$. (Note that the same asymptotic complexity is obtained for the general case where the spectrum of A may include negative values too.)

The quantum phase estimation subroutine must be set up to such that the second register can correctly contain E_{N-1} – the eigenvalue with largest magnitude – and so the relative error if this eigenvalue were to be estimated is upper-bounded by 2^{-t}. It follows that for the eigenvalue with the smallest magnitude, E_0, the relative error will be upper-bounded by $\kappa 2^{-t}$ (as the eigenvalues and singular values are the same in the case considered, $\kappa = |E_{N-1}|/|E_0|$). Thus, $\kappa 2^{-t}$ is a maximum for the relative error of *any* eigenvalue.

This bound on the error can be substituted into Eqn. (13.10), for the extremal case where the relative error is maximum for each term in the superposition, giving:

$$C' \sum_{i=0}^{N-1} \frac{1}{E_i(1 \pm \kappa 2^{-t})} \beta_i |\lambda_i\rangle |0^t\rangle |1\rangle \approx C' \sum_{i=0}^{N-1} \frac{1 \mp \kappa 2^{-t}}{E_i} \beta_i |\lambda_i\rangle |0^t\rangle |1\rangle$$

$$= (1 \mp \kappa 2^{-t}) C' |x\rangle \otimes (|0^t\rangle |1\rangle) \quad (13.11)$$

and thus the relative error (implicitly defined accordingly) is upper-bounded by $\kappa 2^{-t}$.

If the aim is to select t such that the error is restricted to some worst case value, ϵ, rearranging the bounding case, $\epsilon = \kappa 2^{-t}$ gives $t = \log_2(\kappa/\epsilon)$. We also have that an $N \times N$ s-sparse Hamiltonian can be simulated for time T using $\mathcal{O}(\log(N)s^2 T)$ operations and so the total Hamiltonian simulation time can be found by summing this up for each of the blocks, $U_A^{2^0}$, $U_A^{2^1}$, $U_A^{2^{t-1}}$, which gives total time $2^t - 1 < 2^t$. Substituting $t = \log_2(\kappa/\epsilon)$, i.e., $2^t = \kappa/\epsilon$ – as is required to restrict the error to ϵ (recall that error is sub-additive) – the run-time is $\mathcal{O}(\log(N)s^2\kappa/\epsilon)$.

The circuit block U_1 consists of both Hamiltonian evolution, as accounted for above, and the quantum Fourier transform; the latter runs in time polynomial in t and, therefore, poly-logarithmic in κ/ϵ. Rather than explicitly including such poly-logarithmic terms, $\mathcal{O}$ is replaced with $\widetilde{\mathcal{O}}$ to signify that 'lower-order' terms have been suppressed. Turning to the other circuit blocks, the presence of $U_1^\dagger$ (which has the same complexity as U_1) clearly only introduces a constant factor of 2 – which does not appear in the asymptotic expression; and, as already stated, U_2 is simply a rotation, which can be efficiently (i.e., in time polynomial in t and thus poly-logarithmic in κ/ϵ) implemented in the corresponding reversible circuit block, and so inclusion does not change the asymptotic expression for a successful circuit execution:

$$\widetilde{\mathcal{O}}(\log(N)s^2\kappa/\epsilon) \qquad (13.12)$$

Finally, it is necessary to consider the post-selected condition, which occurs with probability proportional to C^2, by simple application of the Born rule to Eqn. (13.9), and as identified $C = \mathcal{O}(1/\kappa)$, meaning that, on average, $\mathcal{O}(\kappa^2)$ repeats is sufficient – which can be reduced to $\mathcal{O}(\kappa)$ by employing amplitude amplification. Multiplying the single circuit runtime in Eqn. (13.12) by the expected number of repeats gives an overall complexity:

$$\widetilde{\mathcal{O}}(\log(N)s^2\kappa^2/\epsilon) \qquad (13.13)$$

This computational complexity is quite remarkable, as the linear system has been solved in a number of operations that grows only logarithmically with the system size. HHL derives its quantum advantage by manipulating the eigenvalues of a large matrix in superposition, which hints at the deeper nature of quantum computation. One of the most important emerging pictures of quantum computers is that they are machines that efficiently manipulate the eigenvalues (or more generally, singular values) of exponentially large matrices, such that matrix transformations are executed. This idea is at the heart of the quantum singular value transformation, in Chapter 14.

13.3 HHL for non-Hermitian matrix inversion and pseudo-inversion

Whilst HHL is most easily analysed for the case where A is already Hermitian – and this in turn provides a suitable way to explain the essence of the algorithm – it is important to note that HHL is not actually restricted just to this setting. For general A, instead the system

$$\begin{bmatrix} b \\ 0 \end{bmatrix} = \begin{bmatrix} 0 & A \\ A^\dagger & 0 \end{bmatrix} \begin{bmatrix} 0 \\ x \end{bmatrix} \tag{13.14}$$

is solved, where $H = \begin{bmatrix} 0 & A \\ A^\dagger & 0 \end{bmatrix}$ is Hermitian. For this generalisation, it is necessary to check:

1. that H is invertible;
2. that the overall complexity is not adversely affected by this variation.

As an intermediate step, between the matrix being inverted being Hermitian and full generality, there is merit in considering the adjusted system of the form in Eqn. (13.14), with A being square and invertible (but not necessarily Hermitian). To check that the resultant matrix H is invertible, it is enough to show that the fact that A has all non-zero eigenvalues implies that H also has all non-zero eigenvalues. For this, we use

$$H^2 = \begin{bmatrix} AA^\dagger & 0 \\ 0 & A^\dagger A \end{bmatrix} \tag{13.15}$$

which is a block diagonal matrix, and so its eigenvalues are the eigenvalues of its blocks ($AA^\dagger$ and $A^\dagger A$).

A few elementary facts about singular values and eigenvalues can now be invoked to obtain the desired result that the invertibility of A implies that of H. First, any invertible A has all non-zero singular values; furthermore, the eigenvalues of $AA^\dagger$ are equal to those of $A^\dagger A$ and also equal to the squares of the singular values of A. From these properties, it follows that the eigenvalues of $AA^\dagger$ as well as those of $A^\dagger A$ are all non-zero; hence, the eigenvalues of H^2 are non-zero, and so are the eigenvalues of H. This suffices to show that invertible A leads to invertible H.

Similar analysis can be employed to show that if A is well conditioned, so is H. In particular, as H^2 is Hermitian, its singular values are simply the magnitudes of its eigenvalues – however, by the above argument, the eigenvalues of H^2 are equal to the squares of the singular values of A, and so the singular values of H are equal to the singular values of A. Thus, the condition number of H is equal to that of A.

The condition number is one of the four terms that appear in the statement of the asymptotic complexity of HHL, along with the sparsity, s, the size of the matrix, N, and the acceptable error, ϵ. The last of these is user-defined and does not directly pertain to any properties of the linear system that is being solved. As shown above, the condition number of a matrix, H, is equal to that of the matrix, A, from which it was constructed, as defined below Eqn (13.14) – therefore, this does not increase the complexity. Furthermore, the size of H is at most twice that of A, and so only causes a (small) constant factor increase in the complexity. Finally, in order for H to be row-sparse, it is necessary for A to be both row- and column-sparse (i.e., in order for the blocks of H, A, and $A^\dagger$ to both be row-sparse) and so in the case of non-Hermitian A, the term s should be taken as the maximum of the row- and column-sparsity.

Following this argument that HHL applied to non-Hermitian, but still invertible, square matrices still works with a small adaptation to the linear system and with essentially the same computational resource requirements, the final generalisation is to *any* matrix. For, perhaps surprisingly, HHL works even for *non-invertible* matrices, when the goal is to find the matrix *pseudo*-inverse rather than inverse. In this case, A need not be full-rank or square – although it can always be zero-padded to be square and of dimension that is an integer power of two. Doing so makes no material difference, as the zero-padded matrix can be thought of as a block diagonal matrix, $\begin{bmatrix} A & 0 \\ 0 & 0 \end{bmatrix}$, which, therefore, has pseudo-inverse $\begin{bmatrix} A^+ & 0 \\ 0 & 0 \end{bmatrix}$, where A^+ is the pseudo-inverse of A. As the zero-padded part of the matrix corresponds to a vector subspace transformation with singular value equal to zero, which is thus left unchanged by matrix pseudo-inversion (apart from swapping the left and right subspaces such that the total dimension is the same for the input and output vectors – see Section 2.4.8 for details), this affects nothing essential in the following analysis. Henceforth, we shall, therefore, use A for the matrix being pseudo-inverted (and not a block therein)

As detailed in Section 2.4.6, every matrix has a singular value decomposition (SVD), and HHL can be analysed by separating out the part of the SVD summation corresponding to singular value equal to zero:

$$A = \sum_{i=0}^{N'-1} \sigma_i |u_i\rangle \langle v_i| + \sum_{i=N'}^{N-1} 0|u_i\rangle \langle v_i| \qquad (13.16)$$

Therefore, the matrix pseudo-inverse of A is:

$$A^+ = \sum_{i=0}^{N'-1} \sigma_i^{-1} |v_i\rangle \langle u_i| + \sum_{i=N'}^{N-1} 0|v_i\rangle \langle u_i| \qquad (13.17)$$

which is such that the result of the pseudo-inverse applied an arbitrary input $|b\rangle = \sum_i \beta_i |u_i\rangle$ is:

$$A^+|b\rangle = \sum_{i=0}^{N'-1} \sigma_i^{-1} |v_i\rangle \langle u_i|b\rangle + \sum_{i=N'}^{N-1} 0|v_i\rangle \langle u_i|b\rangle = \sum_{i=0}^{N'-1} \sigma_i^{-1} \beta_i |v_i\rangle \qquad (13.18)$$

From the matrix A, a corresponding Hermitian matrix, $H = \begin{bmatrix} 0 & A \\ A^\dagger & 0 \end{bmatrix}$ is defined as above, and by the premise that A is $2^n \times 2^n$ for some integer n, the input state $\begin{bmatrix} b \\ 0 \end{bmatrix} = \begin{bmatrix} 1 \\ 0 \end{bmatrix} \otimes b$ corresponds to the state $|0\rangle |b\rangle$.

Using HHL exactly as described for matrix inversion also achieves matrix pseudo-inversion, and to see this, first note that H is Hermitian, with eigenvalues $\{\pm \sigma_i\}$, and corresponding eigenvectors (see chapter problems),

$$|w_i^\pm\rangle = \frac{1}{\sqrt{2}}(|0\rangle|u_i\rangle \pm |1\rangle|v_i\rangle) \quad (13.19)$$

(there are also further eigenvectors that are orthogonal to all of the above and correspond to the eigenvalue zero). It is further useful to note:

$$|0\rangle|u_i\rangle = \frac{1}{\sqrt{2}}(|w_i^+\rangle + |w_i^-\rangle) \quad (13.20)$$

$$|1\rangle|v_i\rangle = \frac{1}{\sqrt{2}}(|w_i^+\rangle - |w_i^-\rangle) \quad (13.21)$$

which enables the input state $|0\rangle|b\rangle$ to be suitably expressed,

$$|0\rangle|b\rangle = \sum_{i=0}^{N'-1} \beta_i \frac{1}{\sqrt{2}}(|w_i^+\rangle + |w_i^-\rangle) + \beta^\perp |w^\perp\rangle \quad (13.22)$$

where $|w^\perp\rangle$ is orthogonal to $|w_i\rangle$ (for all $i = 0 \ldots N' - 1$). Attaching the 'first register' – initially set in the zero state – as well as the qubit that will subsequently be used for the rotation (thus an additional $t + 1$ qubits in the state $|0\rangle$) and stepping through the circuit of HHL applied to the initial state gives:

$$\left(\frac{1}{\sqrt{2}} \sum_{i=0}^{N'-1} \beta_i(|w_i^+\rangle + |w_i^-\rangle) + \beta^\perp|w^\perp\rangle\right)|0^{t+1}\rangle$$

$$\xrightarrow{U_1} \left(\frac{1}{\sqrt{2}} \sum_{i=0}^{N'-1} \beta_i(|w_i^+\rangle|\sigma_i\rangle + |w_i^-\rangle|-\sigma_i\rangle) + \beta^\perp|w^\perp\rangle|0^t\rangle\right)|0\rangle$$

$$\xrightarrow{U_2} \frac{1}{\sqrt{2}} \sum_{i=0}^{N'-1} \beta_i \left(|w_i^+\rangle|\sigma_i\rangle\left(\sqrt{1-\frac{C^2}{\sigma_i^2}}|0\rangle + \frac{C}{\sigma_i}|1\rangle\right)\right.$$

$$\left. + |w_i^-\rangle|-\sigma_i\rangle\left(\sqrt{1-\frac{C^2}{\sigma_i^2}}|0\rangle - \frac{C}{\sigma_i}|1\rangle\right)\right) + \beta^\perp|w^\perp\rangle|0^t\rangle|0\rangle$$

$$\xrightarrow{U_1^\dagger} \frac{1}{\sqrt{2}} \sum_{i=0}^{N'-1} \beta_i \left(|w_i^+\rangle|0^t\rangle\left(\sqrt{1-\frac{C^2}{\sigma_i^2}}|0\rangle + \frac{C}{\sigma_i}|1\rangle\right)\right.$$

$$\left. + |w_i^-\rangle|0^t\rangle\left(\sqrt{1-\frac{C^2}{\sigma_i^2}}|0\rangle - \frac{C}{\sigma_i}|1\rangle\right)\right) + \beta^\perp|w^\perp\rangle|0^t\rangle|0\rangle$$

$$\xrightarrow{\text{post selection}} C'' \sum_{i=0}^{N'-1} \beta_i \left(|w_i^+\rangle |0^t\rangle \frac{1}{\sigma_i} |1\rangle - |w_i^-\rangle |0^t\rangle \frac{1}{\sigma_i} |1\rangle \right)$$

$$= C'' \sum_{i=0}^{N'-1} \frac{\beta_i}{\sigma_i} |1\rangle |v_i\rangle |0^t\rangle |1\rangle$$

$$= C'' |1\rangle \otimes \left(\sum_{i=0}^{N'-1} \sigma_i^{-1} \beta_i |v_i\rangle \right) \otimes |0^t\rangle |1\rangle \qquad (13.23)$$

where C'' is a normalisation constant. That is, (a state proportional to) the desired state, as in Eqn. (13.18), has indeed been prepared, and we now see why the unitary U_2 is defined such that it first checks if the eigenvalue is zero and does nothing if so, only applying the rotation if it is non-zero.

13.4 HHL: the caveats

HHL is truly a *quantum* algorithm for solving a linear system, in that it maps a quantum state representing the vector b to a quantum state representing the vector x. For this to be useful for classical data:

1. there must be some efficient way to prepare the input state, $|b\rangle$;
2. as HHL prepares the quantum state $|x\rangle$, it is further necessary that the output is some quantity pertaining to x that can be quickly extracted – such as, for example, an observable expectation value, $\langle x| M |x\rangle$, estimated by a number of projective measurements – rather than the x itself, *per se*.

Along with these caveats relating to the form of data that HHL takes as an input and returns as an output, the statement of asymptotic complexity can also be interpreted as presenting two further caveats:

3. the matrix to be inverted must be sparse;
4. the matrix to be inverted must have low condition number.

Finally, it has been shown that if the matrix to be inverted is low rank, then HHL can be *dequantised* – as discussed in more detail in Section 13.6 – and so there is a final caveat:

5. the matrix to be inverted must be close to full-rank.

These five caveats form a 'check-list' that any quantum algorithm based on HHL must be assessed against to justify any claim of quantum advantage. Of these five, item 1 has attracted considerable attention and is sometimes called 'the data-loading problem' (see Box 13.1).

Box 13.1 Loading classical data onto quantum computers and QRAM

Of all the caveats, the problem of preparing a state, $|b\rangle$, of the form defined in Eqn. (13.3) has perhaps received the most attention. It is conceivable that one day such states may arise in natively quantum form, for example if the quantum computer is attached to quantum metrology apparatus, which directly prepares such states but for the most part, it is more realistic to suppose that $|b\rangle$ would have to be prepared from suitable classical data (i.e., the computational basis state coefficients).

This is the problem of *data loading* and is one of the most pressing in contemporary quantum computing, especially with regard to the significant challenge of applying quantum algorithms to data-intensive applications. Solving this challenge would give what is often described as 'superposition access to the data', and from a purely theoretical stand-point, a quantum circuit to load arbitrary classical data requires exponential quantum resources (i.e., either exponential circuit depth and/or exponential circuit width – that is, exponentially many ancilla qubits to assist the state preparation). However, given the importance of data loading, there is hope that in future quantum computers, there could be dedicated hardware units – known as *quantum random access memory (QRAM)* – to achieve this task in a manner that is similar to the operation of random access memory in classical computation. Even though the existence of QRAM would not strictly change the complexity from a theoretical standpoint, the hope is that, in practice, QRAM could enable sufficiently rapid data loading to bring data-centric applications into play.

It is widely agreed that realising QRAM is a steep challenge that will take time, and in the intervening time, it is more productive to focus on classical data that is *structured* (QRAM, by definition, applies to arbitrary classical data) in the sense that $|b\rangle$ can be prepared by some acceptably shallow quantum circuit.

13.5 Solving differential equations using HHL

The caveats associated with HHL are usually viewed as negative facts about HHL, and, in a sense, this is true (especially in the case of many proposals for applying HHL to machine learning, as detailed in Section 13.6), as they concern all of the ways in which HHL does not generically speed-up the solving of linear systems. However, the caveats can also be seen a more positive light: by knowing the caveats precisely, it is possible to focus on certain applications where there *is* real promise of quantum advantage.

One such application is applying HHL to the problem of solving linear differential equations. (As an aside, it is worth knowing that there are other approaches to designing quantum algorithms to solve differential equations, most notably attempts to massage the equation into a form that can be solved 'directly' by Hamiltonian simulation.) A linear differential equation in the variable x is of the form

$$\dot{x}(t) = A(t)x(t) + b(t) \tag{13.24}$$

where $\dot{x}(t)$ denotes differentiation with respect to time, A is a matrix, and b is a vector.

To illustrate the principle of solving linear differential equations quantumly, we employ the *Euler method* of approximating the differential equation with a system of difference equations:

$$\frac{x(t_{n+1}) - x(t_n)}{h} \approx A(t_n)x(t_n) + b(t_n) \tag{13.25}$$

for some finite constant, h.

For simplicity, we now index the time instants such that x_n means $x(t_n)$ and similarly for A and b, which thus enables the difference equation to be represented as a system of linear equations:

$$\begin{bmatrix} I & 0 & 0 & 0 & \cdots \\ -(I + hA_0) & I & 0 & 0 & \cdots \\ 0 & -(I + hA_1) & I & 0 & \cdots \\ 0 & 0 & -(I + hA_2) & I & \cdots \\ \vdots & \vdots & \vdots & \vdots & \ddots \end{bmatrix} \begin{bmatrix} x_0 \\ x_1 \\ x_2 \\ x_3 \\ \vdots \end{bmatrix} = \begin{bmatrix} x_{init} \\ hb_0 \\ hb_1 \\ hb_2 \\ \vdots \end{bmatrix} \tag{13.26}$$

The matrix in Eqn. (13.26) is a block matrix, where each block has dimension equal to that of the original differential equation, and the number of blocks is equal to the number of time steps considered.

As the linear differential equation has now be approximated by a linear system, it follows that, in principle, HHL can be used to invert the matrix and output (a quantum encoding of) x. It is, therefore, incumbent upon us to ask whether solving linear differential equations in this way yields an overall quantum advantage. This entails working through the caveats (in the order they were listed in Section 13.4):

1. (input is quantum encoding $|b\rangle$ that must be prepared) if there exists an efficient classical algorithm to calculate $b(t)$, then this property will be inherited directly by the quantum algorithm and so there will be a (efficient) quantum circuit to prepare $|b\rangle$;
2. (output is quantum encoding $|x\rangle$) in practice, the quantity to be computed will usually be some property of the state at the final time, and one way to boost the probability of collapsing to a measurement outcome pertaining to the final time step is to append a period of steady state to the linear differential equation evolution, where $x_n = x_{n-1}$;
3. (matrix must be sparse) if the number of time-steps is large compared to the number of variables in the original linear differential equation, then the matrix

will be sparse (alternatively the sparsity of the linear differential equation itself is clearly a sufficient condition for the linear system to have a sparse matrix);
4. (matrix must be well-conditioned) it turns out that it is hard to guarantee that the condition number remains small as the linear differential equation system size grows when using the Euler method;
5. (matrix must be close to full rank) in general, the matrix to be inverted will be full-rank, or close to full rank.

It is, therefore, the fourth of these that poses the greatest problem. The Euler method represents the simplest means of discretising linear differential equations, but it is known to be ineffective classically in many cases of practical interest. 'Quantising' more advanced techniques for solving differential equations (including partial and/or non-linear differential equations – which are often needed to describe practically interesting systems) is, therefore, a very active research direction, and one in which meeting *all* of the caveats is of central importance.

13.6 Quantum linear algebra for machine learning and low rank dequantisation

Many problems in machine learning amount to manipulating large matrices. The advent of HHL, therefore, led to a surge of anticipation that quantum algorithms for linear algebraic tasks may lead to exponential speed-ups in machine learning applications. This, in turn, led to proposals for quantum-enhanced algorithms for recommendation systems, principle component analysis, clustering, and support vector machines, amongst many others. In some cases, the algorithms themselves are little more than applications of HHL, but others, for example the quantum recommendation system algorithm – one of the most celebrated – are similar, but algorithmically distinct.

It turns out that in all cases, these quantum algorithms owe their quantum advantage to the assumptions made about the data access and matrix properties (namely being low-rank), rather than because of the quantum information processing itself. Ewin Tang showed that, with commensurate data access, there exist classical algorithms matching the complexity of the quantum algorithms – at least in low rank cases – a procedure that has become known as 'dequantising' the quantum algorithms in question.

A more positive way to look at the process of dequantisation is that it yields *quantum-inspired* classical algorithms. Central to these (at least for the linear algebra-based algorithms considered here) are randomised approaches to approximate linear algebraic operations that leverage the *low rank* of the matrix in question. In the case of linear systems solving, this amounts to a dequantisation of HHL in the low rank setting, which, in turn, gives the fifth caveat in Section 13.4. More generally, the (approximate) low-rank of high-dimensional datasets is often a central property that enables (classical) machine learning to work at all – in that seemingly complex data can often be explained by a relatively small number of factors. This means that the low-rank dequantisation is often fatal, although optimists point at the possibility of still advantageous 'lower-order' quantum speed-ups (for example, even though the dequantisation has eliminated the possibility of an *exponential* quantum advantage, there may still be a polynomial advantage available).

Chapter problems

1. For each of the following matrices, find an alternative suitable for use in HHL (or justify that no alternative is necessary).

    (a) $\begin{bmatrix} 0 & 1+i \\ 1-i & 15 \end{bmatrix}$

    (b) $\begin{bmatrix} 0 & 1+i & 0 \\ 1-i & 15 & 0 \\ 0 & 0 & 25 \end{bmatrix}$

    (c) $\begin{bmatrix} 1 & 0 \\ 1 & 1 \end{bmatrix}$

2. For each of the matrices found in question 1, give the condition number.

3. Let A be a matrix of size $N = 2^n$ for some integer n, which has SVD $A = \sum_{i=0}^{N'-1} \sigma_i |u_i\rangle \langle v_i|$, where $N' \leq N$ is the number of non-zero singular values. Further, let $H = \begin{bmatrix} 0 & A \\ A^\dagger & 0 \end{bmatrix}$, and show that for all non-zero σ_i, $\pm \sigma_i$ are eigenvalues of H with corresponding eigenvectors $|w_i^\pm\rangle = \frac{1}{\sqrt{2}}(|0\rangle |u_i\rangle \pm |1\rangle |v_i\rangle)$.

4. HHL is to be used to find **x** by solving the linear system:

$$\begin{bmatrix} \frac{1}{\sqrt{5}} \\ -\frac{2}{\sqrt{5}} \end{bmatrix} = \begin{bmatrix} \frac{3}{8} & \frac{1}{8} \\ \frac{1}{8} & \frac{3}{8} \end{bmatrix} \mathbf{x} \qquad (13.27)$$

   (a) Verify that $\begin{bmatrix} \frac{1}{\sqrt{5}} \\ -\frac{2}{\sqrt{5}} \end{bmatrix}$ is already correctly normalised and hence express $|b\rangle$.

   (b) Find the eigenvectors, $|\lambda_0\rangle$ and $|\lambda_1\rangle$ and associated eigenvalues, E_0 and E_1 of the matrix $\begin{bmatrix} \frac{3}{8} & \frac{1}{8} \\ \frac{1}{8} & \frac{3}{8} \end{bmatrix}$.

   (c) Express $|b\rangle$ as a superposition of $|\lambda_0\rangle$ and $|\lambda_1\rangle$.

   (d) Let two qubits be used in the second register, what is the state prepared by HHL immediately before the measurement pertaining to the post-selection condition (i.e., the state $|\psi_3\rangle$)?

   (e) What is the maximum value of the constant, C, which can be chosen in this case without an over-rotation?

(f) If this maximum value of C is used, what is the probability that the post-select condition is satisfied on any single run?

(g) When the post-selection condition is met, verify that the final state is proportional to **x** obtained by direct calculation.

Further reading

The quantum algorithm for solving a linear system is due to Harrow, Hassidim, and Lloyd [2009]. Since the original HHL paper, there have been a number of variations and developments proposed, notably when efficiency in the error is crucial a version running in time polynomial in the logarithm of the reciprocal of the error by Childs *et al* [2017], and conversely, when the error and sparsity are treated as constants, a reduction in complexity from $\widetilde{\mathcal{O}}(\log(N)\kappa^2)$ to $\widetilde{\mathcal{O}}(\log(N)\kappa \log^3 \kappa)$ by Andris Ambainis [2012]. The caveats of HHL are well-known within the community, and Scott Aaronson [2015] does a nice job of summing up the general problems with proposals of quantum advantage of this type.

QRAM was proposed by Giovannetti *et al* [2008]; using quantum computers to solve differential equations has been proposed in myriad articles, but the version presented here is based on that of Dominic Berry [2014]. The most famous quantum machine learning algorithm mentioned here is the *quantum recommendation system* by Kerenidis and Prakash [2016], which was dequantised by Ewin Tang [2019]; Tang and her co-authors [2020] then presented a general framework for dequantising 'quantum' machine learning algorithms of this type in a wide range of settings.

14

Quantum signal processing and the quantum singular value transformation

The quantum algorithms presented in the previous chapters are underpinned by three fundamental features of quantum computation. First, it is possible to construct Grover iterates that amplify amplitude and concentrate probability mass on desired outcomes quadratically faster than their classical counterparts; second, quantum phase estimation is a subroutine that (in the version presented) uses the quantum Fourier transform to interfere a quantum state such that wrong answers cancel out, leaving a state from which the solution to the problem can be extracted; and finally, using a digital quantum computer, equipped with a standard finite universal gateset, Hamiltonian evolution can be simulated.

Remarkably, these three phenomena can be seen to be instances of the same underlying computational effect: the *quantum singular value transformation*. This chapter introduces the quantum singular value transformation, starting from *quantum signal processing* for a single-qubit instance, before moving on to show how *block encoding* and *qubitisation* allow states in arbitrary-sized Hilbert spaces to benefit from this extremely flexible means of transformation. It is then shown how each of search, quantum phase estimation, and Hamiltonian simulation is, as claimed, achieved by the quantum singular value transformation. This impressive property of the quantum singular value transformation has led to it being thought of not as a single algorithm, but rather a general framework for developing quantum algorithms, and indeed has led to it being dubbed the 'grand unification of quantum algorithms'.

14.1 Quantum signal processing

The powerhouse of the quantum singular value transformation is the method of applying a polynomial transformation to a unitary matrix (in a sense that is implicitly defined in what follows) known as *quantum signal processing*. Quantum signal processing is a technique that applies to a 2×2 (single-qubit) matrix, W, parameterised by some real variable x,

$$W(x) = \begin{bmatrix} x & i\sqrt{1-x^2} \\ i\sqrt{1-x^2} & x \end{bmatrix} \tag{14.1}$$

which can easily be seen to be unitary for $-1 \leq x \leq 1$. The goal of quantum signal processing is to construct a sequence of single-qubit gates such that their product is:

$$W_\Phi(x) = \begin{bmatrix} P(x) & iQ(x)\sqrt{1-x^2} \\ iQ^*(x)\sqrt{1-x^2} & P^*(x) \end{bmatrix} \quad (14.2)$$

where $P(x)$ and $Q(x)$ are polynomials. In applications of the quantum singular value transformation, it is usually the case that $P(x)$ is designed to have certain properties.

The quantum signal processing theorem. $W_\Phi(x)$ can be constructed by the following sequence of gates:

$$W_\Phi(x) = R_z(\phi_0) \prod_{i=1}^{d} W(x) R_z(\phi_i) \quad (14.3)$$

where $\Phi = [\phi_0, \phi_1, \ldots, \phi_d]$, if and only if:

1. $\deg(P) \leq d$ and $\deg(Q) \leq d - 1$
2. If d is odd, P is an odd function and Q is an even function; if d is even, P is an even function and Q is an odd function.
3. $\forall x \in [-1, 1], |P(x)|^2 + (1 - x^2)|Q(x)|^2 = 1$

What these three conditions say is that: (i) the number of unitary matrices needed in the product grows proportionally with the degree of the polynomial (as one may intuitively expect); (ii) this only works for purely odd or purely even functions, $P(x)$ (in practice, this can be made to hold for general functions: one way is to use a linear combination of unitaries in conjunction with quantum signal processing, noting that every function can be decomposed as a purely even part plus a purely odd part); and (iii) the polynomials $P(x)$ and $Q(x)$ must be such that unitarity is preserved in the final matrix, $W_\Phi(x)$.

Conditions (i) and (ii) can be shown by induction to be implied by any unitary of the form given in Eqn. (14.2) that has been constructed according to Eqn. (14.3). For the base case, $d = 0$, $W_\Phi = R_z(\phi_0)$, and so $P(x) = e^{-i\phi_0/2}$ and $Q(x) = 0$. Thus, $P(x)$ is a constant (degree $0 \leq d$ as claimed) and $Q(x)$ is equal to zero (which is commonly, albeit not universally defined as a polynomial of degree -1 which is thus equal to $d - 1$ as claimed); furthermore, as $d = 0$, P should be even, as a non-zero constant indeed is, and Q should be odd, which it is (a function that is equal to the constant zero is both even and odd). Next is the inductive step, where (i) and (ii) are assumed to hold for $W_{[\phi_0,\ldots,\phi_d]}(x)$ and so we consider $W_{[\phi_0,\ldots,\phi_{d+1}]}(x)$:

$$W_{[\phi_0,\ldots,\phi_{d+1}]}(x) = \begin{bmatrix} P(x) & iQ(x)\sqrt{1-x^2} \\ iQ^*(x)\sqrt{1-x^2} & P^*(x) \end{bmatrix} W(x) R_z(\phi_{d+1})$$

$$= \begin{bmatrix} P(x) & iQ(x)\sqrt{1-x^2} \\ iQ^*(x)\sqrt{1-x^2} & P^*(x) \end{bmatrix} \begin{bmatrix} xe^{-i\phi_{d+1}/2} & ie^{i\phi_{d+1}/2}\sqrt{1-x^2} \\ ie^{-i\phi_{d+1}/2}\sqrt{1-x^2} & xe^{i\phi_{d+1}/2} \end{bmatrix}$$

$$= \begin{bmatrix} e^{-i\phi_{d+1}/2}(xP(x) - (1-x^2)Q(x)) & i\sqrt{1-x^2}e^{i\phi_{d+1}/2}(P(x) + xQ(x)) \\ i\sqrt{1-x^2}e^{-i\phi_{d+1}/2}(P^*(x) + xQ^*(x)) & e^{i\phi_{d+1}/2}(xP^*(x) - (1-x^2)Q^*(x)) \end{bmatrix}$$

$$= \begin{bmatrix} \tilde{P}(x) & i\tilde{Q}(x)\sqrt{1-x^2} \\ i\tilde{Q}^*(x)\sqrt{1-x^2} & \tilde{P}^*(x) \end{bmatrix} \tag{14.4}$$

where:

$$\tilde{P}(x) = e^{-i\phi_{d+1}/2}(xP(x) - (1-x^2)Q(x)) \tag{14.5}$$

$$\tilde{Q}(x) = e^{i\phi_{d+1}/2}(P(x) + xQ(x)) \tag{14.6}$$

from which it can be seen that, if the degree of Q is $d-1$ and the degree of P is d, then the maximum degree of $\tilde{P}$ is $d+1$; also the maximum degree of $\tilde{Q}$ is d, thus verifying condition (i); Eqns. (14.5) and (14.6) also show that if P and Q are odd and even, respectively, then $\tilde{P}$ and $\tilde{Q}$ will be even and odd respectively (and vice-versa), thus implying condition (ii). Condition (iii) immediately follows too, as a product of unitary matrices is itself necessarily unitary.

Having shown that the construction given in Eqns. (14.2) and (14.3) implies conditions (i)–(iii), the converse is to show that every pair of polynomials, P, Q, that satisfy conditions (i)–(iii) can be encoded in a unitary matrix as in Eqn. (14.2). This can be proven by induction: the base case $d = 0$ is satisfied as $W_\Phi = R_z(\phi_0)$ trivially captures every degree 0 polynomial P and degree -1 polynomial Q that meets the normalisation condition. For the inductive step, we begin by unpacking condition (iii), in particular by substituting $|\tilde{P}(x)|^2 = \tilde{P}(x)\tilde{P}^*(x)$ and $|\tilde{Q}(x)|^2 = \tilde{Q}(x)\tilde{Q}^*(x)$ into condition (iii):

$$1 = \tilde{P}(x)\tilde{P}^*(x) + (1-x^2)\tilde{Q}(x)\tilde{Q}^*(x) \tag{14.7}$$

Now we observe that in order for this to hold, if P is a polynomial of degree $d+1$, then Q is a polynomial of degree d. Moreover, if $P(x)$ is written $p_{d+1}x^{d+1} + \ldots$ and similarly, $Q(x)$ is written $q_d x^d + \ldots$, then in order for all of the terms to cancel such that Eqn. (14.7) holds, it is further necessary that $|p_{d+1}| = |q_d|$. We choose ϕ_{d+1} such that $p_{d+1}/q_d = e^{-i\phi_{d+1}}$, and consider:

$$\begin{bmatrix} P(x) & iQ(x)\sqrt{1-x^2} \\ iQ^*(x)\sqrt{1-x^2} & P^*(x) \end{bmatrix} R_z^\dagger(\phi_{d+1})W^\dagger(x)$$

$$= \begin{bmatrix} P(x) & iQ(x)\sqrt{1-x^2} \\ iQ^*(x)\sqrt{1-x^2} & P^*(x) \end{bmatrix} \begin{bmatrix} xe^{i\phi_{d+1}/2} & -ie^{i\phi_{d+1}/2}\sqrt{1-x^2} \\ -ie^{-i\phi_{d+1}/2}\sqrt{1-x^2} & xe^{-i\phi_{d+1}/2} \end{bmatrix}$$

$$= \begin{bmatrix} xP(x)e^{i\phi_{d+1}/2} + (1-x^2)Q(x)e^{-i\phi_{d+1}/2} & i\sqrt{1-x^2}(-P(x)e^{i\phi_{d+1}/2} + xQ(x)e^{-i\phi_{d+1}/2}) \\ i\sqrt{1-x^2}(-P^*(x)e^{-i\phi_{d+1}/2} + xQ^*(x)e^{i\phi_{d+1}/2}) & xP^*(x)e^{-i\phi_{d+1}/2} + (1-x^2)Q^*(x)e^{i\phi_{d+1}/2} \end{bmatrix}$$

$$= \begin{bmatrix} \tilde{P}(x) & i\tilde{Q}(x)\sqrt{1-x^2} \\ i\tilde{Q}^*(x)\sqrt{1-x^2} & \tilde{P}^*(x) \end{bmatrix} \tag{14.8}$$

where

$$\tilde{P}(x) = xP(x)e^{i\phi_{d+1}/2} + (1-x^2)Q(x)e^{-i\phi_{d+1}/2} \tag{14.9}$$

$$\tilde{Q}(x) = -P(x)e^{i\phi_{d+1}/2} + xQ(x)e^{-i\phi_{d+1}/2} \tag{14.10}$$

Notably, owing to the choice $p_{d+1}/q_d = e^{-i\phi_{d+1}}$, we get that $\tilde{P}$ is a polynomial of degree d and $\tilde{Q}$ is a polynomial of degree $d-1$. Rearranging, Eqn. (14.8) gives:

$$\begin{bmatrix} P(x) & iQ(x)\sqrt{1-x^2} \\ iQ^*(x)\sqrt{1-x^2} & P^*(x) \end{bmatrix} = \begin{bmatrix} \tilde{P}(x) & i\tilde{Q}(x)\sqrt{1-x^2} \\ i\tilde{Q}^*(x)\sqrt{1-x^2} & \tilde{P}^*(x) \end{bmatrix} W(x)R_z(\phi_{d+1}) \tag{14.11}$$

By the inductive hypothesis, as $\tilde{P}$ and $\tilde{Q}$ have degree d and $d-1$, respectively, the following construction holds:

$$\begin{bmatrix} \tilde{P}(x) & i\tilde{Q}(x)\sqrt{1-x^2} \\ i\tilde{Q}^*(x)\sqrt{1-x^2} & \tilde{P}^*(x) \end{bmatrix} = R_z(\phi_0)\prod_{i=1}^{d} W(x)R_z(\phi_i) \tag{14.12}$$

and substituting this into Eqn. (14.11) gives:

$$\begin{bmatrix} P(x) & iQ(x)\sqrt{1-x^2} \\ iQ^*(x)\sqrt{1-x^2} & P^*(x) \end{bmatrix} = R_z(\phi_0)\left(\prod_{i=1}^{d} W(x)R_z(\phi_i)\right)W(x)R_z(\phi_{d+1})$$

$$= R_z(\phi_0)\prod_{i=1}^{d+1} W(x)R_z(\phi_i) \tag{14.13}$$

thus completing the inductive proof.

Although an interesting construction in its own right, it is not immediately clear how quantum signal processing provides a means of Hamiltonian simulation, or indeed of performing any other quantum algorithm. To see how quantum signal processing can be applied to transform states in larger spaces, it is first helpful to decompose $W(x)$,

$$W(x) = iR_z(\pi/2)R(x)R_z(\pi/2) \tag{14.14}$$

where the leading i is a global phase factor that can be neglected, and

$$R(x) = \begin{bmatrix} x & \sqrt{1-x^2} \\ \sqrt{1-x^2} & -x \end{bmatrix} \tag{14.15}$$

Thus, the $R_z(\pi/2)$ gates can be absorbed into the parameterised angles Φ, and so quantum signal processing can instead be stated in terms of the product:

$$R_z(\phi_0)R(x)R_z(\phi_1)R(x)\ldots R(x)R_z(\phi_d) \qquad (14.16)$$

To see how this is practically useful, it is necessary to introduce the notion of a *block encoding* of a matrix.

14.2 Block encoding and the quantum eigenvalue transformation

Let H be some general $N \times N$ Hermitian matrix, which is (pre-)scaled such that $||H|| \leq 1$. In this case, it is possible to embed H in a larger unitary matrix, U. This is known as *block encoding*, and for simplicity, we (initially) consider the case where H appears in the top-left corner of U, i.e.,

$$U = \begin{bmatrix} H & \cdot \\ \cdot & \cdot \end{bmatrix} \qquad (14.17)$$

This embedding is such that the other three blocks of U are of size at least $N \times N$. If U is such that the other blocks are exactly $N \times N$, to 'access' the encoded block, applied to some input state $|\psi\rangle$, it is necessary to prepare the initial state $|0\rangle|\psi\rangle$. If ψ is the column vector of $|\psi\rangle$, then $|0\rangle|\psi\rangle$ is a column vector of ψ followed by zeros and so (mixing notation slightly):

$$U(|0\rangle|\psi\rangle) = \begin{bmatrix} H & \cdot \\ \cdot & \cdot \end{bmatrix}\begin{bmatrix} \psi \\ 0 \end{bmatrix} = \begin{bmatrix} H\psi \\ \vdots \end{bmatrix} = |0\rangle(H|\psi\rangle) + |1\rangle|\psi'\rangle \qquad (14.18)$$

where $|\psi'\rangle$ is irrelevant (and is not necessarily a properly normalised state as expressed in Eqn. (14.18)). Measuring $|0\rangle$ on the first qubit then collapses the superposition to a state proportional to $H|\psi\rangle$ as desired. Therefore, to 'pick out' the encoded block, the first qubit is prepared and post-selected in the $|0\rangle$ state, as is the case when a linear combination of unitaries is used to simulate Hamiltonian evolution with a truncated Taylor series in Section 11.2.2. Moreover, as is the case with linear combination of unitaries, if the block encoding is such that U is larger than $2N \times 2N$, then the top-left block is 'picked' out by preparing and measuring multiple qubits in the $|0\rangle$ state.

It is convenient to make the further assumption (in the first instance) that U has the specific form:

$$U = \begin{bmatrix} H & \sqrt{1-H^2} \\ \sqrt{1-H^2} & -H \end{bmatrix} \qquad (14.19)$$

(i.e., the other blocks, therefore, have size exactly $N \times N$) where:

$$\sqrt{1 - H^2} = \sum_i \sqrt{1 - \lambda_i^2} |\lambda_i\rangle\langle\lambda_i| \tag{14.20}$$

where $\{|\lambda_*\rangle\}$ are the eigenvectors of H, with associated eigenvalues $\{\lambda_*\}$, i.e., this follows from the standard definition of a function applied to a diagonalisable matrix in Eqn. (2.43). (Note that, even though H is Hermitian, we use $\{\lambda_*\}$ for its eigenvalues rather than $\{E_*\}$ as this makes the analysis considerably easier to follow, and as there is no need for the eigendecomposition of the unitary matrix for exponentiated iH in this chapter, no ambiguity is introduced in doing so.) Eqn. (14.19) can be expressed:

$$\begin{aligned} U &= Z \otimes H + X \otimes \sqrt{1 - H^2} \\ &= \sum_i \lambda_i \begin{bmatrix} 1 & 0 \\ 0 & -1 \end{bmatrix} \otimes |\lambda_i\rangle\langle\lambda_i| + \sum_i \sqrt{1 - \lambda_i^2} \begin{bmatrix} 0 & 1 \\ 1 & 0 \end{bmatrix} \otimes |\lambda_i\rangle\langle\lambda_i| \\ &= \sum_i \begin{bmatrix} \lambda_i & \sqrt{1 - \lambda_i^2} \\ \sqrt{1 - \lambda_i^2} & -\lambda_i \end{bmatrix} \otimes |\lambda_i\rangle\langle\lambda_i| \end{aligned} \tag{14.21}$$

A second operator, which is simply a $R_z(\phi)$ gate applied to the first qubit, is next needed, and this performs the operation on the entire Hilbert space:

$$R_z(\phi) \otimes I_N = \sum_i R_z(\phi) \otimes |\lambda_i\rangle\langle\lambda_i| \tag{14.22}$$

using the fact that the identity can be decomposed into any orthonormal basis (of the appropriate dimension), as in Eqn. (2.39).

The *quantum eigenvalue transformation* is (in some sense) the simplest generalisation of quantum signal processing to a multi-qubit system and, analogously with quantum signal processing, the quantum eigenvalue transformation circuit consists of R_z gates applied to the first qubit alternated with the block encoded operator, U, i.e.,

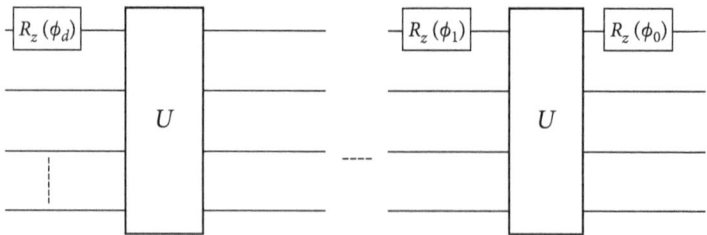

Figure 14.1

which performs the operation

$$(R_z(\phi_0) \otimes I_N) \prod_{k=1}^{d} U(R_z(\phi_k) \otimes I_N)$$

$$= \left(\sum_i R_z(\phi_0) \otimes |\lambda_i\rangle\langle\lambda_i|\right)\left(\sum_i \begin{bmatrix} \lambda_i & \sqrt{1-\lambda_i^2} \\ \sqrt{1-\lambda_i^2} & -\lambda_i \end{bmatrix} \otimes |\lambda_i\rangle\langle\lambda_i|\right)$$

$$\times \left(\sum_i R_z(\phi_1) \otimes |\lambda_i\rangle\langle\lambda_i|\right) \cdots \left(\sum_i \begin{bmatrix} \lambda_i & \sqrt{1-\lambda_i^2} \\ \sqrt{1-\lambda_i^2} & -\lambda_i \end{bmatrix} \otimes |\lambda_i\rangle\langle\lambda_i|\right)$$

$$\times \left(\sum_i R_z(\phi_d) \otimes |\lambda_i\rangle\langle\lambda_i|\right)$$

$$= \sum_i \left(R_z(\phi_0) \begin{bmatrix} \lambda_i & \sqrt{1-\lambda_i^2} \\ \sqrt{1-\lambda_i^2} & -\lambda_i \end{bmatrix} R_z(\phi_1) \right.$$

$$\left. \cdots \begin{bmatrix} \lambda_i & \sqrt{1-\lambda_i^2} \\ \sqrt{1-\lambda_i^2} & -\lambda_i \end{bmatrix} R_z(\phi_d) \right) \otimes |\lambda_i\rangle\langle\lambda_i|$$

$$= \sum_i \left(R_z(\phi_0) R(\lambda_i) R_z(\phi_1) \ldots R(\lambda_i) R_z(\phi_d) \right) \otimes |\lambda_i\rangle\langle\lambda_i|$$

$$= \sum_i \left(R_z(\phi_0) \prod_{k=0}^{d} R(\lambda_i) R_z(\phi_k) \right) \otimes |\lambda_i\rangle\langle\lambda_i| \qquad (14.23)$$

where the (mutual) orthonormality of the basis states means that

$$\langle\lambda_i|\lambda_j\rangle = \begin{cases} 1 & \text{if } i = j \\ 0 & \text{otherwise} \end{cases}$$

and has been used to combine the sums.

This series of operations amounts to manipulating each of the N eigenvalues as if they are each being operated on by a single-qubit operation (the same operation for each eigenvalue). Choosing a suitable series of angles $[\phi_0, \phi_1, \ldots, \phi_d]$ as described in quantum signal processing, polynomial transformations can be applied to the eigenvalues:

$$(R_z(\phi_0) \otimes I_N) \prod_{k=1}^{d} U(R_z(\phi_k) \otimes I_N) = \sum_i \begin{bmatrix} P(\lambda_i) & iQ(\lambda_i)\sqrt{1-\lambda_i^2} \\ iQ^*(\lambda_i)\sqrt{1-\lambda_i^2} & P^*(\lambda_i) \end{bmatrix} \otimes |\lambda_i\rangle\langle\lambda_i|$$

$$(14.24)$$

Using the fact that a function applied to matrix is simply the function applied to its eigenvalues when spectrally decomposed, it can equivalently be written that:

$$(R_z(\phi_0) \otimes I_N) \prod_{k=1}^{d} U(R_z(\phi_k) \otimes I_N) = \begin{bmatrix} P(H) & iQ(H)\sqrt{1-H^2} \\ iQ^*(H)\sqrt{1-H^2} & P^*(H) \end{bmatrix} \quad (14.25)$$

where the quantum signal processing conditions apply to $P(\cdot)$ and $Q(\cdot)$.

14.3 Quantum signal processing as a means of Hamiltonian simulation

One of the original applications envisaged for quantum signal processing, when applied to block-encoded Hermitian matrices, is Hamiltonian simulation. Each of sine and cosine (which are odd and even functions, respectively) can be approximated by polynomials using the Jacobi–Anger expansion:

$$\cos(xt) = J_0(t) + 2\sum_{k=1}^{\infty} (-1)^k J_{2k}(t) T_{2k}(x) \quad (14.26)$$

$$\sin(xt) = 2\sum_{k=0}^{\infty} (-1)^k J_{2k+1}(t) T_{2k+1}(x) \quad (14.27)$$

where $J_k(t)$ is a Bessel function of order k and $T_k(x)$ is a Chebyshev polynomial of order k. The quantum eigenvalue transformation can be used to apply a suitable truncation of these functions to a suitable block encoding of the Hamiltonian. Taking first the polynomial approximation of sine, as the Hermitian matrix is encoded in the top-left block of a unitary matrix, this requires the initial state to be set as $|0\rangle|\psi\rangle$, and the measurement outcome 0 to be post-selected on the first qubit, and so the circuit is:

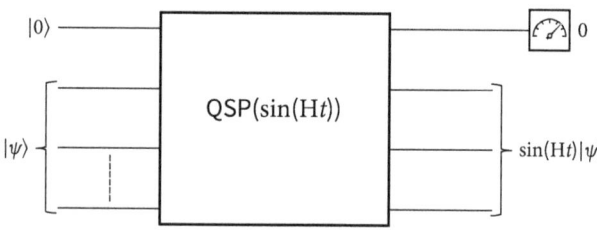

Figure 14.2

Taking the corresponding circuit to apply a polynomial approximation of cosine, together with the familiar expression:

$$e^{iHt} = \cos(Ht) + i\sin(Ht) \quad (14.28)$$

an exponentiated Hamiltonian can be simulated using a simple linear combination of unitaries (see Chapter 11) for cos(Ht) and i sin(Ht), where the linear combination of unitaries essentially acts as a further block encoding:

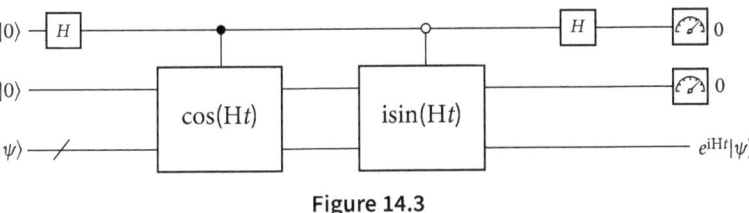

Figure 14.3

Complexity analysis

As a general rule when using the quantum eigenvalue transformation and quantum singular value transformation, the overall computational complexity is found by considering:

1. The number of operations (gates) required to the prepare the block encoding (in some presentations, the complexity will simply be given in terms of the number of queries to some black-box block encoding). Generic techniques for block encoding typically have complexity that grows with the sparsity of the matrix in question – which is also the case for other means of Hamiltonian simulation.
2. The degree of the polynomial approximation of the desired function to be applied (higher degree means closer approximation). As the degree is the number of uses of the block encoding, this translates directly into the computational complexity.
3. The probability of the required post-selection condition occurring, which therefore dictates the expected number of runs in a repeat until success strategy.

Complexity analysis of specific applications of the quantum eigenvalue and singular value transformations will often largely be concerned with the second of these – that is, the efficiency of the polynomial approximation of some desired function – an important topic, but not a particularly 'quantum' one. So it follows that, departing from the convention established in much of the rest of this book, in this chapter, we shall focus on explaining how certain algorithmic effects can be achieved by the quantum eigenvalue/singular value transformation and then simply state the overall complexity. In the case of using quantum signal processing for sparse Hamiltonian simulation, the overall complexity is:

$$\mathcal{O}\left(t + \log\left(\frac{1}{\epsilon}\right)\right) \tag{14.29}$$

where t is the simulation time, and ϵ is the error. This is complexity asymptotically optimal: no means of Hamiltonian simulation can do better.

14.4 The quantum singular value transformation

The quantum eigenvalue transformation applies primarily to (block-encoded) Hermitian matrices. As quantum mechanical Hamiltonians *are* Hermitian, this gives it great power – as exemplified in the application to Hamiltonian simulation. However, going beyond the quantum *eigenvalue* transformation to the quantum *singular value* transformation allows the same essential framework to be applied to a wide range of further computational problems.

Recall from Chapter 2 that not all matrices are diagonalisable, but all matrices *do* have a singular value decomposition:

$$A = \sum_i \sigma_i |u_i\rangle \langle v_i| \quad (14.30)$$

where A is any $N \times M$ matrix, $\{\sigma_*\}$ are singular values (which are positive real numbers), and $\{|u_*\rangle\}$ and $\{|v_*\rangle\}$ are orthonormal bases.

The quantum singular value transformation applies to matrices decomposed according to the singular value decomposition, and to see the quantum singular value transformation in action, we initially assume that A is square (but now not necessarily diagonalisable) and encoded in the top-left block of a unitary of the form:

$$U = \begin{bmatrix} A & \sqrt{1-A^2} \\ \sqrt{1-A^2} & -A \end{bmatrix} \quad (14.31)$$

where:

$$\sqrt{1-A^2} = \sum_i \sqrt{1-\sigma_i^2} |u_i\rangle \langle v_i| \quad (14.32)$$

from which a derivation analogous to that deployed when presenting the quantum eigenvalue transform then yields:

$$U = \sum_i \begin{bmatrix} \sigma_i & \sqrt{1-\sigma_i^2} \\ \sqrt{1-\sigma_i^2} & -\sigma_i \end{bmatrix} \otimes |u_i\rangle \langle v_i| \quad (14.33)$$

and:

$$U^\dagger = \sum_i \begin{bmatrix} \sigma_i & \sqrt{1-\sigma_i^2} \\ \sqrt{1-\sigma_i^2} & -\sigma_i \end{bmatrix} \otimes |v_i\rangle \langle u_i| \quad (14.34)$$

As for the quantum eigenvalue transformation, the sequence of operations deployed to achieve a quantum *singular value* transformation derives from quantum signal processing and, owing to the absence of symmetry in a quantum singular value transformation (i.e., the left- and right-singular spaces are not in general the same), it is necessary to separate out the cases of odd and even d. For odd polynomials, the series of unitaries is

$$(R_z(\phi_0) \otimes I_N)U(R_z(\phi_1) \otimes I_N) \prod_{k=1}^{\frac{d-1}{2}} U^\dagger(R_z(\phi_{2k}) \otimes I_N)U(R_z(\phi_{2k+1}) \otimes I_N)$$

$$= \left(\sum_i R_z(\phi_0) \otimes |u_i\rangle\langle u_i|\right)\left(\sum_i \begin{bmatrix} \sigma_i & \sqrt{1-\sigma_i^2} \\ \sqrt{1-\sigma_i^2} & -\sigma_i \end{bmatrix} \otimes |u_i\rangle\langle v_i|\right)$$

$$\times \left(\sum_i R_z(\phi_1) \otimes |v_i\rangle\langle v_i|\right) \prod_{k=1}^{\frac{d-1}{2}} \left(\sum_i \begin{bmatrix} \sigma_i & \sqrt{1-\sigma_i^2} \\ \sqrt{1-\sigma_i^2} & -\sigma_i \end{bmatrix} \otimes |v_i\rangle\langle u_i|\right)$$

$$\times \left(\sum_i R_z(\phi_{2k}) \otimes |u_i\rangle\langle u_i|\right)\left(\sum_i \begin{bmatrix} \sigma_i & \sqrt{1-\sigma_i^2} \\ \sqrt{1-\sigma_i^2} & -\sigma_i \end{bmatrix} \otimes |u_i\rangle\langle v_i|\right)$$

$$\times \left(\sum_i R_z(\phi_{2k+1}) \otimes |v_i\rangle\langle v_i|\right)$$

$$= \sum_i \left(R_z(\phi_0) \begin{bmatrix} \sigma_i & \sqrt{1-\sigma_i^2} \\ \sqrt{1-\sigma_i^2} & -\sigma_i \end{bmatrix} R_z(\phi_1) \right.$$

$$\left. \cdots \begin{bmatrix} \sigma_i & \sqrt{1-\sigma_i^2} \\ \sqrt{1-\sigma_i^2} & -\sigma_i \end{bmatrix} R_z(\phi_d) \right) \otimes |u_i\rangle\langle v_i|$$

$$= \sum_i \begin{bmatrix} P(\sigma_i) & iQ(\sigma_i)\sqrt{1-\sigma_i^2} \\ iQ^*(\sigma_i)\sqrt{1-\sigma_i^2} & P^*(\sigma_i) \end{bmatrix} \otimes |u_i\rangle\langle v_i| \qquad (14.35)$$

which has been simplified using the fact that as $\sum_i |u_i\rangle\langle u_i| = \sum_i |v_i\rangle\langle v_i| = I_N$, and so the action of a rotation, $R_z(\phi)$ on the first qubit, i.e., $R_z(\phi) \otimes I_N$ can be expressed as either $\sum_i R_z(\phi) \otimes |u_i\rangle\langle u_i|$ or $\sum_i R_z(\phi) \otimes |v_i\rangle\langle v_i|$).

For even d, the sequence of operations is:

$$(R_z(\phi_0) \otimes I_N) \prod_{k=1}^{\frac{d}{2}} U^\dagger(R_z(\phi_{2k-1}) \otimes I_N)U(R_z(\phi_{2k}) \otimes I_N)$$

$$= \left(\sum_i R_z(\phi_0) \otimes |v_i\rangle\langle v_i|\right) \prod_{k=1}^{\frac{d}{2}} \left(\sum_i \begin{bmatrix} \sigma_i & \sqrt{1-\sigma_i^2} \\ \sqrt{1-\sigma_i^2} & -\sigma_i \end{bmatrix} \otimes |v_i\rangle\langle u_i|\right)$$

$$\times \left(\sum_i R_z(\phi_{2k-1}) \otimes |u_i\rangle\langle u_i| \right) \left(\sum_i \begin{bmatrix} \sigma_i & \sqrt{1-\sigma_i^2} \\ \sqrt{1-\sigma_i^2} & -\sigma_i \end{bmatrix} \otimes |u_i\rangle\langle v_i| \right)$$

$$\times \left(\sum_i R_z(\phi_{2k}) \otimes |v_i\rangle\langle v_i| \right)$$

$$= \sum_i \left(R_z(\phi_0) \begin{bmatrix} \sigma_i & \sqrt{1-\sigma_i^2} \\ \sqrt{1-\sigma_i^2} & -\sigma_i \end{bmatrix} R_z(\phi_1) \right.$$

$$\cdots \left. \begin{bmatrix} \sigma_i & \sqrt{1-\sigma_i^2} \\ \sqrt{1-\sigma_i^2} & -\sigma_i \end{bmatrix} R_z(\phi_d) \right) \otimes |v_i\rangle\langle v_i|$$

$$= \sum_i \begin{bmatrix} P(\sigma_i) & iQ(\sigma_i)\sqrt{1-\sigma_i^2} \\ iQ^*(\sigma_i)\sqrt{1-\sigma_i^2} & P^*(\sigma_i) \end{bmatrix} \otimes |v_i\rangle\langle v_i| \qquad (14.36)$$

The former (i.e., the case of odd d) may be thought of as a more genuine *singular value transformation* in the sense that the overall action is to map a state represented as a superposition in the basis $\{|v_*\rangle\}$ to one represented as a superposition in the basis $\{|u_*\rangle\}$ – this follows from the fact that the tensor product term on the right-hand side of Eqn. (14.35) is $|u_i\rangle\langle v_i|$.

14.5 Matrix inversion by the quantum singular value transformation

One of the most obvious matrix transformations that can be represented as a function of its eigenvalues/singular values is matrix (pseudo)-inversion. As the pseudo-inverse is equal to the inverse for invertible matrices, we can immediately focus just on the former, which is a function of the singular values as originally given in Section 2.4.8. If A has SVD $A = \sum_i \sigma_i |u_i\rangle\langle v_i|$, then the pseudo-inverse is given by:

$$A^+ = \sum_i f(\sigma_i) |v_i\rangle\langle u_i| \qquad (14.37)$$

where

$$f(x) = \begin{cases} \dfrac{1}{x} & \text{if } x \neq 0 \\ 0 & \text{otherwise} \end{cases} \qquad (14.38)$$

Thus, we can see that the function $f(x)$ is odd, which is consistent with the fact that the overall effect is to map from the right into the left singular space. A polynomial approximation of $f(x)$ is therefore required, and this can be achieved with $\mathcal{O}(\kappa \log(\kappa/\epsilon))$ queries to the block encoding of the matrix to be inverted, where κ is the condition number and ϵ is the error. The complexity of the circuit performing the block encoding will vary depending on its precise nature, but in general will certainly be Poly($s \log N$) for an s-sparse matrix of size N, thus representing a slight asymptotic improvement over the Harrow–Hassidim–Lloyd algorithm (HHL) in the condition number, and a significantly improved scaling with error. Indeed, the dependence on the condition number is especially apparent when using the quantum singular value transformation to achieve matrix inversion, as can be illustrated by considering a case with relatively low condition number (on the left-hand side) and versus one relatively high condition number (on the right-hand side):

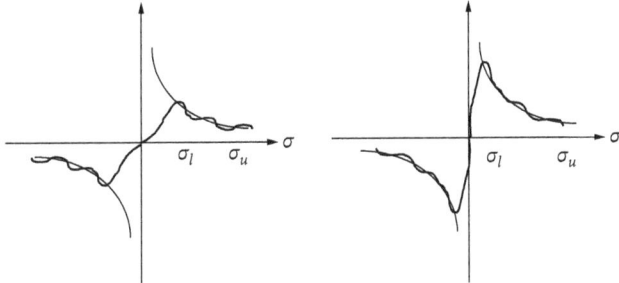

Figure 14.4

As a general principle, the 'sharper' the function is, the higher the degree needed for a good polynomial approximation. In the case of high condition number, the smallest singular value is relatively close to zero, and hence, a steep rise either side of zero is needed in order for the approximation to be good.

14.6 Search by the quantum singular value transformation

In all of the above analysis, it has been convenient to assume that (a scaled version of) the matrix of interest appears in the top-left block of a unitary matrix that is exactly twice as large as the matrix being transformed. This has meant that the rotation can be performed by applying a rotation to the first qubit. However, this simplification – although useful for pedagogical clarity – hides the fact that *any* block encoding suffices for the quantum eigenvalue/singular value transformations. This is particularly important when the quantum singular value transformation is used for unstructured search. In general, a unitary matrix U block encodes a matrix, A, if

$$A = (\langle \pi_l | \otimes I) U (|\pi_r\rangle \otimes I) \tag{14.39}$$

for some states $|\pi_r\rangle$ and $|\pi_l\rangle$ acting on some or all of the qubits. This is sometimes expressed in terms reminiscent of the top-left block encoding:

$$U = \begin{array}{c} |\pi_l\rangle \\ |\pi_l^\perp\rangle \end{array} \begin{array}{cc} |\pi_r\rangle & |\pi_r^\perp\rangle \\ \begin{bmatrix} A & \cdot \\ \cdot & \cdot \end{bmatrix} \end{array} \tag{14.40}$$

where the annotations around the matrix indicate that the encoding block is not actually found around in the top left, but rather in the part of the matrix obtained by Eqn. (14.39). This strictly generalises the case of top-left block encoding, where $|\pi_l\rangle = |\pi_r\rangle = |0\rangle$.

To explain how quantum signal processing can be applied to a general block encoding, it is instructive to first introduce an intermediate case where the block encoded matrix is Hermitian (so we can restrict our attention to the quantum eigenvalue transformation – although we shall continue to use A rather than H to denote the block-encoded matrix to facilitate subsequent generalisation to the non-Hermitian case) and $|\pi\rangle = |\pi_l\rangle = |\pi_r\rangle \neq |0\rangle$. The operation U can then be described as the sum of operations acting on parallel two-dimensional subspaces. In particular, each of the subspaces is spanned by a different eigenvector of block-encoded A, $|\pi\rangle \otimes |\lambda_i\rangle$ and some orthogonal vector to this that we denote $|\lambda_i^\perp\rangle \neq |\pi\rangle \otimes |\lambda_j\rangle$ for any j (that is, the not equals condition precludes the orthogonal vector being one of the other block-encoded eigenvectors of A). The action of U can thus be expressed

$$U = \bigoplus_{\lambda_*} \begin{bmatrix} \lambda_* & \sqrt{1-\lambda_*^2} \\ \sqrt{1-\lambda_*^2} & -\lambda_* \end{bmatrix} \tag{14.41}$$

where $\bigoplus$ is the standard notation for the direct sum of linear operations applied to the subspaces defined above for each $|\lambda_i\rangle$. In the case of the top-left block encoding, $|\pi\rangle = |0\rangle$ and $|\lambda_i^\perp\rangle = |1\rangle \otimes |\lambda_i\rangle$ so Eqn. (14.41) is exactly equivalent to Eqn. (14.21).

Every block encoding can be converted into the form of Eqn. (14.41) by a process known as *qubitisation*, and any block encoding of the form of Eqn. (14.19) is naturally qubitised, as the derivation of the form of Eqn. (14.21) therefrom shows. The consequence of a general qubitised block encoding is that the entire quantum eigenvalue transformation can be expressed:

$$\bigoplus_{\lambda_*} \prod_{k=1}^{d} \begin{bmatrix} e^{-i\phi_k/2} & 0 \\ 0 & e^{i\phi_k/2} \end{bmatrix} \begin{bmatrix} \lambda_* & \sqrt{1-\lambda_*^2} \\ \sqrt{1-\lambda_*^2} & -\lambda_* \end{bmatrix} \tag{14.42}$$

To remain consistent with the literature, the definition has very slightly changed as the right-most operation is now $\begin{bmatrix} \lambda_* & \sqrt{1-\lambda_*^2} \\ \sqrt{1-\lambda_*^2} & -\lambda_* \end{bmatrix} = R(\lambda_*)$ (as defined in Eqn. (14.15)). In practice, this makes no difference to the validity of quantum signal processing achieving the desired transformation, and the important thing to remember is that the number of occurrences of $R(\lambda_*)$ is what dictates whether the polynomial is even or odd (even number of occurrences means even P and likewise for odd). This setting also means that the 'R_z' gate is no longer simply that applied to a single qubit, but rather one that performs the rotation on the two dimensional subspace in question. This can be achieved by the projective rotation circuit:

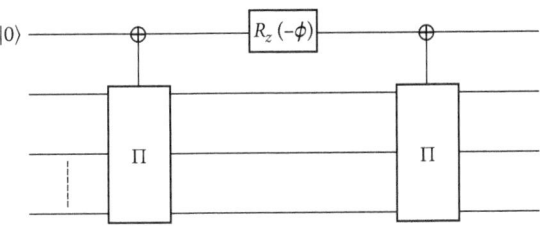

Figure 14.5

where Π is the projector $|\pi\rangle\langle\pi| \otimes I$. (In the chapter problems, you are asked to verify that this circuit has the desired effect for the simple top-left block encoding.) We denote this circuit $R_\pi(\phi)$, this means that the quantum eigenvalue transformation in Eqn. (14.42) can be expressed in a form that is not decomposed by the direct sum operator. For even d

$$\prod_{k=1}^{\frac{d}{2}} R_\pi(\phi_{2k-1})U^\dagger R_\pi(\phi_{2k})U \qquad (14.43)$$

and for odd d:

$$R_\pi(\phi_1)U\prod_{k=1}^{\frac{d-1}{2}} R_\pi(\phi_{2k})U^\dagger R_\pi(\phi_{2k+1})U \qquad (14.44)$$

Note that in Eqns. (14.43) and (14.44), the 'canonical' equations for the quantum eigenvalue transformation have been given using both U and $U^\dagger$. This is important when the encoded block appears at a general location within U (i.e., if $|\pi_l\rangle \neq |\pi_r\rangle$); however, it is also worth highlighting that Eqns. (14.43) and (14.44) exactly correspond to the previous definition of the quantum eigenvalue transformation for the top-left block encoding in Eqn. (14.25) (apart from the aforementioned fact that the last term in the product is now U rather than some R or R_π), as U in Eqn. (14.19) is Hermitian (as the eigenvalues are all real for Hermitian, H, and moreover also have magnitudes smaller than one, for H to be block-encodable) and so $U = U^\dagger$ therein.

Turning to the quantum singular value transformation for general block encodings, as already noted in Section 14.4, a general statement using both U and $U^\dagger$ is clearly needed when applying the quantum *singular value* rather than *eigenvalue* transformation. We also now consider the generalisation to any block encoding, where the location of the matrix of interest within the larger unitary is determined by $|\pi_l\rangle$ and $|\pi_r\rangle$ (which may not be equal), and so the correct projective rotation must be chosen. When $|\pi_l\rangle \neq |\pi_r\rangle$, the alternating U and $U^\dagger$ operations means that the state is mapped back and forth between the right and left singular spaces, and so the relevant projector controlled rotation will itself alternate between that controlled by $|\pi_l\rangle\langle\pi_l|$ and $|\pi_r\rangle\langle\pi_r|$, giving the full quantum singular value transformation as, for even d:

$$\prod_{k=1}^{\frac{d}{2}} R_{\pi_r}(\phi_{2k-1}) U^\dagger R_{\pi_l}(\phi_{2k}) U \qquad (14.45)$$

and for odd d:

$$R_{\pi_l}(\phi_1) U \prod_{k=1}^{\frac{d-1}{2}} R_{\pi_r}(\phi_{2k}) U^\dagger R_{\pi_l}(\phi_{2k+1}) U \qquad (14.46)$$

With the general framework defined, we are ready to address the problem of search with the quantum singular value transformation. Recall that the problem of search, as originally addressed by Grover, is to find a single 'marked' N-bit vector, denoted $|z\rangle$ (for ease of exposition, only the case of a single marked element is considered here), where we have access to an n-qubit oracle (where $N = 2^n$), V, which is such that:

$$V(|x'\rangle|x\rangle) = \begin{cases} |x'\rangle|x\rangle, & x \neq z \\ |x' \oplus 1\rangle|x\rangle, & x = z \end{cases} \qquad (14.47)$$

for some computational basis state, $|x\rangle$ and single bit, x'. Notice that the oracle acts on the first rather than (the usual) last qubit as defined in Eqn. (14.47), as this allows the following to be presented slightly more clearly.

The approach when employing the quantum singular value transformation for unstructured search is to treat the $N \times N$ identity matrix as a block encoding of a single number (i.e., a one-element matrix), which is equal to the overlap between $|z\rangle$ and the uniform superposition,

$$|\psi\rangle = \frac{1}{\sqrt{2^n}} \sum_{x \in \{0,1\}^n} |x\rangle \qquad (14.48)$$

More precisely, the goal is to amplify the overlap between the lone vector in the right singular space, $|\pi_r\rangle = |\psi\rangle$, and the lone vector in the left singular space, $|\pi_l\rangle = |z\rangle$, and the identity serves as a suitable block encoding:

$$\langle z| I |\psi\rangle = \frac{1}{\sqrt{N}} \qquad (14.49)$$

(As the vectors $|z\rangle$ and $|\psi\rangle$ already have dimension equal to that of entire space, there is no need for them to be tensored with the identity on any of the qubits.)

In order to use the quantum singular value transformation, four circuits are required:

- U: the block encoding – in this case this is trivial, as U is the identity;
- $U^\dagger$: which is also trivial as $U^\dagger$ is also the identity;

- $|\pi_r\rangle\langle\pi_r|$-controlled NOT;
- $|\pi_l\rangle\langle\pi_l|$-controlled NOT.

$|\pi_l\rangle\langle\pi_l|$ is the projector which picks out the marked element, and so $|\pi_l\rangle\langle\pi_l|$-controlled NOT is a gate that applies a NOT to a qubit when the marked element is in the input register, and applies the identity otherwise, and hence is precisely the oracle, V. $|\pi_r\rangle\langle\pi_r|$ is the projector which picks out the state $|+^n\rangle$ when the state is expressed as a superposition in the $|+\rangle, |-\rangle$ basis, and therefore, $|\pi_r\rangle\langle\pi_r|$-controlled NOT can be realised by the following circuit:

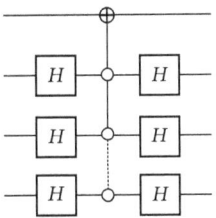

Figure 14.6

These circuits can then be deployed in the sequence specified in Eqn. (14.46) to give the circuit (omitting $U = U^\dagger = I$):

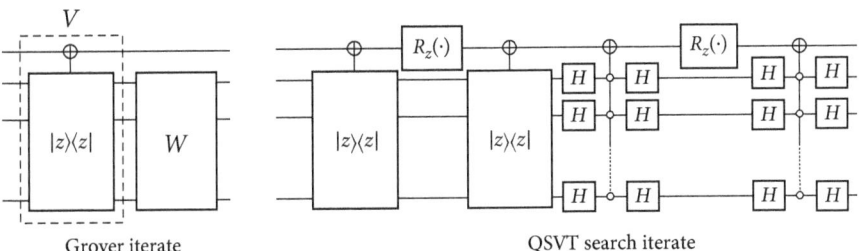

Figure 14.7

As always, applying the quantum singular value transformation to the application of unstructured search requires a suitable polynomial transformation to be defined. In this case, the goal is to transform the initial singular value, $1/\sqrt{N}$, to (approximately) 1, such that the final state is equal to (or dominated by) the marked state. Thus, an ostensibly suitable transformation would be:

$$W_{QSVT} = \begin{matrix} |z\rangle \\ |z^\perp\rangle \end{matrix} \begin{bmatrix} |\psi\rangle & |\psi^\perp\rangle \\ \approx 1 & \cdot \\ \cdot & \cdot \end{bmatrix} \qquad (14.50)$$

However, as the overall effect is to map from the right singular space to the left singular space, the function must be purely odd, and this is not so for the constant function $P(x) = 1$, as above. Instead, a suitable alternative is the sign function with $c = 0$:

$$\Theta(x - c) = \begin{cases} -1, & x < c \\ 0, & x = c \\ 1, & x > c \end{cases} \tag{14.51}$$

which is purely odd, but is not polynomial – and so must be approximated by a polynomial for quantum signal processing to be used. An infinite degree polynomial would be necessary in principle; however, all that is actually required is that the polynomial approximation rises steeply in the region $[0, \frac{1}{\sqrt{N}}]$ such that it reaches a value of approximately one (and then stays close to one). That is, a polynomial approximation that appears:

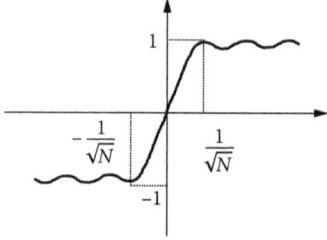

Figure 14.8

To achieve this, it turns out that a polynomial approximation of degree $d = \Theta(\sqrt{N})$ suffices, and hence, $\Theta(\sqrt{N})$ oracle calls are used, matching the asymptotically optimal performance of Grover search.

It is worth further remarking on the fact that Grover's algorithm is itself naturally qubitised, in that the action of the Grover iterate may be viewed in terms of an evolution in the subspace spanned by $|y\rangle|-\rangle$ and $|z\rangle|-\rangle$ and in particular each Grover iterate is a rotation by a constant angle in the great circle of the Bloch sphere corresponding to real quantum states. In the case of unstructured search by the quantum singular value transformation, as the encoded block is a single element, there is only one 'parallel' subspace, i.e., the overall action may also be represented by a single Bloch sphere. Unlike Grover search, the evolution of the state when the polynomial function approximation of the sign function is applied is not contained with any single great circle, and instead 'swirls' over the surface of the Bloch sphere to transform the state from $|\psi\rangle$ to one close to $|z\rangle$.

14.7 A grand unification of quantum algorithms?

In this chapter, we have seen how the technique of quantum signal processing, when applied to a qubitised matrix block encoding, can transform the singular or eigenvalues of said matrix to achieve a desired computational effect. In particular, we have seen how this technique enables (i) Hamiltonian simulation; (ii) inversion of sparse well-conditioned matrices; and (iii) unstructured search to be achieved with asymptotic complexity equal to, or better than the canonical quantum algorithms for each of these.

There is, however, a further more fundamental reason why the quantum singular value transformation has garnered widespread interest, and that is that it has been labelled the *grand unification of quantum algorithms*, in the sense that every major quantum algorithm may be thought of as a special case of the quantum singular value transformation. Notably, as well as the applications given above, the quantum singular value transformation can be used to perform quantum phase estimation. Full exposition of how the quantum singular value transformation achieves quantum phase estimation is out of scope, as it is built on a form of quantum phase estimation that is slightly different to the one studied in this book; however, even from a simplified exposition, it is possible to gain an intuition for why quantum phase estimation can be achieved by a process of extracting singular values.

Recall that the problem of quantum phase estimation is to find the phase, ϕ, of an eigenvalue, $e^{2\pi i \phi}$ associated with a certain eigenvector, $|u\rangle$ of a unitary matrix, U – where $|u\rangle$ is provided (in the simplest instance) and the quantum algorithm has access to the circuit for U (and controlled versions thereof). To use the quantum singular value transformation for quantum phase estimation, consider the matrix

$$A_j(\theta) = \frac{1}{2}\left(I + e^{-2\pi i \theta} U^{2^j}\right) \tag{14.52}$$

which has a right-singular vector $|u\rangle$ associated with singular values:

$$\sigma^{(j)} = |\cos(\pi(2^j \phi - \theta))| \tag{14.53}$$

and may be block encoded as $\begin{bmatrix} A_j(\theta) & \cdot \\ \cdot & \cdot \end{bmatrix}$ by the circuit:

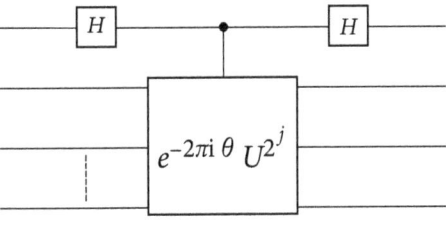

Figure 14.9

The phase may be thought of as a binary fraction, $\phi = 0 \cdot \phi_1 \phi_2 \ldots \phi_m$, and the method then proceeds by 'picking out' the bits of ϕ from the least to the most significant. If ϕ has some m bits (in the case where ϕ is not exactly a binary fraction then the following algorithm can be shown still to work to the desired precision), then first $\theta \leftarrow 0$ and $j \leftarrow m-1$ such that:

$$\sigma^{(m-1)} = |\cos(\pi(2^{m-1}(0 \cdot \phi_1 \phi_2 \ldots \phi_m) - \theta))|$$
$$= |\cos(\pi(\phi_1 \phi_2 \ldots \phi_{m-1} \cdot \phi_m) - 0))|$$
$$= |\cos(\pi(0 \cdot \phi_m)))|$$
$$= \begin{cases} 1, & \phi_m = 0 \\ 0, & \phi_m = 1 \end{cases} \tag{14.54}$$

(using the fact that $|\cos(\cdot)|$ has period π). θ is a semi-classical parameter, saving the least significant bits of θ such that to extract the $(j+1)$th bit, the block encoding $A_j(\theta)$ is prepared with $\theta = 0 \cdot 0 \phi_{j+2} \ldots \phi_m$, giving

$$\sigma^{(j)} = |\cos(\pi(2^j(0 \cdot \phi_1 \phi_2 \ldots \phi_j \phi_{j+1} \ldots \phi_m) - \theta))|$$
$$= |\cos(\pi((\phi_1 \phi_2 \ldots \phi_j \cdot \phi_{j+1} \ldots \phi_m) - \theta))|$$
$$= |\cos(\pi((0 \cdot \phi_{j+1} \ldots \phi_m) - (0 \cdot 0 \phi_{j+2} \ldots \phi_m)))|$$
$$= |\cos(\pi(0 \cdot \phi_{j+1}))|$$
$$= \begin{cases} 1, & \phi_{j+1} = 0 \\ 0, & \phi_{j+1} = 1 \end{cases} \tag{14.55}$$

And so the singular value is the complement of ϕ_{j+1}, which can be efficiently extracted by measurement (the quantum singular value transformation achieves this by eigenvalue thresholding, using the sign function), thus meaning that a recursive approach indeed extracts each bit in turn. This procedure leads to an overall complexity that is comparable with other techniques for quantum phase estimation. Whilst θ is a *semi*-classical parameter, the whole algorithm can be conducted coherently (as one quantum circuit), which is indeed necessary when a superposition of eigenvectors is used, rather than a single eigenvector (as indeed is the case in factoring) to ensure that each measurement-induced collapse corresponds to the same eigenvector.

Thus, we have sketched out how the singular values of a block encoded matrix can be used to perform quantum phase estimation, confirming the role of the quantum singular value transformation in unifying the major quantum algorithms. It is worth stepping back and asking the significance of this. The term *unification* suggests that all *possible* quantum algorithms may be seen as instances of the quantum singular value transformation. In one sense, this is unquestionably the case: the quantum singular value transformation can be used to simulate Hamiltonian evolution, and every quantum algorithm may be cast of as the evolution of *some* Hamiltonian. However, many quantum computational frameworks are known to be 'complete' in this way, including quantum walk and adiabatic quantum computation, but it does not follow that every computational effect that may be achieved using quantum mechanics is naturally thought of as a quantum walk, or an (adiabatic) evolution of some Hamiltonian, and so is also the case with the quantum singular value transformation.

Instead, the real benefit of the quantum singular value transformation as a unified structure for various quantum algorithms is that it *does* give a powerful framework for thinking up new quantum algorithms. Indeed, treating quantum computers as machines that manipulate the eigenvalues and singular values of very large matrices, and investigating how this can be put to useful effect may be thought of as the *modern* approach to quantum algorithm design. One way that this can be leveraged is when some large matrix 'hides' a block encoding of a smaller matrix, which holds the solution to the problem—as in unstructured search by the quantum singular value transformation.

Alternatively, there are quantum algorithms where the block encoding is explicit and the problem requires the eigenvalues or singular values of some large, sparse matrix to be transformed (all eigenvalues/singular values in the same way). Hamiltonian simulation and matrix inversion are examples of this, and another nice example concerns Markov chain fast forwarding, as outlined in Box 14.1. (Note that the explanation in Box 14.1 has been deliberately framed in terms of the quantum singular value transformation and Szegedy originally proposed a very slightly different quantum walk operator.)

Box 14.1 Markov chain fast-forwarding by transition matrix block encoding

A Markov chain is succinctly defined by its transition matrix, P, which such that $p = e_i P$, where e_i is a unit vector of all zeros, except for a single one in the ith element, and p is the probability distribution after a single step of the Markov chain starting from the ith vertex. As P will not in general be unitary, some work is needed to 'quantise' the stochastic matrix, that is, to define a related matrix that can be used as a quantum operator. In 2004, Mario Szegedy devised a general way to achieve this, which although many years before the quantum singular value transformation, can be thought of as a block encoding not of the transition matrix itself, but of a closely related matrix, the Markov chain's *discriminant*, D,

$$D_{i,j} = \sqrt{P_{i,j} P_{j,i}} \qquad (14.56)$$

In the case where the Markov chain is symmetric, that is $P^T = P$, then $D = P$. For simplicity, we shall assume that the Markov chain is over a graph with some $N = 2^n$

vertices; to block encode the discriminant, first consider an operator, U_P that operates on an N^2 element space (i.e., a 2n-qubit operator) and encodes the transitions:

$$U_P|0\rangle|i\rangle = \sqrt{P_{i,j}}|j\rangle|i\rangle \qquad (14.57)$$

Access to such an operator is essentially commensurate with classical access to look up (or compute) transition probabilities – even when the Markov chains is too large to explicitly express the transition matrix. The N^2 element space can thus be thought of as a first register of n qubits representing the vertex transitioned *to* followed by a second register of n qubits representing the vertex transitioned *from*. A second operator is the SHIFT operator, which is the permutation matrix that swaps the two registers:

$$\text{SHIFT}|j,i\rangle = |i,j\rangle \qquad (14.58)$$

We then get that $U_P^\dagger \text{SHIFT} U_P$ block encodes D:

$$\langle 0|\langle i|(U_P^\dagger \text{SHIFT} U_P)|0\rangle|j\rangle = \left(\sqrt{P_{i,k}}\langle k,i|\right)\text{SHIFT}\left(\sqrt{P_{j,l}}|l,j\rangle\right)$$

$$= \sqrt{P_{i,k}P_{j,l}}\langle k,i|j,l\rangle$$

$$= \sqrt{P_{i,j}P_{j,i}} = D_{i,j} \qquad (14.59)$$

Recalling that for symmetric Markov chains $D = P$, an immediate application of the block encoded transition matrix is to *fast forward* the Markov chain. In general, a random process may be defined as a number of steps, k, of the Markov chain, when applied to some initial state e_i, which amounts to sampling from the distribution prepared by $e_i P^k$. Classically, this means applying P some k times, however with a suitable block encoding of P, it is a useful fact that the *monomial* x^k can be well-approximated by some *polynomial* of degree $\Theta(\sqrt{k})$. This means that a quantum encoding of the probability distribution after a given number of Markov chain transitions (steps) can be prepared with quadratically fewer uses of P.

Chapter problems

1. What are $|\pi_l\rangle$ and $|\pi_r\rangle$ for the block encoding $\begin{bmatrix} \cdot & \cdot \\ \cdot & A \end{bmatrix}$?
2. Verify that the projector controlled rotation has the same effect as an $R_z(\phi)$ directly applied to the first qubit when a top-left block encoding is used.
3. From the complexity of quantum singular value transformation for unstructured search, what can you conclude about how well a polynomial of a given degree can approximate the sign function?

4. For the application of quantum singular value transformation to quantum phase estimation:
   (a) Let U be a unitary matrix with eigenvector $|u\rangle$ and associated eigenvalue, $e^{2\pi i \phi}$. Show that the matrix:
   $$A_j(\theta) = \frac{1}{2}\left(I + e^{-2\pi i \theta} U^{2^j}\right) \qquad (14.60)$$
   has a right-singular vector $|u\rangle$ associated with singular value:
   $$\sigma^{(j)} = |\cos(\pi(2^j \phi - \theta))| \qquad (14.61)$$
   (b) Show that A can be block-encoded as $\begin{bmatrix} A_j(\theta) & \cdot \\ \cdot & \cdot \end{bmatrix}$ by the circuit:

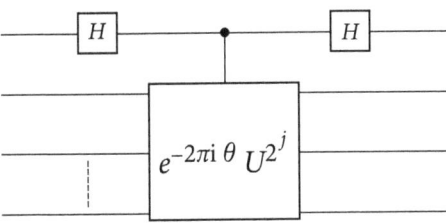

Figure 14.10

Further reading

Quantum signal processing is due to Low et al [2016; 2019], and the quantum singular value transformation due to Gilyén et al [2019] in which Section 5.1 shows that this is optimal for Hamiltonian simulation. Martyn et al [2021] published a wonderful pedagogical paper, *a grand unification of quantum algorithms* celebrating and explaining the quantum singular value transformation on which many of the explanations in this chapter were based. Martyn et al [2023] also discuss ways to relax the rather strict conditions of quantum signal processing (in its original form). Szegedy [2004] proposed the method of quantising Markov chains, and quantum fast forwarding of Markov chains is due to Apers and Sarlette [2019].

15
An introduction to quantum error correction

Quantum computation is conventionally presented, in this book and in most of the relevant literature, as an abstract model of computation derived from the postulates of quantum mechanics as they apply to *closed* quantum systems. This approach, however, misses one crucial factor: real computational devices are *open* quantum systems exposed to environmental effects. These environmental effects lead to deviations in the quantum state from that predicted by the postulates of quantum mechanics, and these deviations are variously referred to as *quantum noise*, *quantum error* and/or *decoherence* of the quantum state.

This, in turn, means that the 'correct' outcome of a quantum computation (when performed on real quantum hardware) may only occur with a certain probability. Of course, the essential principle that environmental effects can adversely interfere with a computational state is not peculiar to quantum computation; however, modern (classical) digital computers are, to all intents and purposes, error-free. The same cannot, however, be said of quantum computers, and the remedy for this is *quantum error correction*. Quantum error correction is a deep and wide-reaching subject in its own right, and we cannot hope to cover all or even most of it here. Instead, what this chapter aims to do is to present the essential principles that show how quantum errors can be corrected, and thus to shine a light on how it is possible to simulate noiseless quantum computation on real hardware.

15.1 Classical error correction

Before getting into the correction of *quantum* errors, it is helpful to briefly introduce *classical* error correction. To motivate quantum error correction from classical first principles, all that is required is a very basic error model and a correspondingly trivial error-correction code. One of the simplest models for single-bit (classical) errors is the *binary symmetric channel*, in which each possible state of the bit, 0 and 1 'flips' to the other with some probability p_e:

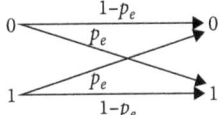

Figure 15.1

Without loss of generality, it can be assumed that $p_e \leq 0.5$, because if $p_e > 0.5$, then it is more likely than not that a bit flip has occurred, and so a received '0' can be interpreted instead as a '1' and vice versa. In the case where $p_e = 0.5$, it is not possible to recover any information from the channel.

One simple way to counter error in a binary symmetric channel is to *encode* the bit by repeating it (say) three times: this is known as a *repetition code*. That is, if the bit to be transmitted is '0', then three bits (sequentially) in the state 0 are transmitted (and likewise for 1). This can be denoted as:

$$0 \to 000 \tag{15.1}$$

$$1 \to 111 \tag{15.2}$$

Once the three bits have been received, the *decoding* is done by a 'majority vote'. That is, any three-bit string will have a majority (i.e., either two or all three) of the bits with the same value, and this is taken as the 'error-corrected' value. This simple error-correction protocol fails when at least two of the bits have been flipped, in which case the majority of the received bits are erroneous, and the protocol therefore makes a false 'correction'.

For the purposes of explaining quantum error correction, it is helpful to define an alternative (but equivalent) way to do the decoding step. Suppose that the received three bits are stored in some 'fenced-off' part of the system, and the only information that is forwarded to the error-correction controller is two bits: the first indicating the parity of the first two received bits (parity means the indicator bit is set to zero if the two bits are the same, and one if they are different), and a second bit indicating the parity of the second and third bits. For example, if the received bit string is 001, then the parity bits would be 01.

The error-correction controller receives these two parity bits and then returns an error-correction instruction. It turns out that such parity information is sufficient for error correction when there is at most one bit flip, as the error-correction controller can adopt the following strategy: if the parity bits are 00, the instruction is to do nothing; if the parity bits are 10, the instruction is to flip the first bit; if the parity bits are 01, the instruction is to flip the third bit; if the parity bits are 11, the instruction is to flip the second bit. This strategy can be shown, as in Table 15.1 for all cases where there is at most one bit flip.

Thus, this (somewhat convoluted) protocol for correcting the error without the error-correcting controller having direct visibility of the received bits performs the same as error correction by majority vote.

With either approach, the error-correction procedure only fails if two of the three bits have been flipped (which can occur in three different ways), or all three have been. Denoting the probability of an error on the bit encoded with the three-bit repetition code p'_e gives:

$$p'_e = 3p_e^2(1 - p_e) + p_e^3 = 3p_e^2 - 2p_e^3 \tag{15.3}$$

which is smaller than p_e if $p_e < 0.5$. Typically, p_e is small, and this may be described as suppressing the error to $\mathcal{O}(p_e^2)$.

Quantum error correction

Table 15.1 Illustration of classical error correction with the three-bit repetition code.

Data bit	Transmitted	Received	Parity bits	Recovery	Corrected
0	000	000	00	None	000
0	000	100	10	Flip bit 1	000
0	000	010	11	Flip bit 2	000
0	000	001	01	Flip bit 3	000
1	111	111	00	None	111
1	111	011	10	Flip bit 1	111
1	111	101	11	Flip bit 2	111
1	111	110	01	Flip bit 3	111

15.2 Quantum error correction

An obvious starting point for quantum error correction would be to naively attempt to port classical error-correction techniques directly. However, this approach immediately runs into three fundamental problems:

1. the no-cloning principle forbids the copying of quantum states, which seems to rule out repetition codes;
2. measurement destroys quantum information, which seems to rule out the possibility of decoding by majority votes – although the alternative strategy (above) for classical error correction is deliberately suggestive as a way to corrects errors without revealing the state;
3. quantum states are continuous: for any single-qubit state, $\alpha|0\rangle + \beta|1\rangle$, a quantum error channel in general has the effect $\alpha|0\rangle + \beta|1\rangle \to (\alpha + \epsilon_0)|0\rangle + (\beta + \epsilon_1)|1\rangle$ (for some error terms ϵ_0 and ϵ_1). So it follows that simply correcting (qu)bit flips would seem to be insufficient to correct all possible quantum errors.

Nevertheless, it turns out that, with some ingenuity, it *is* possible to correct quantum errors.

15.2.1 The three-qubit bit-flip code

To see how quantum errors can be corrected, a suitable starting point is to disregard (for now) the third item listed above, and assume that the only errors affecting qubits are bit flips. With this simplification, the first two items can be resolved:

- entanglement, rather than cloning, can be used to achieve repetition;
- errors can be detected using parity-check measurements: these do not destroy the quantum information, but reveal only whether an error has occurred or not.

The three-bit repetition code guarantees to return the correct value, so long as at most one of the bits in the code is flipped. This provides inspiration for the *three-qubit bit-flip code*, which is prepared by the circuit applied to an arbitrary single-qubit state:

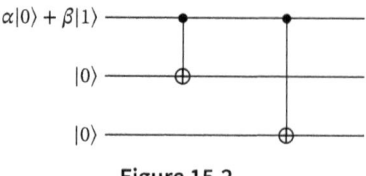

Figure 15.2

and encodes the state as

$$(\alpha|0\rangle + \beta|1\rangle)|0\rangle^{\otimes 2} \to \alpha|000\rangle + \beta|111\rangle \tag{15.4}$$

To correct errors that may occur in a 'noisy channel' (as stated above, in the first instance, this can only impart bit flips), the circuit is supplemented with two ancillas, which are used for error *detection*:

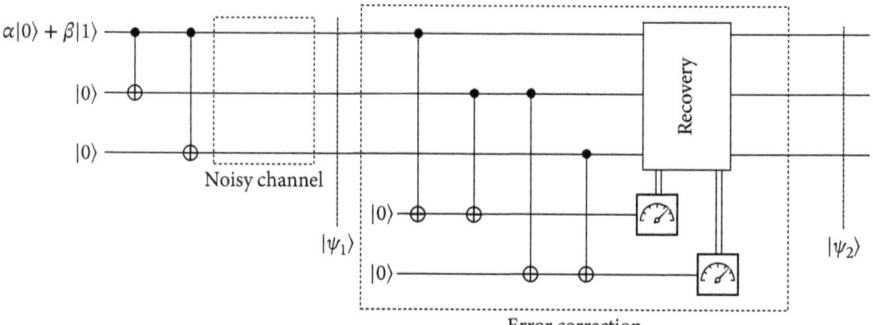

Figure 15.3

The error *detection* ancillas are then measured, with the measurement outcomes used for error *correction*. (Note that the box marked 'error correction' in Fig. 15.3 includes both error detection and correction, and indeed in general the process of error correction is treated as including error detection.) To see how this works, first consider the case in which the noisy channel actually imparts no error, in which case:

$$|\psi_1\rangle = (\alpha|000\rangle + \beta|111\rangle)|00\rangle \tag{15.5}$$

where the two error-detection ancillas, each initially in the state $|0\rangle$, have been explicitly included. 'Simulating' by hand the effect of the first two error-correction CNOTs, we can (by linearity) take the two parts of $|\psi_1\rangle$ in turn. For the first part, $\alpha|000\rangle$, the

Quantum error correction

CNOTs clearly have no effect on the ancilla, as in each case the control of the CNOT is zero; for the second part $\beta|111\rangle$, the ancilla is 'flipped' by the CNOT controlled by the first qubit, but then flipped back by the CNOT controlled by the second qubit. Thus, the state is:

$$(\alpha|000\rangle|0\rangle + \beta|111\rangle|0\rangle)|0\rangle = (\alpha|000\rangle + \beta|111\rangle)|00\rangle = |\psi_1\rangle \quad (15.6)$$

and similarly for the third and fourth CNOTs in the error-correction block.

On the other hand if, for example, the noisy channel is such that the first qubit experiences a bit flip, then

$$|\psi_1\rangle = (\alpha|100\rangle + \beta|011\rangle)|00\rangle \quad (15.7)$$

In this case, in each of the superposed terms, the first ancilla is always flipped by exactly one of the first and second CNOTs, whilst the third and fourth CNOTs still have no overall effect, and so the state before the measurements is:

$$(\alpha|100\rangle + \beta|011\rangle)|10\rangle \quad (15.8)$$

Thus, the two ancillas play the same role as the parity bits in the second decoder for the (classical) three-bit repetition code (indeed these are often called 'parity-check ancillas'). It follows that the first ancilla being flipped to $|1\rangle$ indicates that a bit flip has occurred on exactly one of the first and second encoded qubits; if the second ancilla remains $|0\rangle$, then either both or neither of the second and third qubits have experienced a bit flip. This information can be decoded to infer that, assuming at most one bit flip has occurred, the first qubit has suffered a bit flip whilst the second and third have remained error free. The same process could be stepped through for a sole bit flip suffered by (i) the second qubit and (ii) the third qubit, showing that if there is at most one bit flip, then it can be detected. Furthermore, this detection can be used to *correct* the error (where M_1 and M_2 are the measurement outcomes), as shown in Table 15.2.

Note that, by design, the measurement outcomes M_1 and M_2 merely detect the bit flip and are 'unaware' of the quantum state itself – so these measurements have not destroyed the quantum state.

Table 15.2 Illustration of quantum error correction with the three-qubit bit-flip code.

(Qu)bit flip	$|\psi_1\rangle$	M_1	M_2	Recovery	$|\psi_2\rangle$
-	$\alpha|000\rangle + \beta|111\rangle$	0	0	$I \otimes I \otimes I$	$\alpha|000\rangle + \beta|111\rangle$
1	$\alpha|100\rangle + \beta|011\rangle$	1	0	$X \otimes I \otimes I$	$\alpha|000\rangle + \beta|111\rangle$
2	$\alpha|010\rangle + \beta|101\rangle$	1	1	$I \otimes X \otimes I$	$\alpha|000\rangle + \beta|111\rangle$
3	$\alpha|001\rangle + \beta|110\rangle$	0	1	$I \otimes I \otimes X$	$\alpha|000\rangle + \beta|111\rangle$

15.2.2 The three-qubit phase-flip code

Turning now to the (temporarily) disregarded third item: quantum errors are actually continuous – a good starting point is to show how a different type of error, namely a phase flip, can be corrected. By definition, a phase flip does not affect the computational basis state $|0\rangle$, but multiplies the coefficient of the computational basis state $|1\rangle$ by -1. Protection against phase flips can be achieved by the *three-qubit phase-flip code*, which encodes an arbitrary single-qubit state, $\alpha|0\rangle + \beta|1\rangle$, according to:

$$(\alpha|0\rangle + \beta|1\rangle)|0\rangle^{\otimes 2} \rightarrow \alpha|+++\rangle + \beta|---\rangle \tag{15.9}$$

which is achieved by the following circuit:

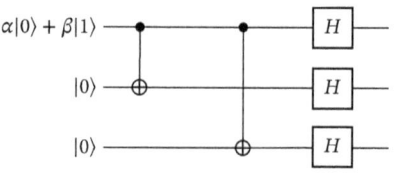

Figure 15.4

Noting that, by definition, a phase flip sends:

$$|+\rangle = \frac{1}{\sqrt{2}}(|0\rangle + |1\rangle) \rightarrow \frac{1}{\sqrt{2}}(|0\rangle - |1\rangle) = |-\rangle \tag{15.10}$$

$$|-\rangle = \frac{1}{\sqrt{2}}(|0\rangle - |1\rangle) \rightarrow \frac{1}{\sqrt{2}}(|0\rangle + |1\rangle) = |+\rangle \tag{15.11}$$

the phase-flip code is essentially identical to the bit-flip code, just working in the $|+\rangle, |-\rangle$ basis. This means that the whole protocol is:

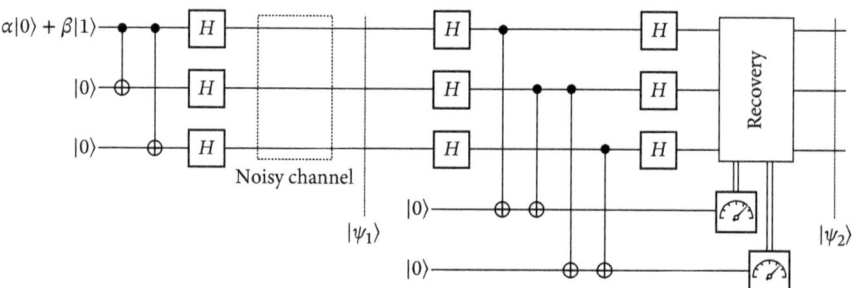

Figure 15.5

Note that the column of Hadamards before the CNOT gates send $|+\rangle \rightarrow |0\rangle$ and $|-\rangle \rightarrow |1\rangle$, so the parity checks have equivalent action of detecting phase flips, and the

Table 15.3 Illustration of quantum error correction with the three-qubit phase-flip code.

Phase flip	$\|\psi_1\rangle$	M_1	M_2	Recovery	$\|\psi_2\rangle$
-	$\alpha\|+++\rangle + \beta\|---\rangle$	0	0	$I \otimes I \otimes I$	$\alpha\|+++\rangle + \beta\|---\rangle$
1	$\alpha\|-++\rangle + \beta\|+--\rangle$	1	0	$Z \otimes I \otimes I$	$\alpha\|+++\rangle + \beta\|---\rangle$
2	$\alpha\|+-+\rangle + \beta\|-+-\rangle$	1	1	$I \otimes Z \otimes I$	$\alpha\|+++\rangle + \beta\|---\rangle$
3	$\alpha\|++-\rangle + \beta\|--+\rangle$	0	1	$I \otimes I \otimes Z$	$\alpha\|+++\rangle + \beta\|---\rangle$

second bank of Hadamards returns the state to its original encoding. Thus, the overall effect of the three-qubit phase-flip code can be summarised, as shown in Table 15.3.

15.3 The Shor code

The Shor code is a nine-qubit code, which is constructed by *concatenating* the three-qubit bit-flip and three-qubit phase-flip codes:

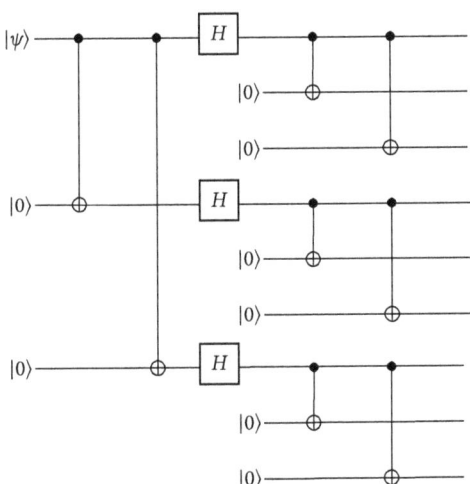

Figure 15.6

This encodes the computational basis states as follows:

$$|0\rangle \rightarrow |0_L\rangle = \frac{1}{2\sqrt{2}}(|000\rangle + |111\rangle)(|000\rangle + |111\rangle)(|000\rangle + |111\rangle) \quad (15.12)$$

$$|1\rangle \rightarrow |1_L\rangle = \frac{1}{2\sqrt{2}}(|000\rangle - |111\rangle)(|000\rangle - |111\rangle)(|000\rangle - |111\rangle) \quad (15.13)$$

and has decoding circuit:

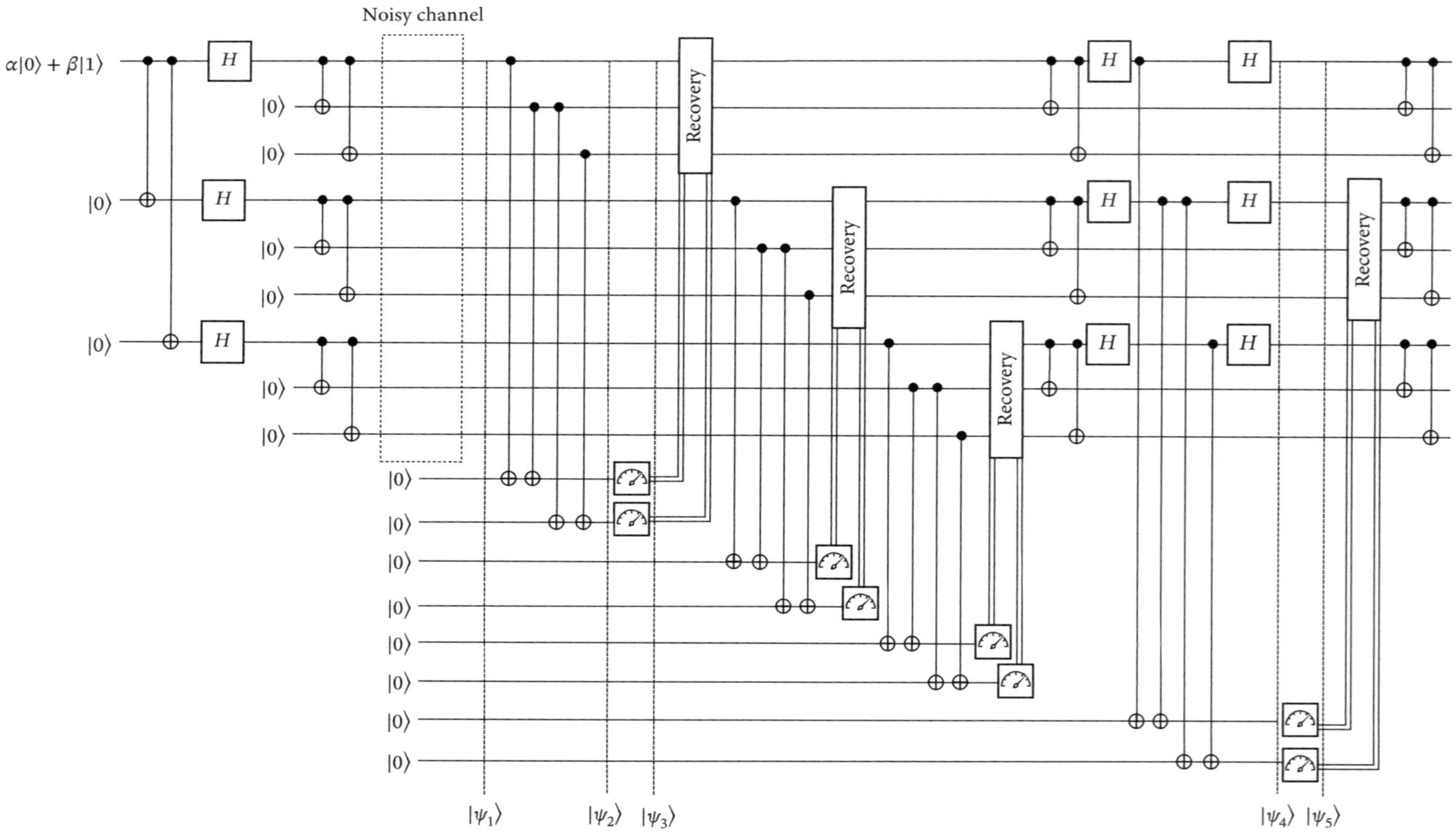

Figure 15.7

where the first three 'recovery' blocks correct bit flips, and thus, the Shor code can detect and correct a bit flip on any single qubit. For example, suppose that we have an arbitrary quantum state $\alpha|0\rangle + \beta|1\rangle$, which encoded with the Shor code as:

$$\frac{1}{2\sqrt{2}}\Big(\alpha(|000\rangle + |111\rangle)(|000\rangle + |111\rangle)(|000\rangle + |111\rangle)$$
$$+ \beta(|000\rangle - |111\rangle)(|000\rangle - |111\rangle)(|000\rangle - |111\rangle)\Big) \quad (15.14)$$

If a bit flip occurs on the first qubit, the state becomes (now explicitly including the eight parity-check ancillas in the order they appear in the circuit diagram):

$$|\psi_2\rangle = \frac{1}{2\sqrt{2}}\Big(\alpha(|100\rangle + |011\rangle)(|000\rangle + |111\rangle)(|000\rangle + |111\rangle)$$
$$+ \beta(|100\rangle - |011\rangle)(|000\rangle - |111\rangle)(|000\rangle - |111\rangle)\Big)|10000000\rangle \quad (15.15)$$

which can be detected (and subsequently corrected) by parity-check measurements. The same principle applies to a single bit flip occurring on any of the nine qubits.

The Shor code can also detect and correct a phase flip on any single qubit. If a phase flip occurs on the first qubit, the state becomes:

$$|\psi_4\rangle = \frac{1}{2\sqrt{2}}\Big(\alpha(|000\rangle - |111\rangle)(|000\rangle + |111\rangle)(|000\rangle + |111\rangle)$$
$$+ \beta(|000\rangle + |111\rangle)(|000\rangle - |111\rangle)(|000\rangle - |111\rangle)\Big)|00000010\rangle$$
$$(15.16)$$

In the chapter problems, you are asked to verify that the Shor code detects and corrects this error.

15.3.1 Correcting any single-qubit error with the Shor code

It turns out that the Shor code can not only correct a single bit flip or phase flip, but can correct an error where one of the nine qubits has suffered both bit and phase flips. In fact, it can do something much more powerful than this: it can correct any single-qubit error.

Noting the symmetry of the Shor code, without loss of generality, the qubits can be re-ordered such that it is the first qubit that experiences the error. Suppose that it encounters a noisy channel that sends $|0\rangle \to a|0\rangle + b|1\rangle$ and $|1\rangle \to c|0\rangle + d|1\rangle$, then the subsequent state is (omitting the 8 parity-check qubits):

$$|\psi_1\rangle = \frac{1}{2\sqrt{2}}\Big(\alpha(a|000\rangle + b|100\rangle + c|011\rangle + d|111\rangle)(|000\rangle + |111\rangle)(|000\rangle + |111\rangle)$$
$$+ \beta(a|000\rangle + b|100\rangle - c|011\rangle - d|111\rangle)(|000\rangle - |111\rangle)(|000\rangle - |111\rangle)\Big)$$
$$(15.17)$$

Letting $k+m=a$, $k-m=d$, $l+n=b$, and $l-n=c$, Eqn. (15.17) can be expanded as

$$|\psi_1\rangle = \frac{1}{2\sqrt{2}}\Big(k\big(\alpha(|000\rangle+|111\rangle)(|000\rangle+|111\rangle)(|000\rangle+|111\rangle)$$
$$+\beta(|000\rangle-|111\rangle)(|000\rangle-|111\rangle)(|000\rangle-|111\rangle)\big)$$
$$+l\big(\alpha(|100\rangle+|011\rangle)(|000\rangle+|111\rangle)(|000\rangle+|111\rangle)$$
$$+\beta(|100\rangle-|011\rangle)(|000\rangle-|111\rangle)(|000\rangle-|111\rangle)\big)$$
$$+m\big(\alpha(|000\rangle-|111\rangle)(|000\rangle+|111\rangle)(|000\rangle+|111\rangle)$$
$$+\beta(|000\rangle+|111\rangle)(|000\rangle-|111\rangle)(|000\rangle-|111\rangle)\big)$$
$$+n\big(\alpha(|100\rangle-|011\rangle)(|000\rangle+|111\rangle)(|000\rangle+|111\rangle)$$
$$+\beta(|100\rangle+|011\rangle)(|000\rangle-|111\rangle)(|000\rangle-|111\rangle)\big)\Big) \quad (15.18)$$

Including the parity-check ancillas, the state $|\psi_2\rangle$ can be expressed:

$$|\psi_2\rangle = \frac{1}{2\sqrt{2}}\Big(k\big(\alpha(|000\rangle+|111\rangle)(|000\rangle+|111\rangle)(|000\rangle+|111\rangle)$$
$$+\beta(|000\rangle-|111\rangle)(|000\rangle-|111\rangle)(|000\rangle-|111\rangle)\big)|00000000\rangle$$
$$+l\big(\alpha(|100\rangle+|011\rangle)(|000\rangle+|111\rangle)(|000\rangle+|111\rangle)$$
$$+\beta(|100\rangle-|011\rangle)(|000\rangle-|111\rangle)(|000\rangle-|111\rangle)\big)|10000000\rangle$$
$$+m\big(\alpha(|000\rangle-|111\rangle)(|000\rangle+|111\rangle)(|000\rangle+|111\rangle)$$
$$+\beta(|000\rangle+|111\rangle)(|000\rangle-|111\rangle)(|000\rangle-|111\rangle)\big)|00000000\rangle$$
$$+n\big(\alpha(|100\rangle-|011\rangle)(|000\rangle+|111\rangle)(|000\rangle+|111\rangle)$$
$$+\beta(|100\rangle+|011\rangle)(|000\rangle-|111\rangle)(|000\rangle-|111\rangle)\big)|10000000\rangle\Big) \quad (15.19)$$

If the parity-check measurement outcome (for the measurement of the first two ancillas) is 00, the state collapses to (note that in the following, normalisation has been omitted for simplicity, so the states are expressed up to a constant of proportionality):

$$|\psi_3\rangle \propto \frac{k}{2\sqrt{2}}\big(\alpha(|000\rangle + |111\rangle)(|000\rangle + |111\rangle)(|000\rangle + |111\rangle)$$
$$+ \beta(|000\rangle - |111\rangle)(|000\rangle - |111\rangle)(|000\rangle - |111\rangle)|00000000\rangle\big)$$
$$+ \frac{m}{2\sqrt{2}}\big(\alpha(|000\rangle - |111\rangle)(|000\rangle + |111\rangle)(|000\rangle + |111\rangle)$$
$$+ \beta(|000\rangle + |111\rangle)(|000\rangle - |111\rangle)(|000\rangle - |111\rangle)|00000000\rangle\big) \quad (15.20)$$

in which case there is no bit flip. Or if the measurement outcome is 10:

$$|\psi_3\rangle \propto \frac{l}{2\sqrt{2}}\big(\alpha(|100\rangle + |011\rangle)(|000\rangle + |111\rangle)(|000\rangle + |111\rangle)$$
$$+ \beta(|100\rangle - |011\rangle)(|000\rangle - |111\rangle)(|000\rangle - |111\rangle)|10000000\rangle\big)$$
$$+ \frac{n}{2\sqrt{2}}\big(\alpha(|100\rangle - |011\rangle)(|000\rangle + |111\rangle)(|000\rangle + |111\rangle)$$
$$+ \beta(|100\rangle + |011\rangle)(|000\rangle - |111\rangle)(|000\rangle - |111\rangle)|10000000\rangle\big) \quad (15.21)$$

i.e., a bit flip has occurred, which is thus correctable.

Following the bit-flip parity-check measurements (and correction if necessary – plus resetting the ancilla to $|0\rangle$), next the parity check is performed to test for a phase flip. Suppose for simplicity that there was no bit flip; then, the state parity-check bits:

$$|\psi_4\rangle \propto k\big(\alpha|+00\rangle^{\otimes 3} + \beta|-00\rangle^{\otimes 3}\big)|00000000\rangle$$
$$+ m\big(\alpha|-00\rangle^{\otimes 3} + \beta|+00\rangle^{\otimes 3}\big)|00000010\rangle \quad (15.22)$$

And so upon measurement, the state collapses to either:

$$|\psi_5\rangle = \big(\alpha|+00\rangle^{\otimes 3} + \beta|-00\rangle^{\otimes 3}\big)|00000000\rangle \quad (15.23)$$

which is thus correct, and the recovery is just the identity. Alternatively, if the measurement collapses the state to:

$$|\psi_5\rangle = \big(\alpha|-00\rangle^{\otimes 3} + \beta|+00\rangle^{\otimes 3}\big)|00000010\rangle \quad (15.24)$$

then a detectable (and correctable) phase flip has occurred.

Therefore, performing bit- and phase-flip parity-check measurements collapses a general state into the case where the error is exactly (i) a bit flip; (ii) a phase flip; (iii) both bit and phase flips; or (iv) no error. Furthermore, which of (i)–(iv) occurs is consistent with the error-detection measurement outcomes. This remarkable property allows a continuum of errors to be corrected by performing only bit- and phase-flip checks.

15.3.2 Digitisation of errors

Even though quantum errors are continuous, quantum error correction can be described as having the effect of *digitising* the errors, and thus, the continuous errors may be treated as digital (bit and/or phase flip) errors occurring with some probability. Moreover, the probability of there being a bit/phase flip is directly connected to the magnitude of the continuous error. In the above example, when detecting a bit flip, the probability of measuring $|00\rangle$ (i.e., no bit flip) on the parity-check bits is $|k|^2 + |m|^2$. By definition, the noise sends $|0\rangle \to a|0\rangle + b|1\rangle$ and $|1\rangle \to c|0\rangle + d|1\rangle$, and also by definition $k = \frac{1}{2}(a+d)$. Thus, for low noise levels ($a, d \approx 1$), there is a correspondingly high probability that no bit flip occurs. Similarly, the probability of there being no phase flip (if there is no bit flip) is close to one if $|k| \gg |m|$ (note $m = \frac{1}{2}(a-d)$ so when $a, d \approx 1$, then $m \approx 0$).

The same principle generalises to the case where noise means that all of the qubit states are subject to distortion. Even if all of the qubits in the nine-qubit code are subject to some unwanted perturbation in a noisy channel, the act of performing the parity-check CNOT gates means that the final state can always be factorising into a form similar to Eqns. (15.19) and (15.22). That is, a superposition of terms with digital (bit or phase flip) errors and parity-check qubits consistent with the corresponding error. This in turn motivates a wholly digitised approach to error channel modelling, at least for the purposes of quantum error correction.

15.3.3 The depolarising channel

A popular choice of digital error channel is the *depolarising channel*, which has the property that a physical qubit is left unchanged with probability $1 - p_e$; experiences a bit flip with probability $\frac{p_e}{3}$; experiences a phase flip with probability $\frac{p_e}{3}$; or experiences both bit and phase flips with probability $\frac{p_e}{3}$.

In reality, some noisy channels are such that bit or phase flips are more likely, but it is always possible to make a conservative choice of error probability. (It is, however, worth noting that some successful practical error-correction techniques leverage asymmetry in error probability between bit and phase flips to achieve the desired error suppression with minimal overhead.) Notably, the fact that quantum error correction has a digitising effect on errors themselves means that general noisy channels can be modelled as depolarising channels, given some assumptions. The most important, and physically contentious of these assumptions, is that the physical qubits experience errors *independently*. In reality, the physical proximity of the qubits in the device makes this somewhat dubious, but for the purposes of designing error-correction codes, and proving the properties thereof, it is nevertheless an assumption that is commonly made.

If a single logical qubit is encoded using the nine (physical) qubit Shor code, each of which experiences an independent depolarising channel (with error probability p_e), then, because the Shor code guarantees to correct any single-qubit error, it is useful to think of it as a code that suppresses the error from p_e to $\mathcal{O}(p_e^2)$. (Note the analogy with the three-bit repetition code applied to the binary symmetric channel.)

15.4 Storing a quantum state indefinitely

The discussion thus far has been framed as a quantum state undergoing a single noisy channel – analogously, for example, to the exposure of a radio signal to noise in a classical wireless communication channel – before being corrected. In reality, a quantum state existing in a quantum computer is exposed continuously to noise, and the remedy therefore is to have error-correction blocks repeating at regular intervals. In this way, each interval between error correction may be modelled as the state undergoing a noisy channel. As above, if the channel is a depolarising channel, then encoding (with the Shor code) a single logical qubit using nine physical qubits to suppress the error from p_e to $\mathcal{O}(p_e^2)$ is clearly beneficial; however, in the long run (after sufficiently many such intervals), an error will still probably occur. This in turn begs the question of how to bulk up the error correction such that the error is suppressed further. One way of doing this is to use *concatenated codes*.

As the Shor code suppresses the error in the depolarising channel from p_e to $\mathcal{O}(p_e^2)$, then a constant c can be defined such that:

$$p'_e \leq c p_e^2 \quad (15.25)$$

where p'_e is the error probability for the encoded qubit. The Shor code can be concatenated by encoding a single *logical* qubit not with nine physical qubits, but rather with nine qubits that are themselves encoded using the Shor code. This requires $9^2 = 81$ physical qubits in a *two-level concatenated Shor code* to form one logical qubit. In this case, the error, p''_e, has been suppressed to:

$$p''_e \leq c \times (c p_e^2)^2 = \frac{(c p_e)^{2^2}}{c} \quad (15.26)$$

In general, any code that suppresses a single-qubit error and is concatenated some k times has error, $p_e^{(k)}$:

$$p_e^{(k)} \leq \frac{(c p_e)^{2^k}}{c} \quad (15.27)$$

By increasing k, the error $p_e^{(k)}$ can be reduced to any desired value, as long as the physical qubit error rate, p_e, is such that $p_e < p_{th} = \frac{1}{c}$. For any error correcting code, p_{th} is known as the code's threshold. It is a simple bit of analysis to calculate k such that the error is suppressed below some user-defined ϵ; however, this is better introduced in terms of fault-tolerant computation, that is, where the stored state undergoes operations rather than just being statically stored.

15.4.1 Correcting noisy 'transversal' logical operations

Consider a single-qubit state $\alpha|0\rangle + \beta|1\rangle$ stored in a (single-level) Shor code, that is, stored as the state:

$$\frac{1}{2\sqrt{2}}\Big(\alpha(|000\rangle + |111\rangle)(|000\rangle + |111\rangle)(|000\rangle + |111\rangle)$$
$$+ \beta(|000\rangle - |111\rangle)(|000\rangle - |111\rangle)(|000\rangle - |111\rangle)\Big) \quad (15.28)$$

If a Pauli-X operation is to be applied to the logical state, then this can be achieved by applying a Pauli-Z gate to each of the physical qubits (in fact, it is only necessary to apply the Pauli-Z to one qubit in each block, but it is in many ways simpler to stick to the requirement of applying a Pauli-Z to each qubit), which thus gives:

$$\frac{1}{2\sqrt{2}}\Big(\alpha(|000\rangle - |111\rangle)(|000\rangle - |111\rangle)(|000\rangle - |111\rangle)$$
$$+ \beta(|000\rangle + |111\rangle)(|000\rangle + |111\rangle)(|000\rangle + |111\rangle)\Big) \quad (15.29)$$

Conversely, applying a Pauli-X to each physical qubit has the effect of performing a Pauli-Z on the logical (encoded) state (see chapter problems). These logical gates can be depicted:

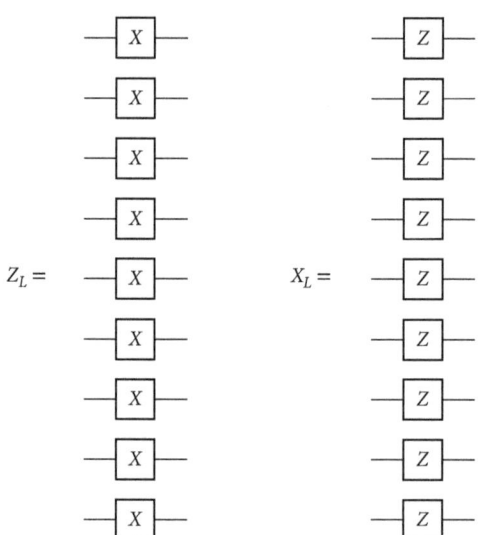

Figure 15.8

Logical gates that can be achieved by applying a physical gate to each qubit individually are termed 'transversal' and play an important role in fault-tolerant quantum computation. A noisy single-qubit operation is generally well-modelled by a perfect execution of the desired operation, followed by a noisy channel, and in this way, it is easy to extend the setting for storing a single-qubit state to correcting errors on a single-qubit state undergoing transversal operations. In particular, if this noise is modelled as independent depolarising noise for each qubit (channel), then the same principle established for storing a quantum state applies to preserving the logical evolution of a quantum state undergoing logical operations – so long as those logical operations are enacted by transversal physical operations.

15.5 Fault-tolerant quantum computation

Showing that single qubits can be protected by error-correction codes when they undergo logical operations that are enacted by transversal physical operations (under the standard depolarising model of noise) is an important step towards showing that general computation can be performed accurately, even in the presence of noise. This is the domain of *fault-tolerant quantum computation*.

Fault-tolerant quantum computation uses logical qubits, each encoded using a number of physical qubits, as the computational primitives. Each logical qubit may be thought of as consisting of a 'block' of physical qubits, and a logical operation is said to be fault-tolerant if a single failure does not propagate to more than one qubit in any one block. Single-qubit transversal operations are trivially fault-tolerant, as a failure (that is, an erroneous physical unitary operation) clearly only affects the corresponding qubit. However, there are two particularly important complicating factors still remaining that need to be addressed:

1. Interesting quantum computations consist not of single-qubit states, but rather of multi-qubit entangled states. This in turn means that (i) the error-correction protocols must still hold for single qubits undergoing unitary operations, when those qubits are components of a multi-qubit entangled state; and (ii) error-correction protocols must hold when the qubits undergo entangling gates.
2. It is necessary either that (i) universal quantum computation can be performed using only transversal logical gates; or (ii) error-correction protocols must work for logical gatesets that include non-transversal gates.

It turns out that the first of these is easy to satisfy, even though it has not explicitly be shown in the above analysis, multi-qubit entangled states *are* protected by error-correction codes, such as the Shor code. It is also the case that two-qubit entangling operations *can* often be performed transversally, at least with many of the most popular error-correction codes. A two-qubit logical gate is transversal if it consists of physical operations such that each operation involves only one qubit in each block. For instance, the Shor code has a transversal logical CNOT, which simply consists of nine physical CNOTS:

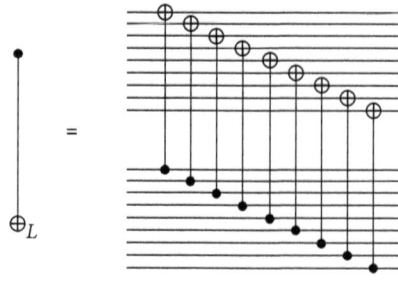

Figure 15.9

Such a two-qubit transversal gate *is* fault-tolerant, as a failure in one of the nine physical CNOTs that constitute the Shor code logical CNOT will cause (at worst) one error in each block of nine qubits.

Conversely, the second item (above) does pose a serious problem: the *Eastin–Knill theorem* states that no quantum error correcting code with a transversal universal gateset exists. There are a number of 'workarounds' to the Eastin–Knill theorem, of which the most prominent is *state injection*. A generic state injection circuit to realise some unitary, U_L, which is not part of the transversal gateset of the code, acting on some logical state $|\psi_L\rangle$ is:

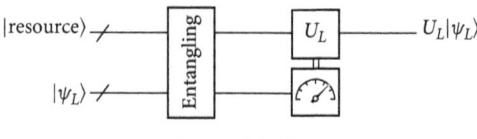

Figure 15.10

Here, '$|\text{resource}\rangle$' is some logical state of the code, which in general may not be preparable using the transversal gateset of the code. In this case, the resource state can be *distilled* such that the error is sufficiently small. The resource state is then entangled with the computational state, $|\psi_L\rangle$, the latter of which is then measured; with the outcome used to (classically) control a unitary, U_L, acting on the first qubit, where U_L is executable using gates from the code's transversal gateset. The overall result of the circuit is that the unmeasured qubit is now in a logical state equal to the operator U_L acting on $|\psi_L\rangle$.

To give a specific example of state injection, it is helpful to use the gateset $\{H, S, T, \text{CNOT}\}$ as a standard universal gateset when assessing various error-correcting codes (note that $\{H, T, \text{CNOT}\}$ is universal as $S = T^2$; however, S is commonly included because it is often much more efficient to execute than two successive T gates) and in this regard the Shor code is not actually particularly satisfactory: of these gates, only CNOT is transversal in the Shor code. A suitable alternative is the *Steane code* – the details of which we do not need here – except to note that it also has the property that it encodes one logical qubit in seven physical qubits such that it protects against an arbitrary single-qubit error (and so can also be thought of as suppressing the error to $\mathcal{O}(p_e^2)$ when each qubit experiences an independent depolarising channel) and that it has transversal gateset $\{H, S, \text{CNOT}\}$.

Thus, to use the Steane code for fault-tolerant quantum computation, all that is required is to synthesise logical T gates using state injection. In particular, to execute a Steane code logical T gate, a *magic state* must be injected. 'Magic state' is the somewhat extravagant name given to any (for our purposes single-qubit) state that cannot be prepared using a product of $\{H, S\}$ applied to $|0\rangle$ – that is, any state other than $|0\rangle, |1\rangle, |+\rangle, |-\rangle, |i\rangle, |-i\rangle$. In particular, when computing with qubits encoded using the Steane code, the logical magic state $(|0\rangle + e^{i\pi/4}|1\rangle)/\sqrt{2}$ should be prepared as a resource, and then entangled by a logical CNOT targeting $|\psi_L\rangle$. Finally, the measurement outcome is used to control the (transversal) gate $U_L = SX$, which achieves the overall effect $T_L|\psi\rangle$.

15.5.1 Elements of fault tolerance

The fundamental characteristic of fault-tolerant quantum computation is that each operation is performed such that a single fault (error) does not propagate to more than one qubit in each encoded block. It has been shown that transversal operations are automatically fault-tolerant, and that magic state injection allows the fault-tolerant execution of non-transversal gates.

In order for the entire computation to be fault-tolerant, it is necessary to further pay attention to three other aspects of the computation:

- The initial preparation of the logical states must be fault-tolerant (that is, the whole enterprise would be somewhat worthless if the initial state had uncorrectable errors).
- Measurement can, in general, introduce errors that may propagate to other qubits, and thus the measurements must be performed fault-tolerantly.
- Finally, the error-correction circuits are *themselves* circuits consisting of physical operations which may be subject to errors.

Fortunately, with appropriate attention, all of these elements of fault tolerance can be successfully addressed; however, detailed exposition of them is beyond the scope of this chapter.

15.5.2 The threshold theorem

It has been shown that concatenation can suppress the error in each encoded (logical) qubit, and moreover that multiple layers of concatenation can suppress the error arbitrarily, as long as the physical error is below the threshold.

Now suppose that we wish to preserve some n-qubit state for a time spanning some $p(n)$ blocks of error correction. (It is important to note that $p(n)$ is defined as the *total* number of blocks of error correction, and so two error-correction blocks on separate qubits that occur in parallel are counted as two and not one.) Recall that $p_e^{(k)}$ is defined as the probability of the failure of a single fault-tolerant gate, when encoded in a k level concatenated code. Using the Union bound, the probability of the computation failing, p_f, can thus be bounded:

$$p_f \le p(n) p_e^{(k)}$$

$$= p(n) \frac{(cp_e)^{2^k}}{c} \quad (15.30)$$

Therefore, in order to achieve a desired maximum error of ϵ, it suffices to choose k such that:

$$\frac{(cp_e)^{2^k}}{c} \le \frac{\epsilon}{p(n)} \quad (15.31)$$

The notation $p(n)$ for the 'number of rounds of error correction' deliberately suggests some function of n and in particular a quantum circuit may be seen as some $p(n)/n$ layers of operations, between each of which is a round of error correction. (Note that in this setting, the identity – that is a wire without a gate – is treated as an operation.). We are most interested in cases where $p(n)$ is upper-bounded by some polynomial, and an important question to ask is how much this number of operations is increased by the error-correction circuitry (that is, what is the quantum error correction *overhead?*). To answer this, first notice that a k-level concatenated code requires d^k operations, where d is a constant. It follows that some $d^k p(n)$ operations are required to perform the fault-tolerant computation, compared to $p(n)$ in the original circuit. To evaluate d^k, consider an encoding that just satisfies the allowable error (i.e., Eqn. (15.31) is taken as an equality). Taking logarithms of each side of Eqn. (15.31), the following can be obtained:

$$2^k \log(p_e c) = \log(c\epsilon/p(n))$$
$$\implies -2^k \log(p_e c) = -\log(c\epsilon/p(n))$$
$$\implies 2^k \log(1/(p_e c)) = \log(p(n)/(c\epsilon))$$
$$\implies 2^k = \frac{\log(p(n)/(c\epsilon))}{\log(1/(p_e c))}$$
$$\implies d^k = \left(\frac{\log(p(n)/(c\epsilon))}{\log(1/(p_e c))} \right)^{\log_2 d}$$
$$\in \mathcal{O}(\text{poly}(\log p(n)/\epsilon)) \quad (15.32)$$

This analysis leads to the threshold theorem:

The threshold theorem. *A quantum circuit containing $p(n)$ gates may be simulated with probability of error at most ϵ using*

$$\mathcal{O}\left(\text{poly}\left(\log p(n)/\epsilon\right) p(n)\right) \quad (15.33)$$

quantum gates on hardware whose gates fail with probability at most p_e, provided p_e is less than some constant threshold $p_e < p_{th}$.

The threshold is a property of each error-correction code, for example the threshold of the Shor code can be estimated. Letting N_e be the number of errors when each qubit of the Shor encoded state is subject to a depolarising channel with error probability p_e, the probability of an uncorrected error is the probability of at least two physical errors occurring, that is:

$$\begin{aligned}\Pr(N_e \geq 2) &= 1 - ((1-p_e)^9 + 9p_e(1-p_e)^8) \\ &= 1 - ((1-p_e + 9p_e)(1-p_e)^8) \\ &= 1 - (1 - 36p_e^2 + \mathcal{O}(p_e^3)) \\ &= 36p_e^2 + \mathcal{O}(p_e^3)\end{aligned} \quad (15.34)$$

So it follows that this simple estimate of the threshold gives a value of $1/36 = 2.78\%$. However, this estimate omits the possibility of error in the error-correction and recovery itself, and the actual threshold of the Shor code is estimated to be 10^{-3}–10^{-4}. As a general principle, it is desirable to find error-correction codes that have as high a threshold as possible, as this makes it easier to suppress the hardware noise below the threshold (such that fault-tolerant computation is possible), and there are other problems with using the Shor code – and indeed any concatenated code – on real hardware with limited qubit connectivity.

15.6 Quantum error correction that respects device layout

One particularly salient point has been omitted altogether in the above discourse, namely that real quantum hardware often has restricted connectivity – in the sense that qubits are laid out in some 'interaction graph' and non-adjacent qubits that must 'interact' (that is to undergo a two-qubit operation together) must therefore first undergo swap operations in order to be adjacent. These swaps decompose into three CNOT gates (see Chapter 4), and so themselves are potentially error-inducing. As the purpose of studying quantum error correction is precisely to show the capability of performing quantum computations (with acceptably suppressed error) on real hardware, it therefore follows that omitting this source of error would render the whole exercise pointless.

It is possible to adjust the approach discussed above to account for the necessity of swap operations, but this is rather contrived and tends to lead to a very poor (low) threshold. A much more successful and widely-researched approach is to design the error-correction protocols around the device connectivity. Of such approaches, the *surface code* and *colour codes* are the most prominent – and have parity-check CNOTs that are 'local' (do not require qubit swap operations). Moreover, such approaches tend to have attractively high thresholds – for example, around 1% is the estimated threshold of the surface code.

Chapter problems

1. If a repetition code is used to encode a single bit, which is then sent through a binary symmetric channel with error probability p_e, following which a majority vote is used for correction, what is the probability of error after this correction when:
   (a) a five-bit repetition code is used;
   (b) a seven-bit repetition code is used;
   (c) an n-bit repetition code is used (for some odd n).
2. If a qubit experiences a bit flip and a phase flip, then show that the order in which these occur does not matter by showing that the Pauli-X (bit flip) and Pauli-Z (phase flip) matrices commute (up to a global phase factor).
3. Let a qubit $\alpha|0\rangle + \beta|1\rangle$ be encoded as $\alpha|000\rangle + \beta|111\rangle$ to protect against a bit flip. For the following, what will the state be after a bit-flip code error correction:
   (a) if all three qubits suffer a bit flip;
   (b) if just the first two qubits suffer a bit flip.
4. Show that the Shor code detects and corrects:
   (a) a phase flip on the first qubit;
   (b) a phase flip *and* bit flip on the first qubit.
5. Show that if a Pauli-X is applied (transversally) to each of the nine physical qubits in the Shor code, then the result is a (logical) Pauli-Z applied to the encoded state.

Further reading

There are many comprehensive books on quantum error correction. For example, Chapter 10 of Nielsen and Chuang [2010] cover the topic in a reasonable amount of detail; *quantum error correction* by Lidar and Brun [2013] is a weighty tome covering all of the essential concepts; and Daniel Gottesman's [2024] textbook is probably the first choice of most in the quantum error correction community. The discovery of quantum error correction is generally attributed to Peter Shor [1995] and Andrew Steane [1996a; 1996b].

Gottesman [1996] invented the stabiliser formalism; the Eastin–Knill theorem is due to Eastin and Knill [2009]; the threshold theorem has been developed by various parties, and Aharonov and Ben-Or [1997; 2008] and Kitaev [1997a; 1997b] give clear original expositions. The surface code is based on the *toric code* proposed by Kitaev [2003]; Litinski [2019] provides a nice discussion of the surface code itself from a practitioners perspective.

APPENDIX A
Simulating the Hadamard gate in the quadratic form expansion

It is possible to maintain the matrix A of the quadratic form expansion in a form, namely *principal row form*, which immediately reveals whether the Hadamard update has led to A no longer being full-rank. A matrix in principal row form has $e_1^T \ldots e_r^T$ as rows, with the remaining rows as general binary row vectors. For instance,

$$A = \begin{bmatrix} 1 & 1 & 0 \\ 1 & 0 & 1 \\ 0 & 1 & 0 \\ 1 & 1 & 1 \\ 1 & 0 & 0 \\ 0 & 0 & 1 \end{bmatrix} \tag{A.1}$$

is in principal row form, with the third, fifth, and sixth rows being the principal rows. As the principal rows are linearly independent, a matrix in principal row form is certainly full-rank. Furthermore, *every* full-rank binary matrix can be converted to principal row form by first performing Gaussian elimination to obtain column-echelon form and then (if necessary) performing some further row subtractions to obtain principal row form. If A is converted to principal row form, then the quadratic form expansion incurs a corresponding change of variable; however (as A is full-rank), this can be accommodated such that the summation remains over the same range of bit strings and the updated Q has the correct form (i.e., the change of variable can be absorbed into Q such that it remains symmetric).

Alternatively, the method of stabiliser simulation with quadratic form expansions can be adjusted to not actually require Gaussian elimination at all, by using slightly elaborated versions of the update procedures (for all of the gates, not just the Hadamard), such that A remains in principal row form. For the remainder of the analysis, we assume that this has been taken care of by one means or another, such that A is indeed in principal row form.

As A is principal row form, it is immediately apparent if the effect of simulating H has updated A such that it is no longer full-rank. This can be seen from Eqn. (6.25), which we repeat here:

$$H_j|\psi\rangle = \frac{(\sqrt{i})^g}{\sqrt{2^{r+1}}} \sum_{\substack{x \in \{0,1\}^r \\ z \in \{0,1\}}} i^{x^T Q x + 2a_{j,\bullet} x z + 2b_j z} |(A'x + ze_j) \oplus b'\rangle \tag{A.2}$$

(Recall $A' = K_j A = (I \oplus e_j e_j^T) A$, i.e., it is the operation that zeroes out the jth row.)

In the itemization below Eqn. (6.25), A is then updated according to $A \leftarrow [K_j A, e_j]$, and we can now break this down into three cases:

1. If the jth row of A is not a principal row, then the effect of the update will be to turn the jth row of A into e_{r+1}^T i.e., a principal row for $r + 1$. As none of the other r principal rows will have been altered, A will have rank $r + 1$ and already be in principal row form.

2. If the jth row of A is a principal row (say there exists c such that $a_{j,*} = e_c^T$), but at least one other element of the cth column of A is equal to 1, then A' will still be full-rank: The jth row of A will now be the principal row e_{r+1}^T; however, there may no longer be a principal row e_c^T and so A may not be in principal row form.

In either of these cases, all that is further required is to update:

- $r \leftarrow r + 1$

(and then put A in principal row form, if necessary). Continuing with the case analysis:

3. If the jth row of A is a principal row (again say $a_{j,*} = e_c^T$) and the update means that the cth column is now all zero, then A is clearly no longer full-rank (in fact, as there are r principal rows, $e_1^T \ldots e_{r+1}^T$ with e_c^T missing), and a non-trivial update is needed as below.

For the third case, following the update, the matrix A will be of the form:

$$A = \begin{bmatrix} & & c & & r+1 \\ & & 0 & & 0 \\ & & \vdots & & \vdots \\ 0 & \cdots & 0 & \cdots & 1 \\ & & \vdots & & \vdots \end{bmatrix} j \qquad (A.3)$$

i.e., the cth column is all zero, the $(r+1)$th column is e_j and the jth row is e_{r+1}^T. The quadratic form expansion for the case where the cth column is all zero can be derived from the general expression for the quadratic form expansion (as obtained by the simulation of the Hadamard):

$$|\psi\rangle = \frac{\left(\sqrt{i}\right)^g}{\sqrt{2^{r+1}}} \sum_{x \in \{0,1\}^{r+1}} i^{x^T Qx} |Ax \oplus b\rangle$$

$$= \frac{\left(\sqrt{i}\right)^g}{\sqrt{2^{r+1}}} \sum_{x \in \{0,1\}^r} i^{x^T Q'x} (1 + i^{2x^T q''_{*,c} + q_{cc}}) |A'x \oplus b\rangle \qquad (A.4)$$

where A' is A with the (all zero) cth column removed (note that this is a new use of A'); Q' is Q with the cth column and row removed; and Q'' is Q with the cth row removed (i.e., $q''_{*,c}$ is the cth column of Q with the cth element removed). This expression has been obtained by explicitly writing out the terms for $x_c \in \{0,1\}$.

The final updates then depend on the value of $q_{c,c} \in \{0,1,2,3\}$. First, we note that for $d \in \{0,1\}$, the following useful identities hold:

$$\frac{1 + (-1)^d i}{\sqrt{2}} = \frac{i^{-d}(1+i)}{\sqrt{2}} = \left(\sqrt{i}\right) i^{-d}$$

$$\frac{1 - (-1)^d i}{\sqrt{2}} = \frac{i^d(1-i)}{\sqrt{2}} = \left(\sqrt{i}\right)^{-1} i^d \qquad (A.5)$$

Using these, and the fact that for any integer $\tilde{d}$, $(-1)^{\tilde{d}} = (-1)^{(\tilde{d}^2)}$, when $q_{c,c} \in \{1,3\}$, Eqn. (A.4) can be expressed:

$$|\psi\rangle = \frac{(\sqrt{i})^g (\sqrt{i})^{2-q_{c,c}}}{\sqrt{2^r}} \sum_{x \in \{0,1\}^r} i^{x^T Q' x_i (q_{c,c}-2) x^T q''_{*,c} (q''_{*,c})^T x} |A'x \oplus b\rangle$$

$$= \frac{(\sqrt{i})^{g+2-q_{c,c}}}{\sqrt{2^r}} \sum_{x \in \{0,1\}^r} i^{x^T Q' x + (q_{c,c}-2) x^T q''_{*,c} (q''_{*,c})^T x} |A'x \oplus b\rangle \quad (A.6)$$

and thus directly gives the updates:

- $g \leftarrow g + 2 - q_{c,c}$
- $Q \leftarrow Q' + (q_{c,c} - 2) q''_{*,c} (q''_{*,c})^T$
- $A \leftarrow A'$

Conversely, in the cases where $q_{c,c} \in \{0, 2\}$, we get that Eqn. (A.4) can be rewritten:

$$|\psi\rangle = \frac{(\sqrt{i})^g}{\sqrt{2^{r+1}}} \sum_{x \in \{0,1\}^r} i^{x^T Q' x} (1 + (-1)^{x^T q''_{*,c} + q_{c,c}/2}) |A'x \oplus b\rangle$$

$$= \frac{(\sqrt{i})^g}{\sqrt{2^{r-1}}} \sum_{x \in \{0,1\}^r} i^{x^T Q' x} \delta_{x^T q''_{*,c} \oplus q_{c,c}/2} |A'x \oplus b\rangle \quad (A.7)$$

By the update $Q \leftarrow \begin{bmatrix} Q & a^T_{j,*} \\ a_{j,*} & 2b_j \end{bmatrix}$ (in Section 6.3.3) and the premise that $a_{j,*} = e_c^T$, the final element of the cth column of Q, i.e., the final (rth) element of $q''_{*,c}$ is equal to 1. This means $\delta_{x^T q''_{*,c} \oplus q_{c,c}/2} = \delta_{x_r \oplus (x')^T q'''_{*,c} \oplus q_{c,c}/2}$, where x' is x with the final element removed, and $q'''_{*,c}$ is $q''_{*,c}$ with the final element removed. This, in turn, provides a way to eliminate x_r such that only non-zero terms remain as $x_r \oplus (x')^T q'''_{*,c} \oplus q_{c,c}/2 = 0 \implies x_r = (x')^T q'''_{*,c} \oplus q_{c,c}/2)$. This can be distilled into a procedure to update A':

- Let A'' be A' with the final column removed; then A'' can be updated such that for every i, if $a'_{i,r} = 1$ (i.e., the last element of the ith row of A' is one), then: $a''_{i,*} \leftarrow a''_{i,*} \oplus (q'''_{*,c})^T$, $b_i \leftarrow b_i \oplus q_{c,c}/2$; and finally $A \leftarrow A''$.

Q' can next be updated using the fact that its final (rth) row and column are zero, expect for possibly $q'_{r,r}$. Therefore, the update is to remove the final column and row of Q', and multiply each term in the summation by

$$i^{q_{r,r}((x')^T q'''_{*,c} \oplus q_{c,c}/2)} = i^{q_{r,r}((x')^T q'''_{*,c} (q'''_{*,c})^T x' + q_{c,c}/2 - (x')^T q'''_{*,c} (q'''_{*,c})^T x' q_{c,c})} \quad (A.8)$$

which uses the identity in Eqn. (6.2.1) along with the above. Putting this together gives the updates:

- $g \leftarrow g + q_{r,r} q_{c,c}$;
- Remove the last row and column of Q'; $q'''_{*,c}$ is $q''_{*,c}$ with the last element removed; then $Q \leftarrow Q' + q_{r,r}(1 - q_{c,c}) q'''_{*,c} (q'''_{*,c})^T$.

As the last element of x has now been eliminated from all terms, the final update is:

- $r \leftarrow r - 1$.

and the matrix A is now full rank. Even though we have omitted a full complexity analysis, all of the above updates are elementary matrix and vector operations on matrices and vectors whose sizes are $\mathcal{O}(n^2)$ and $\mathcal{O}(n)$ (respectively) and so this has been achieved in an efficient manner.

Answers to chapter problems

Chapter 2

1. (a) $I \otimes H = \frac{1}{\sqrt{2}} \begin{bmatrix} 1 & 1 & 0 & 0 \\ 1 & -1 & 0 & 0 \\ 0 & 0 & 1 & 1 \\ 0 & 0 & 1 & -1 \end{bmatrix}$

   (b) $H \otimes I = \frac{1}{\sqrt{2}} \begin{bmatrix} 1 & 0 & 1 & 0 \\ 0 & 1 & 0 & 1 \\ 1 & 0 & -1 & 0 \\ 0 & 1 & 0 & -1 \end{bmatrix}$

2. -
3. -
4. (a) • $\lambda = \frac{1}{\sqrt{2}} + \frac{i}{\sqrt{2}}$, with eigenvector $\begin{bmatrix} 1 \\ -i \end{bmatrix}$.

   • $\lambda = \frac{1}{\sqrt{2}} - \frac{i}{\sqrt{2}}$, with eigenvector $\begin{bmatrix} 1 \\ i \end{bmatrix}$.

   (b) • $\lambda = 1$, with eigenvector $\begin{bmatrix} 1 \\ i \end{bmatrix}$.

   • $\lambda = -1$, with eigenvector $\begin{bmatrix} 1 \\ -i \end{bmatrix}$.

5. The trace is 5, the determinant is −2.
6. (a)
$$\frac{1}{\sqrt{2}}\begin{bmatrix} 1 & -1 \\ 1 & 1 \end{bmatrix} = \frac{1}{\sqrt{2}}(|0\rangle\langle 0| - |0\rangle\langle 1| + |1\rangle\langle 0| + |1\rangle\langle 1|) \qquad (A.9)$$

   (b)
$$\begin{bmatrix} 0 & -i \\ i & 0 \end{bmatrix} = i(-|0\rangle\langle 1| + |1\rangle\langle 0|) \qquad (A.10)$$

7. -
8. (a) No inverse.

   (b)
$$A^+ = \begin{bmatrix} -0.7778 & 0.6111 & -0.1667 \\ -0.1111 & 0.1111 & 0.0000 \\ 0.5556 & -0.3889 & 0.1667 \end{bmatrix} \qquad (A.11)$$

Chapter 3

1. (a), (b), (d), (e), and (g) are valid qubit states.

The probabilities of measuring $|0\rangle$ and $|1\rangle$ are (respectively): (a) 0.5 and 0.5; (b) 0.25 and 0.75; (d) 0.64 and 0.36; (e) $\cos^2 \theta$ and $\sin^2 \theta$; (g) 0.5 and 0.5.

The probabilities of measuring $|+\rangle$ and $|-\rangle$ (respectively): (a) 1 and 0; (b) $\sin^2 15°$ and $\cos^2 15°$; (d) 0.98 and 0.02; (e) 0.5 and 0.5; (g) 0.5 and 0.5.

2. The post-measurement state is $-|1\rangle \otimes \frac{1}{\sqrt{25}}(3|0\rangle + 4i|1\rangle)$ and so the probability of measuring one on the second qubit is $\frac{16}{25}$.

3. -

4. $Z = iYX; X = iZY; Y = iXZ$

5. -

6. -

7. The computational basis.

8. The computational basis, probability of success 15/16.

9. Let the two-qubit state be: $a|00\rangle + b|01\rangle + c|10\rangle + d|11\rangle$:
   (a) At least one of a and b is zero, **and** at least one of c and d is zero.
   (b) One of $a, b, c,$ and d equals zero, and the other three are strictly non-zero.
   (c) The state is unentangled.
   Final part: yes, such a basis can always be chosen.

Chapter 4

1. -

2. $\frac{1}{2}(|00\rangle + |01\rangle + |10\rangle - |11\rangle)$ which is entangled.

3. The circuit:

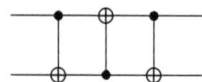

4. The circuit:

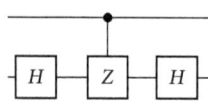

5.

$$\begin{bmatrix} 1 & 0 & 0 & 0 & 0 & 0 & 0 & 0 \\ 0 & 1 & 0 & 0 & 0 & 0 & 0 & 0 \\ 0 & 0 & 1 & 0 & 0 & 0 & 0 & 0 \\ 0 & 0 & 0 & 1 & 0 & 0 & 0 & 0 \\ 0 & 0 & 0 & 0 & 0 & 1 & 0 & 0 \\ 0 & 0 & 0 & 0 & 1 & 0 & 0 & 0 \\ 0 & 0 & 0 & 0 & 0 & 0 & 0 & 1 \\ 0 & 0 & 0 & 0 & 0 & 0 & 1 & 0 \end{bmatrix} \quad (A.12)$$

6. -

7. -

8. -

Answers to chapter problems 303

Chapter 5

1. -
2. (a) Yes, (without loss of generality) let Alice hold the first qubit and Bob the second and third. Then if Bob applies a CNOT controlled by the second qubit, targeting the third, this transforms the state to: $\frac{1}{\sqrt{2}}(|000\rangle + |110\rangle) = \frac{1}{\sqrt{2}}(|00\rangle + |11\rangle)|0\rangle$. Bob can then discard the third qubit, and Alice and Bob share $\frac{1}{\sqrt{2}}(|00\rangle + |11\rangle)$ as is required for teleportation.
   (b) After Charlie applies the Hadamard gate, the state will be $\frac{1}{2}(|000\rangle + |001\rangle + |110\rangle - |111\rangle)$.
   If Charlie measures 0, the state will be $\frac{1}{\sqrt{2}}(|00\rangle + |11\rangle)$ as is required for teleportation between Alice and Bob, so they need do nothing; if Charlie measures 1, then the state will be $\frac{1}{\sqrt{2}}(|00\rangle - |11\rangle)$, in which case (exactly) one of Alice and Bob should apply a Z gate to their qubit, to return the state to $\frac{1}{\sqrt{2}}(|00\rangle + |11\rangle)$, and again it is then possible to do teleportation.
3. As the qubit encodes four bits, a single measurement made by Eve cannot reveal all of the information.
4. 1/2.
5. (a) $U_1 = I$, $U_2 = H$.
   (b) It is always entangled. Single-qubit unitaries applied to a two-qubit state cannot change its entanglement.
   (c) Alice applies X to her qubit; Bob applies HZ to his qubit and then the state is such that the superdense coding protocol can be used.
   (d) Yes, they can: Let U_{XZ} be the unitary that is either X or Z (unknown to Alice, but known to Bob). If Alice just applies the operation as she would for standard superdense coding without trying to invert U_{XZ}, which we let be U_A, and Bob again applies HZ, then after Alice transmits the qubit, the state is: $((U_A U_{XZ}) \otimes I)\left(\frac{1}{\sqrt{2}}(|00\rangle + |11\rangle)\right)$. However, as U_A is one of $\{I, X, Z, XZ\}$ and U_{XZ} is either X or Z then noting that X and Z commute (up to global phase), then we can rewrite this as $((U_{XZ} U_A) \otimes I)\left(\frac{1}{\sqrt{2}}(|00\rangle + |11\rangle)\right)$ to which Bob can apply U_{XZ} (as he knows whether it is X or Z – and each is self-inverse), and then the state is in the standard form for superdense coding.
6. (a) The two bases are orthonormal and mutually unbiased, and so BB84 can be used.
   (b) The two bases are orthonormal but are not mutually unbiased, and so BB84 will not work exactly as described.

Chapter 6

1. -
2. (a) In each case: (i) when the accepted language consists of strings accepted with certainty; and (ii) the accepted language consists of strings accepted with probability at least $\frac{2}{3}$, the accepted language is all strings of length l such that $l \mod 4 = 2$.
   (b) -

(c)

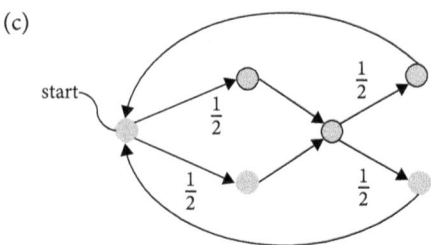

where the black circled states are accepted.

3. (a) -
   (b) ccdd
   (c) Any string containing four of each of c and d, such that one letter occurs four times consecutively.

4. Probability of accepting a string in L is at least $\frac{20}{27}$; the probability of accepting a string not in L is at most $\frac{7}{27}$.

5. (a)
$$|\psi\rangle = \frac{1}{2} \sum_{x\in\{0,1\}^2} i^{[x_1\ x_2]\begin{bmatrix}0&1\\1&0\end{bmatrix}\begin{bmatrix}x_1\\x_2\end{bmatrix}}|x_1,x_2\rangle = \frac{1}{2}\sum_{x\in\{0,1\}^2} i^{[x_1\ x_2]\begin{bmatrix}0&1\\1&0\end{bmatrix}\begin{bmatrix}x_1\\x_2\end{bmatrix}}|\begin{bmatrix}1&0\\0&1\end{bmatrix}\begin{bmatrix}x_1\\x_2\end{bmatrix}\rangle$$
(A.13)

(b)
$$\frac{1}{2}\sum_{x\in\{0,1\}^2} i^{[x_1\ x_2]\begin{bmatrix}0&1\\1&0\end{bmatrix}\begin{bmatrix}x_1\\x_2\end{bmatrix}}|\begin{bmatrix}1&0\\1&1\end{bmatrix}\begin{bmatrix}x_1\\x_2\end{bmatrix}\rangle$$
(A.14)

(c) $\frac{1}{2}(|00\rangle + |01\rangle + |11\rangle - |10\rangle)$

Chapter 7

1. -
2. -
3. (a) 01 is obtained with certainty.
   (b) 00 is obtained with certainty.
4. (a) Toffoli is neither constant or balanced.
   (b) There is a 1/4 chance that 00 will be measured, in which case the Deutsch–Jozsa algorithm decides that the function is constant; the other 3/4 of the time one of $\{01, 10, 11\}$ will be measured, in which case the Deutsch–Jozsa algorithm decides that the function is balanced.
   (c) The simplest solution is to repeat the self-inverse Toffoli such that the unknown unitary is the identity in this case. This corresponds to the constant function $f(x) = 0$.
5. -
6. -

Chapter 8

1. The probability of measuring the marked state after 0, 1, 2, and 3 iterations is, respectively: $\frac{1}{8}$, 0.77, 0.94, 0.32.
2. (a) The simplest solution is the six-qubit circuit $(X \otimes X \otimes I \otimes I \otimes X \otimes I)C^5 X(X \otimes X \otimes I \otimes I \otimes X \otimes I)$.
   (b) The initial state should be $|+\rangle^{\otimes 5}|-\rangle$. After a single Grover iterate the state is:

$$\left(0.8612\left(\sum_{x \in \{0,1\}^5/00110} \frac{1}{\sqrt{31}}|x\rangle\right) + 0.5082|00110\rangle\right)|-\rangle \qquad (A.15)$$

   (c) N = 4, the probability of success is 0.9992.
   (d) The probability of success is 0.929.
3. For example, run Grover's algorithm with search oracle $V' = (I \otimes X)V$ or $V' = V(I \otimes X)$.
4. This is equivalent to running Grover's search algorithm with the usual Grover iterate $(W \otimes I)V$ applied one time fewer.

Chapter 9

1. (a) The eigenvalues are ±1.
   (b)

$$\frac{1}{4\sqrt{5}}\Big(\big(|0\rangle + |1\rangle\big)\big(|0\rangle + |1\rangle\big)\big((1+\sqrt{5})|0\rangle + 2|1\rangle\big)$$
$$- \big(|0\rangle + |1\rangle\big)\big(|0\rangle - |1\rangle\big)\big((1-\sqrt{5})|0\rangle + 2|1\rangle\big)\Big) \qquad (A.16)$$

   (c) 00 with probability 0.7236; 01 with probability 0.2764.
2. (a) The eigenvectors and eigenvalues of U are:

   $[0.3827, -0.9239, 0, 0]^T$ with eigenvalue -1 and hence with phase 1/2

   $[0.9239, 0.3827, 0, 0]^T$ with eigenvalue 1 and hence with phase 0

   $[0, 0, 0.3827, -0.9239]^T$ with eigenvalue $-e^{2\pi i \times (1/8)} = e^{2\pi i \times (5/8)}$
   and hence with phase 5/8

   $[0, 0, 0.9239, 0.3827]^T$ with eigenvalue $e^{2\pi i \times (1/8)}$ and hence with phase 1/8

   (b) Phase = 0 with probability 0.8536; Phase = 1/2 with probability 0.1465. Three bits of precision suffice for any input.
3. (a) 2 bits.
   (b) $0.924|0\rangle + 0.383|1\rangle$ with eigenvalue 1; $0.383|0\rangle - 0.924|1\rangle$ with eigenvalue -1.

(c)

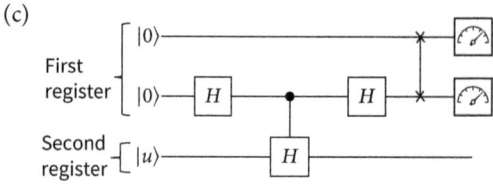

(d)
$$|\psi\rangle = \frac{1}{\sqrt{2}}(|0\rangle + |1\rangle)((a/\sqrt{2})|00\rangle + (b/\sqrt{2})|01\rangle + (a/2 + b/2)|10\rangle + (a/2 - b/2)|11\rangle) \tag{A.17}$$

(e) 00 with probability 0.146; 10 with probability 0.854.

4. In this case, $Q = S_\chi = I \otimes Z$; hence, the amplitude would not be amplified and QAE would fail.
5. If $|\psi\rangle = A|0\rangle$, then perform QAE with $A' = (I \otimes X)A$.
6. (a)
$$|\psi\rangle = \cos\frac{\pi}{10}|+\rangle|0\rangle + \sin\frac{\pi}{10}|1\rangle|1\rangle \tag{A.18}$$

hence $\theta = \frac{\pi}{5}$, $|\Phi_0\rangle = |+\rangle$, and $|\Phi_1\rangle = |1\rangle$.

(b) The circuit $Q^2 A|0\rangle$ has the desired effect.

Chapter 10

1. -
2. (a)
$$1 + \frac{1}{2 + \frac{1}{8}} \tag{A.19}$$

(b)
$$15 + \frac{1}{1 + \frac{1}{3}} \tag{A.20}$$

3. -
4. -
5. (a) The order is 6.
   (b) The eigenvalue is $e^{\pi i/3}$ and

$$|u_1\rangle = \frac{1}{\sqrt{6}}\left(|1\rangle + e^{-\pi i/3}|10\rangle + e^{-2\pi i/3}|16\rangle + e^{-\pi i}|13\rangle + e^{-4\pi i/3}|4\rangle + e^{-5\pi i/3}|19\rangle\right) \tag{A.21}$$

Chapter 11

1. -
2. (a) -
   (b) $\mathcal{O}(\Delta t^3)$.
3. 2^{-n}
4. (a) The Hadamard gate.
   (b) $\tilde{b} = \frac{1}{2}$
   (c) $\frac{1}{2}$
   (d) For example: $(I \otimes X)\text{CNOT}(H \otimes I)$
5. -
6. (a)

$$H = \begin{bmatrix} 1 & 0 & 0 & 0 & 0 & 0 & 0 & 0 \\ 0 & -1 & 0 & 0 & 0 & 0 & 0 & 0 \\ 0 & 0 & 1 & 0 & 0 & 0 & 0 & 0 \\ 0 & 0 & 0 & -1 & 0 & 0 & 0 & 0 \\ 0 & 0 & 0 & 0 & -1 & 0 & 0 & 0 \\ 0 & 0 & 0 & 0 & 0 & 1 & 0 & 0 \\ 0 & 0 & 0 & 0 & 0 & 0 & -1 & 0 \\ 0 & 0 & 0 & 0 & 0 & 0 & 0 & 1 \end{bmatrix} \quad (A.22)$$

(b) The eigenvectors are the computational basis states.
Eigenvalue = 1: $\{|000\rangle, |010\rangle, |101\rangle, |111\rangle\}$; eigenvalue = -1: $\{|001\rangle, |011\rangle, |100\rangle, |110\rangle\}$. That is, whenever exactly one of the first and third qubits is 1, the eigenvalue is -1, and it is $+1$ otherwise, as claimed.

Chapter 12

1. -
2. Satisfying assignments: (i) $x_1 = 1, x_2 = 0, x_3 = 0$; (ii) $x_1 = 0, x_2 = 1, x_3 = 0$; (iii) $x_1 = 0, x_2 = 0, x_3 = 1$.
   The Hamiltonian is:

$$H_y = \begin{bmatrix} 1 & 0 & 0 & 0 & 0 & 0 & 0 & 0 \\ 0 & 0 & 0 & 0 & 0 & 0 & 0 & 0 \\ 0 & 0 & 0 & 0 & 0 & 0 & 0 & 0 \\ 0 & 0 & 0 & 1 & 0 & 0 & 0 & 0 \\ 0 & 0 & 0 & 0 & 0 & 0 & 0 & 0 \\ 0 & 0 & 0 & 0 & 0 & 1 & 0 & 0 \\ 0 & 0 & 0 & 0 & 0 & 0 & 1 & 0 \\ 0 & 0 & 0 & 0 & 0 & 0 & 0 & 4 \end{bmatrix} \quad (A.23)$$

3. $|000\rangle$.
4. (a) 7
   (b) 0 and 1 in one subset; 2 and 3 in the other subset.

(c)

$\tilde{V}$	$\sum_{i,j \in E} 2x_i x_j - x_i - x_j$
$\{0\}$	-2
$\{1\}$	-1
$\{2\}$	-1
$\{3\}$	-2
$\{0,1\}$	-3
$\{0,2\}$	-1
$\{0,3\}$	-2

Chapter 13

1. (a) This is already Hermitian and its size is a power of 2, so no alternative is necessary.

   (b) $\begin{bmatrix} 0 & 1+i & 0 & 0 \\ 1-i & 15 & 0 & 0 \\ 0 & 0 & 25 & 0 \\ 0 & 0 & 0 & \pm x \end{bmatrix}$, where $0.1322 \le x \le 25$

   (c)
   $$\begin{bmatrix} 0 & 0 & 1 & 0 \\ 0 & 0 & 1 & 1 \\ 1 & 1 & 0 & 0 \\ 0 & 1 & 0 & 0 \end{bmatrix} \tag{A.24}$$

2. (a) 114.5
   (b) 189.1
   (c) 2.618

3. -

4. (a) -
   (b) $\lambda_0 = |-\rangle$, $E_0 = 0.25$ and $\lambda_1 = |+\rangle$, $E_1 = 0.5$.
   (c)
   $$|b\rangle = \frac{1}{\sqrt{10}} (3|-\rangle - |+\rangle) \tag{A.25}$$

   (d)
   $$|\psi_3\rangle = \frac{1}{\sqrt{10}} \left(3|-\rangle |00\rangle \left(\sqrt{1 - \frac{C^2}{0.25^2}} |0\rangle + \frac{C}{0.25}|1\rangle \right) \right.$$
   $$\left. -|+\rangle |00\rangle \left(\sqrt{1 - \frac{C^2}{0.5^2}} |0\rangle + \frac{C}{0.5}|1\rangle \right) \right) \tag{A.26}$$

   (e) $C = 0.25$.

(f) $\frac{37}{40}$.

(g) -

Chapter 14

1. $|\pi_l\rangle = |\pi_r\rangle = |1\rangle$ applied to the first qubit.
2. -
3. If we let $\Delta = 1/\sqrt{N}$, then the 'rising part' of the polynomial approximation occurs between $\pm\Delta$ and the polynomial has degree $\Theta(\Delta)$. We can immediately see that no approximation of the sign function can have degree that grows sub-linearly in Δ as this would allow the unstructured search lower-bound to be violated.
4. -

Chapter 15

1. (a)
$$\frac{5!}{3!2!}p_e^3(1-p_e)^2 + \frac{5!}{4!1!}p_e^4(1-p_e) + p_e^5 \tag{A.27}$$

(b)
$$\frac{7!}{4!3!}p_e^4(1-p_e)^3 + \frac{7!}{5!2!}p_e^5(1-p_e)^2 + \frac{7!}{6!1!}p_e^6(1-p_e) + p_e^7 \tag{A.28}$$

(c)
$$\sum_{i=\lceil n/2 \rceil} n! \frac{n}{i!(n-i)!} p_e^i (1-p_e)^{n-i} \tag{A.29}$$

2. -
3. (a)
$$\alpha|111\rangle + \beta|000\rangle \tag{A.30}$$

(b)
$$\alpha|111\rangle + \beta|000\rangle \tag{A.31}$$

4. -
5. -

Bibliography

S. Aaronson, and A. Arkhipov. 'The Computational Complexity of Linear Optics'. In: *STOC '11*. San Jose, California, USA: Association for Computing Machinery, 2011, pp. 333–342. ISBN: 9781450306911. DOI: 10.1145/1993636.1993682. URL: https://doi.org/10.1145/1993636.1993682.

S. Aaronson. *Quantum Computing since Democritus* (Cambridge: Cambridge University Press, 2013).

S. Aaronson. 'Why Philosophers Should Care About Computational Complexity'. In: *Computability: Turing, Gödel, Church, and Beyond* (London, Ontario: Centre for Digital Philosophy at the University of Western Ontario, 2013), pp. 261–328.

S. Aaronson. 'Read the Fine Print'. *Nature Physics* 11/4 (2015): 291–293. DOI: 10.1038/nphys3272.

S. Aaronson. *Shtetl-Optimized: MIP* = RE*. 2020. URL: https://scottaaronson.blog/?p=4512.

S. Aaronson. *How Much Structure Is Needed for Huge Quantum Speedups?* 2022. arXiv: 2209.06930 [quant-ph]. URL: https://arxiv.org/abs/2209.06930.

D. Aharonov, and M. Ben-Or. 'Fault-Tolerant Quantum Computation with Constant Error Rate'. *SIAM Journal on Computing* 38/4 (2008), 1207–1282. DOI: 10.1137/S0097539799359385. URL: https://doi.org/10.1137/S0097539799359385.

D. Aharonov, and M. Ben-Or. 'Fault-Tolerant Quantum Computation with Constant Error'. In: *Proceedings of the Twenty-Ninth Annual ACM Symposium on Theory of Computing*. STOC '97. El Paso, Texas, USA: Association for Computing Machinery, 1997, pp. 176–188. ISBN: 0897918886. DOI: 10.1145/258533.258579. URL: https://doi.org/10.1145/258533.258579.

B. Apolloni, N. Cesa-Bianchi, and D. De Falco. *A Numerical Implementation of "Quantum Annealing"*. Tech. rep. Bielefeld: Bielefeld TU. Bielefeld-Bochum-Stochastik, 1988. URL: https://cds.cern.ch/record/192546.

B. Apolloni, C. Carvalho, and D. de Falco. 'Quantum Stochastic Optimization'. *Stochastic Processes and their Applications* 33/2 (1989): 233–244. ISSN: 0304-4149. DOI: https://doi.org/10.1016/0304-4149(89)90040-9. URL: https://www.sciencedirect.com/science/article/pii/0304414989900409.

S. Aaronson, and D. Gottesman. 'Improved Simulation of Stabilizer Circuits'. *Physical Review A* 70/5 (Nov. 2004). ISSN: 1094-1622. DOI: 10.1103/physreva.70.052328. URL: http://dx.doi.org/10.1103/PhysRevA.70.052328.

D. Aharonov et al. 'Adiabatic Quantum Computation is Equivalent to Standard Quantum Computation'. In: *45th Annual IEEE Symposium on Foundations of Computer Science*. 2004, pp. 42–51. DOI: 10.1109/FOCS.2004.8.

T. Albash, and D. A. Lidar. 'Adiabatic Quantum Computation'. *Reviews of Modern Physics* 90/1 (Jan. 2018). ISSN: 1539-0756. DOI: 10.1103/revmodphys.90.015002. URL: http://dx.doi.org/10.1103/RevModPhys.90.015002.

D. S. Abrams, and S. Lloyd. 'Simulation of Many-Body Fermi Systems on a Universal Quantum Computer'. *Physical Review Letters* 79/13 (Sept. 1997): 2586–2589. ISSN: 1079-7114. DOI: 10.1103/physrevlett.79.2586. URL: http://dx.doi.org/10.1103/PhysRevLett.79.2586.

A. Ambainis. 'Quantum Lower Bounds by Quantum Arguments'. *Journal of Computer and System Sciences* 64/4 (2002): 750–767. ISSN: 0022-0000. DOI: https://doi.org/10.1006/jcss.2002.1826. URL: https://www.sciencedirect.com/science/article/pii/S002200000291826X.

A. Ambainis. 'Variable Time Amplitude Amplification and Quantum Algorithms for Linear Algebra Problems'. In: *29th International Symposium on Theoretical Aspects of Computer Science (STACS 2012)*. Vol. 14. Leibniz International Proceedings in Informatics (LIPIcs). Dagstuhl, Germany: Schloss Dagstuhl – Leibniz-Zentrum für Informatik, 2012, pp. 636–647. ISBN: 978-3-939897-35-4. DOI: 10.4230/LIPIcs.STACS.2012.636.

F. T. Arecchi et al. 'Atomic Coherent States in Quantum Optics'. *Physical Review A* 6 (6 Dec. 1972): 2211–2237. DOI: 10.1103/PhysRevA.6.2211. URL: https://link.aps.org/doi/10.1103/PhysRevA.6.2211.

F. Arute et al. 'Quantum Supremacy Using a Programmable Superconducting Processor'. *Nature* 574/7779 (Oct. 2019): 505–510. DOI: 10.1038/s41586-019-1666-5. arXiv: 1910.11333 [quant-ph].

S. Apers, and A. Sarlette. 'Quantum Fast-Forwarding: Markov Chains and Graph Property Testing'. *Quantum Information & Computing* 19/3–4 (Mar. 2019): 181–213. ISSN: 1533-7146.

C. H. Bennett, and G. Brassard. 'Quantum Cryptography: Public Key Distribution and Coin Tossing'. *Theoretical Computer Science* 560 (2014): 7–11. Theoretical Aspects of Quantum Cryptography – Celebrating 30 years of BB84. ISSN: 0304-3975. DOI: https://doi.org/10.1016/j.tcs.2014.05.025. URL: https://www.sciencedirect.com/science/article/pii/S0304397514004241.

J. C Bridgeman, and C. T Chubb. 'Hand-Waving and Interpretive Dance: An Introductory Course on Tensor Networks'. *Journal of Physics A: Mathematical and Theoretical* 50/22 (May 2017): 223001. issn: 1751-8121. DOI: 10.1088/1751-8121/aa6dc3. URL: http://dx.doi.org/10.1088/1751-8121/aa6dc3.

R. Beals et al. 'Quantum Lower Bounds by Polynomials'. *Journal of the ACM* 48/4 (July 2001): 778–797. ISSN: 0004-5411. DOI: 10.1145/502090.502097. URL: https://doi.org/10.1145/502090.502097.

J. S. Bell. 'On the Einstein Podolsky Rosen Paradox'. *Physics Physique Fizika* 1 (3 Nov. 1964): 195–200. DOI: 10.1103/PhysicsPhysiqueFizika.1.195. URL: https://link.aps.org/doi/10.1103/PhysicsPhysiqueFizika.1.195.

M. Benedetti et al. 'A Generative Modeling Approach for Benchmarking and Training Shallow Quantum Circuits'. *npj Quantum Information* 5/1 (May 2019). ISSN: 2056-6387. DOI: 10.1038/s41534-019-0157-8. URL: http://dx.doi.org/10.1038/s41534-019-0157-8.

M. Benedetti et al. 'Parameterized Quantum Circuits as Machine Learning Models'. *Quantum Science and Technology* 4/4 (Nov. 2019): 043001. ISSN: 2058-9565. DOI: 10.1088/2058-9565/ab4eb5. URL: http://dx.doi.org/10.1088/2058-9565/ab4eb5.

C. H. Bennett et al. 'Teleporting an Unknown Quantum State via Dual Classical and Einstein-Podolsky-Rosen Channels'. *Physical Review Letters* 70 (13 Mar. 1993): 1895–1899. DOI: 10.1103/PhysRevLett.70.1895. URL: https://link.aps.org/doi/10.1103/PhysRevLett.70.1895.

D. W. Berry et al. 'Efficient Quantum Algorithms for Simulating Sparse Hamiltonians'. *Communications in Mathematical Physics* 270/2 (Dec. 2006): 359–371. ISSN: 1432-0916. DOI: 10.1007/s00220-006-0150-x. URL: http://dx.doi.org/10.1007/s00220-006-0150-x.

D. W. Berry et al. 'Simulating Hamiltonian Dynamics with a Truncated Taylor Series'. *Physical Review Letters* 114/9 (Mar. 2015). ISSN: 1079-7114. DOI: 10.1103/physrevlett.114.090502. URL: http://dx.doi.org/10.1103/PhysRevLett.114.090502.

D. W Berry. 'High-Order Quantum Algorithm for Solving Linear Differential Equations'. *Journal of Physics A: Mathematical and Theoretical* 47/10 (Feb. 2014): 105301. DOI: 10.1088/1751-8113/47/10/105301. URL: https://dx.doi.org/10.1088/1751-8113/47/10/105301.

N. de Beaudrap, and S. Herbert. 'Fast Stabiliser Simulation with Quadratic Form Expansions'. *Quantum* 6 (Sept. 2022): 803. ISSN: 2521-327X. DOI: 10.22331/q-2022-09-15-803. URL: http://dx.doi.org/10.22331/q-2022-09-15-803.

G. Brassard, P. HØyer, and A. Tapp. 'Quantum Counting'. In: *Lecture Notes in Computer Science* (Springer: Berlin Heidelberg, 1998), pp. 820–831. ISBN: 9783540686811. DOI: 10.1007/bfb0055105. URL: http://dx.doi.org/10.1007/BFb0055105.

M. J. Bremner, R. Jozsa, and D. J. Shepherd. 'Classical Simulation of Commuting Quantum Computations Implies Collapse of the Polynomial Hierarchy'. *Proceedings of the Royal Society A: Mathematical, Physical and Engineering Sciences* 467/2126 (Aug. 2010): 459–472. ISSN: 1471-2946. DOI: 10.1098/rspa.2010.0301. URL: http://dx.doi.org/10.1098/rspa.2010.0301.

L. Bittel, and M. Kliesch. 'Training Variational Quantum Algorithms Is NP-Hard'. *Physical Review Letters* 127/12 (Sept. 2021). ISSN: 1079-7114. DOI: 10.1103/physrevlett.127.120502. URL: http://dx.doi.org/10.1103/PhysRevLett.127.120502.

K. Boothby et al. *Next-Generation Topology of D-Wave Quantum Processors*. 2020. arXiv: 2003.00133 [quant-ph]. URL: https://arxiv.org/abs/2003.00133.

M. Born. 'Zur Quantenmechanik der Stoßvorgänge'. *Zeitschrift fur Physik* 37 (1926): 863–867. DOI: 10.1007/BF01397477.

M. Boyer et al. 'Tight Bounds on Quantum Searching'. *Fortschritte der Physik* 46/4–5 (June 1998): 493–505. ISSN: 1521-3978. DOI: 10.1002/(sici)1521-3978(199806)46:4/5<493::aid-prop493>3.0.co;2-p. URL: http://dx.doi.org/10.1002/(SICI)1521-3978(199806)46:4/5%3C493::AID-PROP493%3E3.0.CO;2-P.

P. Oscar Boykin et al. *On Universal and Fault-Tolerant Quantum Computing*. 1999. arXiv: quant-ph/9906054 [quant-ph]. URL: https://arxiv.org/abs/quant-ph/9906054.

G. Brassard et al. *Quantum Amplitude Amplification and Estimation*. 2002. DOI: 10.1090/conm/305/05215. URL: http://dx.doi.org/10.1090/conm/305/05215.

D. J. Brod. 'Efficient Classical Simulation of Matchgate Circuits with Generalized Inputs and Measurements'. *Physical Review A* 93 (6 June 2016): 062332. DOI: 10.1103/PhysRevA.93.062332. URL: https://link.aps.org/doi/10.1103/PhysRevA.93.062332.

F. L. Brandao and K. M. Svore. 'Quantum Speed-Ups for Solving Semidefinite Programs'. In: *2017 IEEE 58th Annual Symposium on Foundations of Computer Science (FOCS)*. Los Alamitos, CA, USA: IEEE Computer Society, Oct. 2017, pp. 415–426. DOI: 10.1109/FOCS.2017.45. URL: https://doi.ieeecomputersociety.org/10.1109/FOCS.2017.45.

E. Bernstein, and U. Vazirani. 'Quantum Complexity Theory'. *SIAM Journal on Computing* 26/5 (1997): 1411–1473. DOI: 10.1137/S0097539796300921. URL: https://doi.org/10.1137/S0097539796300921.

C. H. Bennett, and S. J. Wiesner. 'Communication via One- and Two-Particle Operators on Einstein-Podolsky-Rosen States'. *Physical Review Letters* 69 (20 Nov. 1992): 2881–2884. DOI: 10.1103/PhysRevLett.69.2881. URL: https://link.aps.org/doi/10.1103/PhysRevLett.69.2881.

B. Coecke, and R. Duncan. 'Interacting Quantum Observables: Categorical Algebra and Diagrammatics'. *New Journal of Physics* 13/4 (Apr. 2011): 043016. ISSN: 1367-2630. DOI: 10.1088/1367-2630/13/4/043016. URL: http://dx.doi.org/10.1088/1367-2630/13/4/043016.

P. R. Chernoff. 'Note on Product Formulas for Operator Semigroups'. *Journal of Functional Analysis* 2/2 (1968): 238–242. ISSN: 0022-1236. DOI: https://doi.org/10.1016/0022-1236(68)90020-7. URL: https://www.sciencedirect.com/science/article/pii/0022123668900207.

A. M. Childs et al. 'Toward the First Quantum Simulation with Quantum Speedup'. *Proceedings of the National Academy of Sciences* 115/38 (Sept. 2018): 9456–9461. ISSN: 1091-6490. DOI: 10.1073/pnas.1801723115. URL: http://dx.doi.org/10.1073/pnas.1801723115.

N.-H. Chia et al. 'Sampling-Based Sublinear Low-rank Matrix Arithmetic Framework For Dequantizing Quantum Machine Learning'. In: *Proceedings of the 52nd Annual ACM SIGACT Symposium on Theory of Computing*. STOC'20. ACM, June 2020. DOI: 10.1145/3357713.3384314. URL: http://dx.doi.org/10.1145/3357713.3384314.

A. Chi-Chih Yao. 'Quantum Circuit Complexity'. In: *Proceedings of 1993 IEEE 34th Annual Foundations of Computer Science*. 1993, pp. 352–361. DOI: 10.1109/SFCS.1993.366852.

B. Coecke, and A. Kissinger. *Picturing Quantum Processes: A First Course in Quantum Theory and Diagrammatic Reasoning* (Cambridge: Cambridge University Press, 2017).

A. M. Childs, R. Kothari, and R. D. Somma. 'Quantum Algorithm for Systems of Linear Equations with Exponentially Improved Dependence on Precision'. *SIAM Journal on Computing* 46/6 (2017): 1920–1950. DOI: 10.1137/16M1087072. URL: https://doi.org/10.1137/16M1087072.

J. W. Cooley, and J. W. Tukey. 'An Algorithm for the Machine Calculation of Complex Fourier Series'. *Mathematics of Computation* 19 (1965): 297–301. URL: https://api.semanticscholar.org/CorpusID:121744946.

A. M. Childs, and N. Wiebe. 'Hamiltonian Simulation Using Linear Combinations of Unitary Operations'. *Quantum Information & Computing* 12/11–12 (Nov. 2012): 901–924. ISSN: 1533-7146.

N. David Mermin. 'What's Wrong with this Pillow?' *Physics Today* 42/4 (Apr. 1989): 9–11. ISSN: 0031-9228. DOI: 10.1063/1.2810963. eprint: https://pubs.aip.org/physicstoday/article-pdf/42/4/9/8301150/9_1_online.pdf. URL: https://doi.org/10.1063/1.2810963.

D. Deutsch. 'Quantum Theory, the Church-Turing Principle and the Universal Quantum Computer'. *Proceedings of the Royal Society of London Series A* 400/1818 (July 1985): 97–117. DOI: 10.1098/rspa.1985.0070.

D. Deutsch. 'Quantum Computational Networks'. *Proceedings of the Royal Society of London. Series A, Mathematical and Physical Sciences* 425/1868 (1989): 73–90. ISSN: 00804630. URL: http://www.jstor.org/stable/2398494 (visited on 02/08/2024).

D. Deutsch. *The Fabric of Reality: The Science of Parallel Universes–And Its Implications* (London: Allen Lane, 1997).

D. Dieks. 'Communication by EPR Devices'. *Physics Letters A* 92/6 (1982): 271–272. ISSN: 0375-9601. DOI: https://doi.org/10.1016/0375-9601(82)90084-6. URL: https://www.sciencedirect.com/science/article/pii/0375960182900846.

P. A. M. Dirac. 'A New Notation for Quantum Mechanics'. *Mathematical Proceedings of the Cambridge Philosophical Society* 35/3 (1939): 416–418. DOI: 10.1017/S0305004100021162.

D. Deutsch, and R. Jozsa. 'Rapid Solution of Problems by Quantum Computation'. *Proceedings of the Royal Society of London Series A* 439/1907 (Dec. 1992): 553–558. DOI: 10.1098/rspa.1992.0167.

A. Edgington. *Simplex: A Fast Simulator for Clifford Circuits*. URL: https://github.com/CQCL/simplex/releases/tag/v1.4.0.

B. Eastin, and E. Knill. 'Restrictions on Transversal Encoded Quantum Gate Sets'. *Physics Review Letters* 102 (11 Mar. 2009): 110502. DOI: 10.1103/PhysRevLett.102.110502. URL: https://link.aps.org/doi/10.1103/PhysRevLett.102.110502.

A. Einstein, B. Podolsky, and N. Rosen. 'Can Quantum-Mechanical Description of Physical Reality Be Considered Complete?' *Physical Review* 47 (10 May 1935): 777–780. DOI: 10.1103/PhysRev.47.777. URL: https://link.aps.org/doi/10.1103/PhysRev.47.777.

H. Everett. '"Relative State" Formulation of Quantum Mechanics'. *Reviews of Modern Physics* 29/3 (July 1957): 454–462. DOI: 10.1103/RevModPhys.29.454.

E. Farhi et al. *Quantum Computation by Adiabatic Evolution*. 2000. arXiv: quant-ph/0001106 [quant-ph]. URL: https://arxiv.org/abs/quant-ph/0001106.

E. Farhi et al. 'A Quantum Adiabatic Evolution Algorithm Applied to Random Instances of an NP-Complete Problem'. *Science* 292/5516 (2001): 472–475. DOI: 10.1126/science.1057726. eprint: https://www.science.org/doi/pdf/10.1126/science.1057726. URL: https://www.science.org/doi/abs/10.1126/science.1057726.

E. Farhi and A. W. Harrow. *Quantum Supremacy through the Quantum Approximate Optimization Algorithm*. 2019. arXiv. URL: https://arxiv.org/abs/1602.07674.

R. P. Feynman. 'Simulating Physics with Computers'. *International Journal of Theoretical Physics* 21/6 (1st June 1982): 467–488. ISSN: 1572-9575. DOI: 10.1007/BF02650179.

E. Farhi, J. Goldstone, and S. Gutmann. *A Quantum Approximate Optimization Algorithm*. 2014. arXiv: 1411.4028 [quant-ph]. URL: https://arxiv.org/abs/1411.4028.

G. C. Ghirardi et al. 'Experiments of the EPR Type Involving CP-Violation Do not Allow Faster-than-Light Communication between Distant Observers'. *Europhysics Letters* 6/2 (May 1988): 95. DOI: 10.1209/0295-5075/6/2/001. URL: https://dx.doi.org/10.1209/0295-5075/6/2/001.

A. Gilyén et al. 'Quantum Singular Value Transformation and Beyond: Exponential Improvements for Quantum Matrix Arithmetics'. In: *Proceedings of the 51st Annual ACM SIGACT Symposium on Theory of Computing. STOC'19*. ACM, June 2019. DOI: 10.1145/3313276.3316366. URL: http://dx.doi.org/10.1145/3313276.3316366.

V. Giovannetti, S. Lloyd, and L. Maccone. 'Quantum Random Access Memory'. *Physical Review Letters* 100/16 (Apr. 2008). ISSN: 1079-7114. DOI: 10.1103/physrevlett.100.160501. URL: http://dx.doi.org/10.1103/PhysRevLett.100.160501.

D. Gottesman. 'Surviving as a Quantum Computer in a Classical World'. In: *Textbook manuscript preprint* (2024). URL: http://www.cs.umd.edu/class/spring2024/cmsc858G/QECCbook-2024-ch1-11.pdf.

D. Gottesman. 'Class of Quantum Error-Correcting Codes Saturating the Quantum Hamming Bound'. *Physical Review A* 54 (3 Sept. 1996): 1862–1868. DOI: 10.1103/PhysRevA.54.1862. URL: https://link.aps.org/doi/10.1103/PhysRevA.54.1862.

D. Gottesman. *Stabilizer Codes and Quantum Error Correction*. 1997. arXiv: quant-ph/9705052 [quant-ph]. URL: https://arxiv.org/abs/quant-ph/9705052.

D. Grinko et al. 'Iterative Quantum Amplitude Estimation'. *npj Quantum Information* 7/1 (2021). ISSN: 2056-6387. DOI: 10.1038/s41534-021-00379-1. URL: http://dx.doi.org/10.1038/s41534-021-00379-1.

L. K. Grover. 'A Fast Quantum Mechanical Algorithm for Database Search'. In: *Proceedings of the Twenty-Eighth Annual ACM Symposium on Theory of Computing. STOC '96*. Philadelphia, Pennsylvania, USA: Association for Computing Machinery, 1996, pp. 212–219. ISBN: 0897917855. DOI: 10.1145/237814.237866. URL: https://doi.org/10.1145/237814.237866.

W. Gerlach, and O. Stern. 'Der experimentelle Nachweis der Richtungsquantelung im Magnetfeld'. *Zeitschrift fur Physik* 9/1 (Dec. 1922): 349–352. DOI: 10.1007/BF01326983.

J. Gacon, C. Zoufal, and S. Woerner. 'Quantum-Enhanced Simulation-Based Optimization'. In: *2020 IEEE International Conference on Quantum Computing and Engineering (QCE)*. 2020, pp. 47–55. DOI: 10.1109/QCE49297.2020.00017.

C. W. Helstrom. 'Quantum Detection and Estimation Theory'. *Journal of Statistical Physics* 1/2 (June 1969): 231–252. DOI: 10.1007/BF01007479.

S. Herbert. *No free lunch theorems for quantum state measurements as resources in classical sampling and generative modelling*. 2024. arXiv: 2309.13967 [quant-ph]. URL: https://arxiv.org/abs/2309.13967.

A. W. Harrow, A. Hassidim, and S. Lloyd. 'Quantum Algorithm for Linear Systems of Equations'. *Physical Review Letters* 103 (15 Oct. 2009): 150502. DOI: 10.1103/PhysRevLett.103.150502. URL: https://link.aps.org/doi/10.1103/PhysRevLett.103.150502.

P. Hoyer, T. Lee, and R. Spalek. 'Negative Weights Make Adversaries Stronger'. In: *Proceedings of the Thirty-Ninth Annual ACM Symposium on Theory of Computing. STOC '07*. San Diego, California, USA: Association for Computing Machinery, 2007, pp. 526–535. ISBN: 9781595936318. DOI: 10.1145/1250790.1250867. URL: https://doi.org/10.1145/1250790.1250867.

J. E. Hopcroft, R. Motwani, and J. D. Ullman. *Introduction to Automata Theory, Languages, and Computation (3rd Edition)*. USA: Addison-Wesley Longman Publishing Co., Inc., 2006. ISBN: 0321455363.

A. S. Holevo. 'Statistical decision theory for quantum systems'. *Journal of Multivariate Analysis* 3/4 (1973): 337–394. ISSN: 0047-259X. DOI: https://doi.org/10.1016/0047-259X(73)90028-6. URL: https://www.sciencedirect.com/science/article/pii/0047259X73900286.

S. Herbert, J. Sorci, and Y. Tang. 'Almost-optimal computationalbasis-state transpositions'. *Physical Review A* 110 (1 July 2024): 012437. DOI: 10.1103/PhysRevA.110.012437. URL: https://link.aps.org/doi/10.1103/PhysRevA.110.012437.

Z. Ji et al. 'MIP* = RE'. *Communications of the ACM* 64/11 (Oct. 2021): 131–138. ISSN: 0001-0782. DOI: 10.1145/3485628. URL: https://doi.org/10.1145/3485628.

R. Jozsa, and A. Miyake. 'Matchgates and classical simulation of quantum circuits'. *Proceedings of the Royal Society A: Mathematical, Physical and Engineering Sciences* 464/2100 (July 2008): 3089–3106. ISSN: 1471-2946. DOI: 10.1098/rspa.2008.0189. URL: http://dx.doi.org/10.1098/rspa.2008.0189.

P. Jordan, and E. Wigner. 'Über das Paulische Äquivalenzverbot'. *Zeitschrift fur Physik* 47: 631–651. DOI: 10.1007/BF01331938.

A. Y. Kitaev. 'Fault-Tolerant Quantum Computation by Anyons'. *Annals of Physics* 303/1 (Jan. 2003): 2–30. ISSN: 0003-4916. DOI: 10.1016/s0003-4916(02)00018-0. URL: http://dx.doi.org/10.1016/S0003-4916(02)00018-0.

A. Y. Kitaev. 'Quantum Measurements and the Abelian Stabilizer Problem'. In: *Electronic Colloquium on Computational Complexity*. TR96 (1995). URL: https://api.semanticscholar.org/CorpusID:17023060.

A. Y. Kitaev. 'Quantum Computations: Algorithms and Error Correction'. *Russian Mathematical Surveys* 52/6 (Dec. 1997): 1191. DOI: 10.1070/RM1997v052n06ABEH002155. URL: https://dx.doi.org/10.1070/RM1997v052n06ABEH002155.

A. Y. Kitaev. 'Quantum Error Correction with Imperfect Gates'. In: *Quantum Communication, Computing, and Measurement* (Springer, 1997), pp. 181–188.

A. Y. Kitaev. 'Quantum Computations: Algorithms and Error Correction'. *Russian Mathematical Surveys* 52/6 (Dec. 1997: 1191–1249. DOI: 10.1070/RM1997v052n06ABEH002155.

I. Kerenidis, and A. Prakash. *Quantum Recommendation Systems*. 2016. arXiv: 1603.08675 [quant-ph]. URL: https://arxiv.org/abs/1603.08675.

R. Landauer. 'Information is Physical'. *Physics Today* 44/5 (May 1991): 23–29. ISSN: 0031-9228. DOI: 10.1063/1.881299. eprint: https://pubs.aip.org/physicstoday/article-pdf/44/5/23/8304036/23_1_online.pdf. URL: https://doi.org/10.1063/1.881299.

D. Lidar, and T. Brun. *Quantum Error Correction* (Cambridge: Cambridge University Press, 2013).

G. Hao Low, and I. L. Chuang. 'Hamiltonian Simulation by Qubitization'. *Quantum* 3 (July 2019): 163. issn: 2521-327X. DOI: 10.22331/q-2019-07-12-163. URL: http://dx.doi.org/10.22331/q-2019-07-12-163.

D. Litinski. 'A Game of Surface Codes: Large-Scale Quantum Computing with Lattice Surgery'. *Quantum* 3 (Mar. 2019): 128. ISSN: 2521-327X. DOI: 10.22331/q-2019-03-05-128. URL: http://dx.doi.org/10.22331/q-2019-03-05-128.

S. Lloyd. 'Universal Quantum Simulators'. *Science* 273/5278 (1996): 1073–1078. DOI: 10.1126/science.273.5278.1073. eprint: https://www.science.org/doi/pdf/10.1126/science.273.5278.1073. URL: https://www.science.org/doi/abs/10.1126/science.273.5278.1073.

J.-G. Liu, and L. Wang. 'Differentiable Learning of Quantum Circuit Born Machines'. *Physical Review A* 98/6 (Dec. 2018). ISSN: 2469-9934. DOI: 10.1103/physreva.98.062324. URL: http://dx.doi.org/10.1103/PhysRevA.98.062324.

G. Hao Low, T. J. Yoder, and I. L. Chuang. 'Methodology of Resonant Equiangular Composite Quantum Gates'. *Physical Review X* 6 (4 Dec. 2016): 041067. DOI: 10.1103/PhysRevX.6.041067. URL: https://link.aps.org/doi/10.1103/PhysRevX.6.041067.

H. Maassen and J. B. M. Uffink. 'Generalized entropic uncertainty relations'. *Physical Review Letters* 60 (1988): 1103–1106. DOI: 10.1103/PhysRevLett.60.1103. URL: https://link.aps.org/doi/10.1103/PhysRevLett.60.1103.

J. M. Martyn et al. 'Grand Unification of Quantum Algorithms'. *PRX Quantum* 2/4 (Dec. 2021). ISSN: 2691-3399. DOI: 10.1103/prxquantum.2.040203. URL: http://dx.doi.org/10.1103/PRXQuantum.2.040203.

J. Martyn et al. 'Efficient Fully-Coherent Quantum Signal Processing Algorithms for Real-Time Dynamics Simulation'. *The Journal of Chemical Physics* 158 (Jan. 2023): 024106. DOI: 10.1063/5.0124385.

Sa. McArdle et al. 'Quantum Computational Chemistry'. *Reviews of Modern Physics* 92/1 (Mar. 2020). ISSN: 1539-0756. DOI: 10.1103/revmodphys.92.015003. URL: http://dx.doi.org/10.1103/RevModPhys.92.015003.

J. R. McClean et al. 'Barren Plateaus in Quantum Neural Network Training Landscapes'. *Nature Communications* 9/1 (Nov. 2018). ISSN: 2041-1723. DOI: 10.1038/s41467-018-07090-4. URL: http://dx.doi.org/10.1038/s41467-018-07090-4.

A. Montanaro. 'Quantum Speedup of Monte Carlo Methods'. *Proceedings of the Royal Society A: Mathematical, Physical and Engineering Sciences* 471/2181 (Sept. 2015): 20150301. ISSN: 1471-2946. DOI: 10.1098/rspa.2015.0301. URL: http://dx.doi.org/10.1098/rspa.2015.0301.

J. C. Napp et al. 'Efficient Classical Simulation of Random Shallow 2D Quantum Circuits'. *Physical Review X* 12 (2 Apr. 2022): 021021. DOI: 10.1103/PhysRevX.12.021021. URL: https://link.aps.org/doi/10.1103/PhysRevX.12.021021.

M. A. Nielsen, and I. L. Chuang. *Quantum Computation and Quantum Information: 10th Anniversary Edition* (Cambridge University Press, 2010).

M. A. Nielsen. 'The Fermionic Canonical Commutation Relations and the Jordan-Wigner Transform'. (2005). Accessed July 3, 2025 URL: https://api.semanticscholar.org/CorpusID:199373281.

J. L. Park. 'The Concept of Transition in Quantum Mechanics'. *Foundations of Physics* 1 (1970): 23–33. URL: https://api.semanticscholar.org/CorpusID:55890485.

A. Pati, and S. Braunstein. 'Impossibility of Deleting an Unknown Quantum State'. In: *Nature* 404 (Apr. 2000), pp. 164–5. DOI: 10.1038/35004532.

A. Peruzzo et al. 'A Variational Eigenvalue Solver on a Photonic Quantum Processor'. *Nature Communications* 5/1 (July 2014). ISSN: 2041-1723. DOI: 10.1038/ncomms5213. URL: http://dx.doi.org/10.1038/ncomms5213.

K. B. Petersen and M. S. Pedersen. *The Matrix Cookbook*. Version 20081110. Oct. 2008. URL: http://www2.imm.dtu.dk/pubdb/p.php?3274.

J. Preskill. *Quantum Computing 40 Years Later*. 2023. arXiv: 2106.10522 [quant-ph]. URL: https://arxiv.org/abs/2106.10522.

M. Reck et al. 'Experimental Realization of Any Discrete Unitary Operator'. *Physical Review Letters* 73 (1 July 1994): 58–61. DOI: 10.1103/PhysRevLett.73.58. URL: https://link.aps.org/doi/10.1103/PhysRevLett.73.58.

R. L. Rivest, A. Shamir, and L. Adleman. 'A Method for Obtaining Digital Signatures and Public-Key Cryptosystems'. In: *Communications of the ACM* 21.2 (Feb. 1978), pp. 120–126. ISSN: 0001-0782. DOI: 10.1145/359340.359342. URL: https://doi.org/10.1145/359340.359342.

E. Schrödinger. 'Die gegenwärtige Situation in der Quantenmechanik'. *Naturwissenschaften* 23 (1935): 807–812.

C. E. Shannon. 'The Synthesis of Two-Terminal Switching Circuits'. *The Bell System Technical Journal* 28/1 (1949): 59–98. DOI: 10.1002/j.1538-7305.1949.tb03624.x.

V. V. Shende et al. 'Synthesis of Reversible Logic Circuits'. *IEEE Transactions on Computer-Aided Design of Integrated Circuits and Systems* 22/6 (2003): 710–722. DOI: 10.1109/TCAD.2003.811448.

P. W. Shor. 'Algorithms for Quantum Computation: Discrete Logarithms and Factoring'. In: *Proceedings 35th Annual Symposium on Foundations of Computer Science*. 1994, pp. 124–134. DOI: 10.1109/SFCS.1994.365700.

P. W. Shor. 'Scheme for Reducing Decoherence in Quantum Computer Memory'. *Physical Review A* 52 (4 Oct. 1995): R2493–R2496. DOI: 10.1103/PhysRevA.52.R2493. URL: https://link.aps.org/doi/10.1103/PhysRevA.52.R2493.

P. W. Shor. 'Polynomial-Time Algorithms for Prime Factorization and Discrete Logarithms on a Quantum Computer'. *SIAM Journal on Computing* 26/5 (1997): 1484–1509. DOI: 10.1137/S0097539795293172. URL: https://doi.org/10.1137/S0097539795293172.

D. R. Simon. 'On the Power of Quantum Computation'. In: *Proceedings 35th Annual Symposium on Foundations of Computer Science*. 1994, pp. 116–123. DOI: 10.1109/SFCS.1994.365701.

D. R. Simon. 'On the Power of Quantum Computation'. *SIAM Journal on Computing* 26/5 (1997): 1474–1483. DOI: 10.1137/S0097539796298637. URL: https://doi.org/10.1137/S0097539796298637.

M. Slutskii et al. 'Analog Nature of Quantum Adiabatic Unstructured Search'. *New Journal of Physics* 21/11 (Nov. 2019): 113025. DOI: 10.1088/1367-2630/ab51f9. URL: https://dx.doi.org/10.1088/1367-2630/ab51f9.

A. M. Steane. 'Error Correcting Codes in Quantum Theory'. *Physical Review Letters* 77 (5 July 1996): 793–797. DOI: 10.1103/PhysRevLett.77.793. URL: https://link.aps.org/doi/10.1103/PhysRevLett.77.793.

A. Steane. 'Multiple Particle Interference and Quantum Error Correction'. *Proceedings of the Royal Society of London. Series A: Mathematical, Physical and Engineering Sciences* 452/1954 (Nov. 1996): 2551–2577. ISSN: 1471-2946. DOI: 10.1098/rspa.1996.0136. URL: http://dx.doi.org/10.1098/rspa.1996.0136.

Y. Suzuki et al. 'Amplitude Estimation without Phase Estimation'. *Quantum Information Processing* 19/2 (2020). ISSN: 1573-1332. DOI: 10.1007/s11128-019-2565-2. URL: http://dx.doi.org/10.1007/s11128-019-2565-2.

M. Szegedy. 'Quantum Speed-Up of Markov Chain based Algorithms'. In: *45th Annual IEEE Symposium on Foundations of Computer Science*. 2004, pp. 32–41. DOI: 10.1109/FOCS.2004.53.

E. Tang. 'A Quantum-Inspired Classical Algorithm for Recommendation Systems'. In: *Proceedings of the 51st Annual ACM SIGACT Symposium on Theory of Computing*. STOC 2019. Phoenix, AZ, USA: Association for Computing Machinery, 2019, pp. 217–228. ISBN: 9781450367059. DOI: 10.1145/3313276.3316310. URL: https://doi.org/10.1145/3313276.3316310.

H. F. Trotter. 'On the Product of Semi-Groups of Operators'. *Proceedings of the American Mathematical Society* 10/4 (1959): 545–551. ISSN: 00029939, 10886826. URL: http://www.jstor.org/stable/2033649 (visited on 06/08/2024).

G. Vidal. 'Efficient Classical Simulation of Slightly Entangled Quantum Computations'. *Physical Review Letters* 91/14 (Oct. 2003). ISSN: 1079-7114. DOI: 10.1103/physrevlett.91.147902. URL: http://dx.doi.org/10.1103/PhysRevLett.91.147902.

W. K. Wootters, W. K. Wootters, and W. H. Zurek. 'A Single Quantum Cannot be Cloned'. *Nature* 299 (1982): 802–803. URL: https://api.semanticscholar.org/CorpusID:4339227.

T. Young. 'The Bakerian Lecture: Experiments and Calculations Relative to Physical Optics'. *Philosophical Transactions of the Royal Society of London* 94 (1804): 1–16. https://doi.org/10.1098/rstl.1804.0001.

T. Yamakawa and M. Zhandry. 'Verifiable Quantum Advantage without Structure'. In: *2022 IEEE 63rd Annual Symposium on Foundations of Computer Science (FOCS)*. Los Alamitos, CA, USA: IEEE Computer Society, Nov. 2022, pp. 69–74. DOI: 10.1109/FOCS54457.2022.00014. URL: https://doi.ieeecomputersociety.org/10.1109/FOCS54457.2022.00014.

Index

1-in-3-SAT, 223
3-SAT, 223

adiabatic condition, 218
ancilla, 64
ansatz, 209
Argand diagram, 9, 179
artificial neural network, 211
associativity, 11
asymptotic notation, 99
authenticated public channel, 89

barren plateau, 211, 214
Bayes' law, 171
Bell (EPR) pair, 46, 94
Bell states, 33, 43
Bell, John Stewart, 42
Bennett, Charles, 89
Bernoulli random variable, 90, 168, 169
binary fractions, 157
binary numbers, 127
binary symmetric channel, 277
bisection search, 226
black box, 128
Bloch sphere, 37, 62, 76, 141, 163, 165, 270
block encoding, 201
block matrix, 73
Boolean function, 68
Boolean literal, 66, 223
Born, Max, 35
BPP, 108, 109, 127, 177
BQP, 109, 127, 177
Brassard, Gilles, 89

Chebyshev polynomial, 260
Church, Alonzo, 107
circuit fragments, 56
Clifford circuits, 116
closed quantum system, 277
clustering, 249
CNOT, 33
colour code, 295
commutation, 12, 58
completeness equation, 34
complex conjugate, 9
computational basis, 15
computationally guaranteed security, 88
concatenated codes, 283, 289
condition number, 237, 246, 265

conditional expectation, 25
conditional probability, 25
conjunctive normal form, 223
constant or balanced function, 128, 131
continued fractions, 180
Copenhagen interpretation, 7

D-wave, 232
data loading, 247
dequantisation, 249
Deutsch, David, 1, 127, 131
digital logic circuit, 66
digital signal processing, 155, 159
Dirac (bra-ket) notation, 6, 10
Dirac, Paul, 10
distributivity, 12
double-slit experiment, 3

Eastin– Knill theorem, 292
eavesdropping, 90
ebit, 86
Einstein, Albert, 42, 46
entanglement as a resource, 83, 86
entropic uncertainty relations, 93
error detection, 280
Euler method, 248
expectation, 25
exploration versus exploitation, 229
extended Church-Turing thesis, 124

Fermionic system, 191
Feynman, Richard, 1, 191
Fourier, Joseph, 155

Gaussian elimination, 121, 297
Gibbs states, 233
global phase factor, 36
Gottesman-Knill theorem, 116
greatest common divisor, 183, 185
Grover iterate, 144, 163, 165, 270
Grover, Lov, 139

Hadamard operation / gate, 31
Hamiltonian, 24, 30
hardware efficient ansatz, 210
Harrow, Aram, 237
Hassidim, Avinatan, 237
Heisenberg, Werner, 93

HHL, 237
hidden string, 134
hidden variable, 42
higher-order product formulas, 206
Holevo-Helstrom bound, 39
hybrid quantum-classical algorithm, 209

identity operation, 11
information entropy, 93
information theoretically guaranteed security, 88
inverse quantum Fourier transform, 159
IQP circuits, 112
Ising model, 191, 225

Jacobi-Anger expansion, 260
Jordan-Wigner transformation, 191

Lambda Calculus, 107
Landauer, Rolf, 1
linear operators / maps, 10
Lloyd, Seth, 237

magic state, 293
many worlds interpretation, 7
Markov chain, 273
matrix inversion, 237
matrix similarity, 22
measurement basis vectors / states, 34
measurement operators, 33
measurement postulate, 33
metaheuristic, 229
mid-circuit measurement, 56
modular exponentiation, 185
mutually unbiased bases, 93

no fast forwarding principle, 206
noisy channel, 280
norm, ℓ^2, 13
norm, ℓ^2 operator, 25
normal matrix, 18
NP, 108, 109
NP-complete, 109, 147, 221
NP-hard, 214
number field sieve, 177

open quantum system, 277
optimisation problems, 111
oracle, 128, 139
orthogonal states, 13
orthonormal bases, 15

$P \stackrel{?}{=} NP$, 110
$P = NP$, 110
P (complexity class), 108, 109
parameterised quantum circuit, 209

parity check measurements, 279
Pauli gadget, 194
Pauli operators, 31
phase (S) gate, 54
phase estimation by QSVT, 271
phase kickback, 129
Planck's constant, 30, 191
Podolsky, Boris, 42
polynomial hierarchy, 113
polynomial method, 148
PREP circuit (LCU), 202
principle component analysis, 249
principle of deferred measurement, 59
principle of deferred measurement, generalisation, 114
principle of implicit measurement, 60
product formula method of Hamiltonian simulation, 197

QRAM, 247
quadratic advantage, 139, 147, 167, 169, 225, 233, 274
quadratic form expansion, 116
quantum gate model, 53
quantum machine learning, 211
quantum recommendation system, 249
quantum supremacy, 113, 124
quantum walk, 273
quantum-inspired algorithms, 249
qubit routing, 80
query complexity, 127

random circuits, 113
Rayleigh-Ritz variational principle, 209
register (of qubits), 68
repetition code, 278
reversibility, quantum circuit, 58, 67, 159
reversible classical circuit, 69
Rosen, Nathan, 42
RSA, 88, 177

sampling problems, 111
Schrödinger equation, 30, 191
Schrödinger, Erwin, 6
search problems, 111
search register, 141
SEL circuit (LCU), 202
semi-definite programming, 233
Shor, Peter, 177
shut up and calculate, 87
simulated annealing, 230
simulation-based optimisation, 233
Solovay-Kitaev theorem, 78
spectral theorem, 22
spin system, 191
standard basis, 16

The manufacturer's authorised representative in the EU for product safety is
Oxford University Press España S.A. of el Parque Empresarial San Fernando de
Henares, Avenida de Castilla, 2 – 28830 Madrid (www.oup.es/en or product.
safety@oup.com). OUP España S.A. also acts as importer into Spain of products
made by the manufacturer.

www.ingramcontent.com/pod-product-compliance
Ingram Content Group UK Ltd.
Pitfield, Milton Keynes, MK11 3LW, UK
UKHW020924230426
470302UK00018B/115